Studies in Computational Intelligence

Volume 1129

Series Editor

Janusz Kacprzyk, Polish Academy of Sciences, Warsaw, Poland

The series "Studies in Computational Intelligence" (SCI) publishes new developments and advances in the various areas of computational intelligence—quickly and with a high quality. The intent is to cover the theory, applications, and design methods of computational intelligence, as embedded in the fields of engineering, computer science, physics and life sciences, as well as the methodologies behind them. The series contains monographs, lecture notes and edited volumes in computational intelligence spanning the areas of neural networks, connectionist systems, genetic algorithms, evolutionary computation, artificial intelligence, cellular automata, self-organizing systems, soft computing, fuzzy systems, and hybrid intelligent systems. Of particular value to both the contributors and the readership are the short publication timeframe and the world-wide distribution, which enable both wide and rapid dissemination of research output.

Indexed by SCOPUS, DBLP, WTI Frankfurt eG, zbMATH, SCImago.

All books published in the series are submitted for consideration in Web of Science.

Ali Kaveh · Hossein Yousefpoor

Chaotic Meta-heuristic Algorithms for Optimal Design of Structures

Springer

Ali Kaveh
Faculty of Civil and Environmental Engineering
Institute of Structural Engineering Research
Unit Mechanics and Structural Dynamics
TU-Wien
Vienna, Austria

Iran University of Science and Technology
Tehran, Iran

Hossein Yousefpoor
Department of Civil Engineering
Islamic Azad University
Maragheh, Iran

ISSN 1860-949X ISSN 1860-9503 (electronic)
Studies in Computational Intelligence
ISBN 978-3-031-48917-4 ISBN 978-3-031-48918-1 (eBook)
https://doi.org/10.1007/978-3-031-48918-1

This Springer imprint is published by the registered company Springer Nature Switzerland AG
The registered company address is: Gewerbestrasse 11, 6330 Cham, Switzerland

Preface

Optimization of non-linear and non-convex modes is one of the most used scientific problems in recent decades. In order to increase the efficiency of meta-heuristic algorithms, the technology related to embedding chaotic systems in the exploration and exploitation parts of these algorithms brings significant progress. By embedding chaotic time series, premature convergence of algorithms is prevented and local optima converge towards global optima. By creating chaotic jumps in the two important phases of exploration and exploitation, chaotic series saves them from the trap of local optima and creates a balance between the phases of exploration and exploitation. According to the inspiration of the algorithms, these series may be needed only in the stages of exploration or exploitation, or in some cases, we will need them in both stages at the same time. Therefore, the existence of chaoite's series with different behaviors and triple scenarios, has brought huge research to introduce the best chaotic series and the best suitable scenario for each meta-heuristic algorithm. In this research, different time series are embedded on a large number of well-known meta-heuristic algorithms and after the competition between local optima, the best of them has been introduced as the sub global optimum. In Chap. 1, there is a brief introduction to the introduction of selected algorithms for embedding chaos series. In Chap. 2, by introducing chaotic systems and examining the Lorenz absorber system, the method of embedding chaotic systems in meta-heuristic algorithms has been investigated. In Chaps. 3–13, chaos series are embedded in a large number of well-known meta-heuristic algorithms and chaotic algorithms are formed. These chaotic meta-heuristic algorithms are:

Chapter 3: Chaotic Cyclical Parthenogenesis Algorithm (CCPA)
Chapter 4: Chaotic Teaching-Learning-Based Optimization (CTLBO)
Chapter 5: Chaotic Biogeography-Based optimization (CBBO)
Chapter 6: Chaotic Differential Evolution (CDE)
Chapter 7: Chaotic Water Evaporation Optimization (CWEO)
Chapter 8: Chaotic Artificial Bees Colony (CABC)
Chapter 9: Chaotic Imperialist Competitive Algorithm (CICA)
Chapter 10: Chaotic Shuffled Frog-Leaping Algorithm (CSFLA)

Chapter 11: Chaotic Particle Swarm Optimization (CPSO)
Chapter 12: Chaotic Tug-of-War Optimization (CTWO)
Chapter 13: Chaotic Thermal Exchange Optimization (CTEO)

In order to validate, determine the amount of improvement of the results and efficiency of chaotic algorithms, benchmarks and well-known structural examples have been investigated. In these studies, by forming a wide range of competition between chaotic modes, the best chaos function of each algorithm is presented.

We would like to take this opportunity to acknowledge a deep sense of gratitude to a number of colleagues and friends who have helped us in different ways in the process of writing this book. Our special thanks are due to Dr. Thomas Ditzinger, the Editorial Director of Interdisciplinary and Applied Sciences and Engineering from Springer, and the Series Editor of Studies in Systems, Decision and Control, Prof. Janusz Kacprzyk for the publication of our new book within Springer and the series for their comments and suggestions during the preparation of this book. Our sincere appreciation is extended to our Springer colleagues, Gowrishankar Ayyasamy, who prepared the layout design of this book.

The first author would likes to take this opportunity to acknowledge a deep sense of gratitude to Prof. Rudolf Heuer for his kind hospitality during the author's sabbatical leave in his department at TU-Wien. The authors are also grateful to Dr. Kiarash Biabani Hamedani for his comments on the results of some of the benchmarks.

We would like to thank the publishers who permitted some of our papers to be utilized in the preparation of this book, consisting of Springer, and Budapest University of Technology and Economics.

Our warmest gratitude is due to our families for their continued support in the course of preparing this book

Every effort has been made to render this book error-free. However, the authors would appreciate any remaining errors being brought to his attention through their email addresses: alikaveh@iust.ac.ir (Ali Kaveh) and hyosefpor@yahoo.com (Hossein Yousefpoor).

Vienna, Austria
Maragheh, Iran
September 2023

Prof. Ali Kaveh
Dr. Hossein Yousefpoor

Contents

Chapter 1
Introduction

1.1 Introduction

Achieving the optimal state is one of the most fundamental issues in the universe. Due to the limited resources accessible to humans, optimization has become a vital matter and its application range is increasing day by day. Mathematics, computer science, civil engineering, industries, mechanics, and economics discuss only a part of these important applications [1]. In this type of critical issues, the goal is to obtain the best decision state among several different states. Today, meta-heuristic algorithms can be considered one of the most important methods for solving optimization problems. In order to optimize structural skeletons that include non-linear and non-convex modes and are often associated with several local optima, meta-heuristic algorithms face problems that cause stagnation in the optimization process. One of the most important of them is premature convergence and falling into the trap of local optima. One of the best ways to improve the efficiency of meta-heuristic algorithms is to combine them with chaos functions. These functions replace the possible choices of exploration, exploitation or both according to the needs of the algorithm in standard mode. By applying chaos functions, a significant improvement is achieved in the optimization results, which is mainly due to creating a balance between the exploration and exploitation stages. Algorithms resulting from the combination of chaos functions are known as chaotic algorithms, but the type of chaos function that causes a significant improvement in a selected algorithm will require research and field investigations. In this book, a number of well-known algorithms whose standard mode research background has been done, combined with chaotic functions and chaotic algorithms are presented.

A. Kaveh and H. Yousefpoor, *Chaotic Meta-heuristic Algorithms for Optimal Design of Structures*, Studies in Computational Intelligence 1129,
https://doi.org/10.1007/978-3-031-48918-1_1

1.2 Traditional Optimization and Meta-heuristic Optimization

In recent decades, many solutions have been proposed for optimization. Traditional gradient-based methods investigate optimal responses by forming the derivative of the objective function. While most engineering problems cannot provide us with an explicit relation of the objective function, and investigations of partial derivatives will be needed. In some cases, the search space is discrete and access to derivatives will be completely impossible. Today, with the complexity of engineering problems and the increase in the number of decision variables, the unresponsiveness of traditional and classical methods has become evident. Therefore, in order to overcome these problems, a special group of optimization methods has been invented. They are based on decisions and the principles of possible and random search [2]. In these algorithms, the value of the objective function itself is evaluated instead of derivatives, and with inspiration from natural phenomena, it obtains an improving process in successive iterations. The main reason for this choice is the characteristics of natural phenomena, and if there is a need for empowerment, nature does it in the best way. The evolution of living organisms is the first choice of researchers for inspiration in meta-heuristic algorithms. In genetic evolution, the characteristics of living organisms are improved by *cross-over* and *mutation* so that they can win in competition with other organisms by adapting to the surrounding environment as much as possible. Examples of these algorithms include Genetic Algorithm (GA), Differential Evolution (DE) and Evolutionary Strategy (ES). In the algorithms of the second group, the source of inspiration is based on the swarm intelligence of animals and their behavior in foraging. In this group of algorithms, components such as *population*, *cooperation*, *communication*, *information exchange*, *information flow* and *self-organization* are considered. Examples of this group include Ant Colony Optimization (ACO), Cyclical Parthenogenesis Algorithm (CPA), Particle Swarm Optimization (PSO) [3], Cuckoo Search (CS), Artificial Bee Colony (ABC), and Gray Wolf Optimization (GWO), which have achieved valuable results in optimizing structural skeletons in their performance. Another group of meta-heuristic algorithms includes inspiration from physical laws. Examples of these algorithms are: Charged System Search (CSS), Colliding Bodies Optimization (CBO), Water Evaporation Optimization (WEO), Vibrating Particle System (VPS), Big Bang-Big Crunch (BB-BC), Ray Optimization (RO) and Harmony Search (HS). In recent years, these algorithms have been chosen on a large scale for the optimization of structures. They are very popular among researchers. The range of inspiration is so wide that the Shuffled Complex Evolution (SCE-UA) was proposed at the University of Arizona based on the traditional Nelder and Mead method. In this algorithm, geometric operators of contraction and reflection are considered. This algorithm is now the basis of the Shuffled Frog-Leaping Algorithm (SFLA), and classified as a memetic algorithm. In this book, by applying chaos maps to it, a significant improvement in the optimization result has been achieved. Other algorithms such as Imperialist Competitive Algorithm (ICA), Teaching–Learning based Optimization (TLBO) and

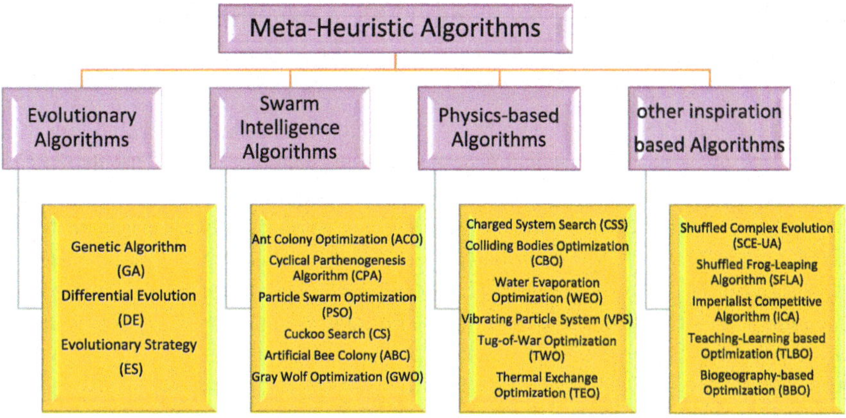

Fig. 1.1 Classification of the metaheuristic algorithms

Biogeography-based Optimization (BBO) are inspired by political behaviors, classroom education process and migration process in wildlife habitats, respectively, with various perspectives proposed. This in most cases have led to improved optimization results. The classification of algorithms is presented in Fig. 1.1.

1.3 Chaotically Enhanced Meta-heuristic Algorithms

Despite the significant progress of meta-exploration algorithms, in engineering optimization problems, pests still threaten them. These pests can be cases such as premature convergence, stopping at local optima and falling into their trap, imbalance between exploration and exploitation, and finally "slowing down" convergence. Research shows that using chaos maps and replacing them in random selection steps largely eliminates the weaknesses of algorithms. Mathematically, chaos refers to the ability of a simple pattern and model, which, although it has no indication of random phenomena, can lead to the emergence of very disordered behaviors in the environment. These dynamic systems have attracted the attention of many scientific societies and are observed in various fields such as engineering, medicine, biology and economics [2]. The salient features of a chaotic system are: (1) They are sensitive to initial conditions. (2) Their alternating rotation is dense. (3) They have quasi-random and non-periodic behavior [2]. The term butterfly effect came from an article by Edward Lawrence. At the 39th Water Summit, he presented an article entitled: "Can a butterfly fly in Brazil cause strong winds in Texas?" Experimental studies show that the use of chaotic signals instead of random signals has yielded very valuable results. Chaos signals are distributed in scattered parts of the decision space, and then a carefully search is done in their neighborhood, and the balance between the exploration and exploitation stages is largely established.

1.4 Chaos Embedded Meta-heuristic Algorithms for Optimal Design and Goals of This Book

In meta-heuristic algorithms, a number of parameters are randomly selected to expand the diversity of investigations and access to various answers [4–6]. The distribution function related to these parameters is different according to the proposed algorithm and can include uniform, normal, logistic, levy, etc. distribution. Although the choice of these parameters is completely "random", but it has significant effects on the performance of the algorithm. Stochastic parameters play a major role in increasing or decreasing the convergence rate, and in some cases these key factors can control the balance between exploration and exploitation. Nowadays, random selection of these parameters is accompanied by skepticism and research shows that some of the factors of inefficiency of meta-heuristic algorithms are the same random parameters, so the dynamic series of turbulence, whose values are deterministic, will be a good alternative to these parameters. Chaos series can accelerate convergence toward general and global responses, and play a major role in preventing local optimization and premature convergence. Series created by chaos maps are similar to random processes, but their values are definite, nonlinear, dynamic, non-repetitive, and non-convergent to a particular boundary. We will not have inverse chaos for the maps and we can consider different situations to apply them. In some cases, these maps are a good alternative to the possible parameters related to the exploration, and in other cases, we will see improved results by replacing the parameters related to the exploitation section [7]. Also, in a small number of meta-heuristic algorithms in both cases simultaneous chaotic maps will lead to improved results. In order to achieve the desired state, we will examine the algorithms in three different scenarios. Which scenario works best for a particular algorithm varies from example to example. In fact, the nonlinear behavior and non-convexity of the target functions in the optimization of truss structures have created these conditions for meta-heuristic algorithms. The main goal of this book is to embedded chaos maps to some well-established and the most recently meta-heuristic algorithms for enhanced optimal design of structures [8]. Schematic of the chapters of the present book in one glance is shown in Fig. 1.2. In order to investigate the effects of chaos maps in improving the optimization results, first these algorithms are examined as standard and then the results are compared with the proposed three scenarios for each chaos map.

1.5 Organization of the Present Book

The remaining chapters of this book are organized as follows: Chap. 2 introduces the stages of exploration and exploitation in meta-heuristic algorithms. If there is no balance between these two steps, we use chaos functions to improve the efficiency of the algorithm. Therefore, by embedding chaotic functions, instead of random numbers, we can save ourselves from premature convergence and converge the local

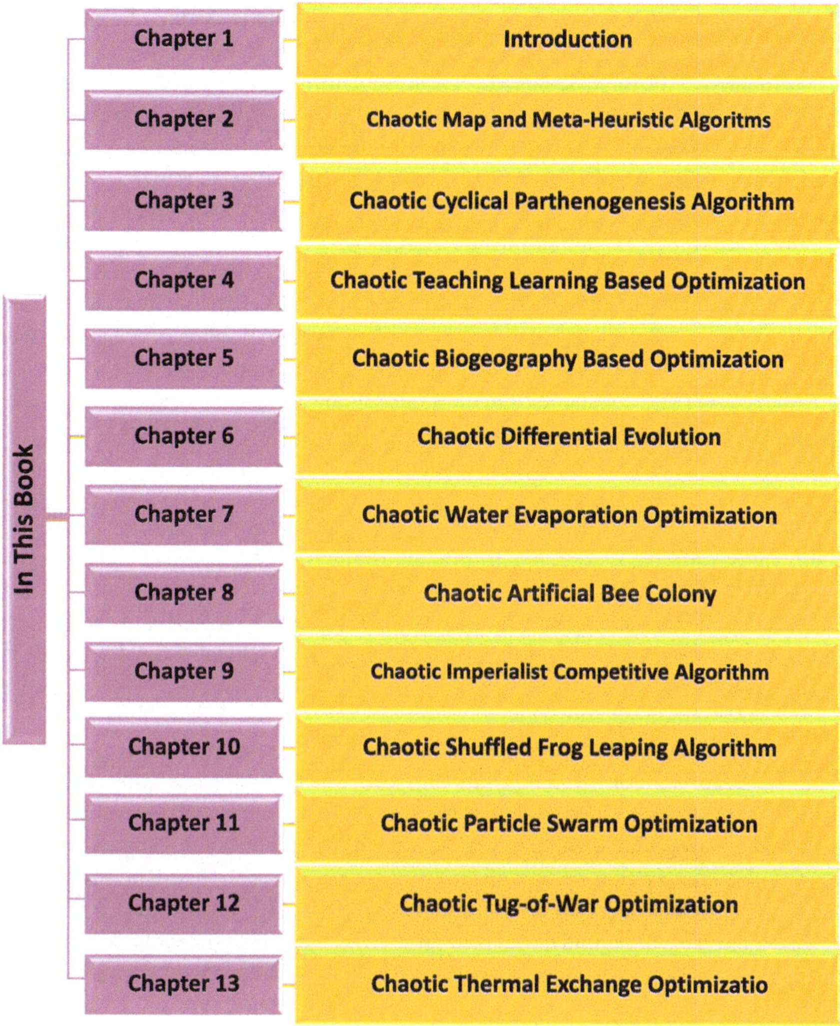

Fig. 1.2 Schematic of the parts and chapters of the present book in one glance

optima to the global optima by creating chaotic jumps. Also, the types of chaos functions, their features and their mathematical relationships are introduced in this chapter. With triple scenarios, it is determined that these functions will be embedded in exploration, exploitation or both at the same time. Chapter 3 introduces a new population-based meta-heuristic optimization method, called Cyclical Parthenogenesis Algorithm (CPA). This algorithm was presented by Kaveh et al. In this algorithm, it was inspired by the social behavior and reproduction of some animal species such as aphids. Mating and without Mating form the two main parts of the exploration and exploitation stage. The statistical distributions used in these two important parts can be replaced with chaos functions. In problems with static loading, Logistic, Tent and Gaussian chaos functions are embedded in the algorithm, and each of them is investigated with three scenarios. These functions improve the parts related to exploration, exploitation or both simultaneously and a significant improvement in the optimization results can be achieved. For multi-frequency constraints, where the problems are strongly non-linear and non-convex, Gaussian, Liebovitch and Piecewise chaos functions are embedded in the algorithm. Chapter 4 introduces the famous Teaching–Learning-Based Optimization (TLBO). This algorithm is inspired by the learning process in the classroom. The algorithm includes two main phases, which are: the influence of the teacher in improving the average academic level of students and the mutual influence of students on each other. Like other population-based algorithms, it starts with a set of initial and random answers, these answers are the knowledge level of the students, and the smartest one is chosen as the teacher in each iteration. In the teacher phase, the average academic level of the students is improved by the teacher, then in the student phase, the students learn from each other. These two phases are exploitation and exploration stage. Chaotic functions are embedded in the algorithm instead of random distributions and the optimization results are significantly improved. As a result, the algorithm is saved from the local optimal trap and with chaotic jumps, the local optimal converges to the global optimal. Also, the problem of premature convergence of the algorithm is solved and the balance between the stages of exploration and exploitation is achieved. Logistic, Tent and Gaussian are embedded for problems with static loading. Also, each chaos function is associated with three scenarios.

In dynamic problems and with multi-frequency constraints that are strongly nonlinear and non-convex, chaos models are formed by embedding Gaussian, Liebovitch and Piecewise functions. In both groups of optimization problems, chaotic modes lead to a significant improvement in the optimization results. Chapter 5 is dedicated to the introduction of one of the most successful meta-heuristic algorithms, namely the Biogeography-Based optimization (BBO). This algorithm is inspired by the distribution of plants and animals in different geographical habitats. Most animal and plant species try to make exclusive use of food resources, but they do not achieve this goal due to the limitation of food resources. Therefore, they try to migrate to less populated settlements. Based on the suitability of each habitat, migration to or from it may take place. In the original version of the algorithm, the coefficients μ and λ are introduced for these two types of migration. Migration of species forms the exploitation phase of the algorithm. In addition, the exploration phase includes the mutation of species to obtain a new settlement. The random distribution used in these phases in the standard version of the algorithm is uniform. Therefore, the imbalance between exploration and exploitation in the standard model is natural. For this reason, in non-linear and non-convex problems, the algorithm falls into the local optimal trap. The random distributions used in these two important phases can be replaced by chaotic dynamic functions and a significant improvement can be achieved. By embedding the chaos functions in the biogeography-based optimization (CBBO), the algorithm is saved from the local optimality trap and converges from the local optimum to the global optimum. As a result, the problem of premature convergence of the algorithm is solved and the balance between the stages of exploration and exploitation is achieved. In this study; *Logistic*, *Tent*, *Gauss*, *Liebovitch* and *Piecewise* chaos functions are considered for embedding in biogeography-based optimization (BBO). Each chaos function is associated with three scenarios. Therefore, we have an extensive mode of competition between the perturbed modes and finally the best optimal response is compared with the standard mode. In Chap. 6, we introduce one of the most famous algorithms of the evolutionary group, that is, Differential Evolution (DE). This algorithm uses a special method to generate the answer. In most evolutionary algorithms, the new generation population is generated using the crossover operator, and then mutation is applied to them, but in the Differential Evolution (DE), a temporary response is generated with the mutation operator and then stabilized by the crossover of the response. During the optimization period, the distance between the community members becomes smaller and smaller and converges towards a global optimum. Therefore, with the distance between the members of the population, the step length and the intensity of the jump can be adjusted. If this distance is longer, we can reach the ideal points faster by choosing big steps, and on the contrary, we need small steps for small distances. In the original version of the algorithm, by defining the difference vector, the algorithm chooses a coefficient of this difference for mutation. To apply the crossover operator, some arrays are selected from the initial population and some from the mutant population. For the mutation, the role of the scale parameter β is very important and the generation of new responses is obtained by the intersection method using a binomial. Considering all the advantages that have been proposed for the algorithm, there are always pests that threaten the algorithm, and that is the random distribution used in the two stages of exploration and exploitation. By embedding chaotic dynamic series with triple scenarios, this weakness is removed and a significant improvement in the optimization results is achieved. Also, the algorithm is saved from the premature convergence problem. In chaotic states, we can get rid of the local optimum trap by making a jump and converge to the global optimum. By embedding *Logistic*, *Tent* and *Gossin* chaos functions with triple scenarios, a great challenge between chaos functions is obtained and the results are significantly improved. In the tests for each model, nine perturbed modes compete with each other, and then the best response is compared to the standard mode to determine the improvement. Chapter 7 is devoted to the review of one of the most successful algorithms imitating physical phenomena, called the Water Evaporation Optimization (WEO). In this algorithm, it is inspired by the surface evaporation of tiny water molecules on a solid surface with different humidity states. Also, in order to increase efficiency, dynamic simulation of water molecules is considered. Based on this simulation, the desired surface changes from the repulsive state to the water absorbing state, but the evaporation rate does not decrease uniformly in this sudden change, contrary to our imagination. Therefore, it first increases and then, after reaching the maximum volume, it starts to decrease. In cases where the humidity of the surface of the desired layer is not enough, the water molecule becomes a spherical droplet. Otherwise, the water molecule spreads as a single layer. Two strategies, single layer and droplet, constitute the discovery and exploitation stages of the algorithm. The imbalance between these strategies is related to the random distribution used in the original version. By embedding *Logistic, Gaussian,* and *Piecewise* chaos functions, the algorithm can be converged from the local optimum to the global optimum. Chaotic functions prevent premature maturation of the algorithm by creating chaotic mutations and a significant improvement in optimization results is achieved. By creating a competition between the nine perturbed models, the best response is selected for comparison with the standard case. In non-linear, non-convex, complex problems, implicit modes, multi-element problems, simultaneous optimization of cross section and geometric shape and multi-frequency constraints, chaotic methods have significant computational speed. In Chap. 8, one of the most famous algorithms from the swarm intelligence group, Artificial Bees Colony (ABC), is examined. The source of inspiration in this algorithm is the group behavior of bees to search for food. Therefore, each proposed answer represents a food source and its quality is checked based on the objective function. The location of food sources is improved by bees in an iterative process. The division of tasks in bees is in three. These groups are: recruited bees, onlooker bees and scout bees. The task of recruited bees is to find better food sources with a specific process, while observer bees search randomly based on the merit of food sites. The strategy of employed and observer bees is considered to carry out exploration and exploitation stages. The existence of random distributions in these strategies is, on the one hand, to create diversity in the search space, but on the other hand, it causes the optimization operation to stagnate and stop at the local optimum. Therefore, chaotic dynamic functions have a high potential to save artificial bee colony algorithm. By embedding *Logistic, Gaussian* and *Piecewise* chaos functions, the local optimum converges to the global optimum and premature convergence is eliminated. As the chaotic models are formed, nine different chaotic modes compete and the best response is compared to the standard mode. For problems with multi-frequency constraints, size and layout optimization, the standard state of the artificial bee colony faces a wide stagnation, which in chaotic states we can improve the results. By examining the state of chaos, the weaknesses and strengths of algorithms in dealing with optimization problems are determined. Chapter 9 belongs to an algorithm in the political, cultural and social field called Imperialist Competitive Algorithm (ICA). The population of this algorithm is made up of countries. At the beginning, some of the best countries are chosen as emperors and think of acquiring colonies based on their superior power. This phase is known as the absorption phase of colonial countries. The states of each empire compete with each other to bring their conditions to the emperor, but the main competition is the competition between the emperors themselves. Therefore, we have two types of competition, including intra-group and extra-group. At this stage, weak countries move towards the emperor to improve their development level. This stage is known as the assimilation policy stage, and the coefficient chosen to move is between 0 and 2. ($0 \leq \beta \leq 2$). In the next step, in order to access unknown places in the search space, the revolution step is implemented and a random change is made in the answers and if the results are successful, it is accepted. Now, in intra-group competition, if a country better than the existing emperor is found, it will replace the emperor and the former country will be considered a colony. In outgroup competition, the empires themselves compete, in this competition the weakest empire loses one of its colonies and the strongest empire takes it. In the process, the empire will fall if it loses all of its colony countries. Finally, if only one empire remains, the operation stops and that empire is introduced as the final answer of this round. In the original version of this algorithm, the strategy of absorption and revolution is formed by random distributions. Although, the main purpose of these random distributions is to increase the variety in the response space, but most of the cases cause the algorithm to fall into the trap of local optima. In these cases, with premature convergence, the optimization operation faces a wide stagnation. Therefore, the chaotic dynamic series by creating jumps, it converges the local optimum to the global optimum. It also achieves a balance between exploration and exploitation stages. By creating a competition between nine chaotic modes, the best response is selected and compared with the standard mode. Chapter 10 is dedicated to the Shuffled Frog-Leaping Algorithm (SFLA). This algorithm is inspired by the behavior of frogs to find food in the wetland. The only known algorithm that has the characteristics of both evolutionary groups and swarm intelligence. At each step, the entire population of frogs is divided into memeplex groups, then each group of frogs' searches based on its best memeplex member. If the results are not improved, the search is performed based on the best member of the whole set. In this step, the algorithm uses the strategy of the distance between the best and the worst answer to determine the corrective steps. Therefore, the method of improving the answers of each memeplex is first the best answer of the same memeplex, and if no improvement is achieved, in the next strategy we go to the best answer of all memeplexes. With these two strategies, exploration and exploitation steps are implemented in the algorithm. In the original version of the algorithm, these two stages of the strategy use random distributions to create diversity. In this case, uniform random distribution causes early convergence in the algorithm. Therefore, the algorithm is stuck in the local optimal trap and stagnation is achieved. Chaotic functions converge the results towards the global optimum by making sudden jumps and create a balance between exploration and exploitation. In the frog jump algorithm, by embedding Logistic, Gaussian and Piecewise chaos functions, a significant progress has been achieved in the optimization of problems. It has been obtained in complex problems with multi-frequency constraints, the standard mode of the algorithm is associated with stagnation and stopping, and by embedding chaos functions, a significant improvement in the optimization results has been achieved. In the competition between 9 chaotic models, the best response is selected and compared with the standard mode. Chapter 11 is dedicated to the investigation of the most famous swarm intelligence algorithm called Particle Swarm Optimization (PSO). In this algorithm, the answers are considered as particles in the search space, and to determine the merit of each particle, its position is evaluated in the objective function. In order to determine the new position in each iteration, the velocity of each particle is added to its previous position. Therefore, all the sensitivity of the algorithm is in determining the speed of the particle in each iteration. The components that play a role in the velocity vector are: a coefficient of the previous velocity, a coefficient of the best position of that particle and a coefficient of the best position of all particles. Two important strategies in this algorithm include the best position of each particle and the best position of all particles. That they are responsible for the exploration and exploitation stages. Although their random selection increases the diversity in the search space, it causes premature convergence and stagnation of the optimization speed. Therefore, getting stuck in the standard mode of the algorithm is a natural thing in the local optimal trap. In order to improve the efficiency of the algorithm, by embedding chaos functions instead of random coefficients, chaotic jumps are formed and local optima converge towards global optima. Another role of chaos functions is to create a balance between exploration and exploitation steps. In this study, by embedding Gaussian, Liebovitch and Piecewise chaos functions, a significant improvement is achieved in the optimization results. With extensive competition between 9 chaotic models, the best response is selected from among them and then compared with the standard case. Also, by embedding chaos functions, we can get to know the strengths and weaknesses of algorithms. In Chap. 12, the Tug-of-War Optimization (TWO), is examined. This algorithm is inspired by the traditional tug-of-war game and is a population-based algorithm. In the initially stage, the algorithm selects a set of primary answers in the search space. In this algorithm, each solution is considered as a team and all of them form a league. In each iteration, the evaluation of the teams is determined based on merit and the results are sorted. In this competition, the best team has the most weight and the worst team has the least weight. Both teams are pulling the rope, the light team loses the competition and goes toward of heavy team. In the algorithm, both exploration and exploitation strategies are applied with a random distribution. The presence of random distributions in these strategies, on the one hand, help to create diversity in the search space, but they cause the optimization operation to stop at the local optimal position with premature maturity. Therefore, chaotic dynamic functions have a high potential to save the algorithm from the trap of local optima. By embedding Logistic and Gaussian chaos functions, the local optima converge to the global optima. In this research, by forming chaotic models with three scenarios, 6 different chaotic modes compete with each other and to determine the best model, a comparison is made with the standard mode. For problem related to multi-frequency optimization constraints, size and layout, the standard Tug-of-War Optimization (TWO), is faced with extensive stagnation, which is balanced between exploration and exploitation by forming chaotic jumps with chaotic functions. Also, in chaos methods, the weaknesses and strengths of algorithms in dealing with optimization problems are determined. In Chap. 13, Thermal Exchange Optimization (TEO), has been investigated. This algorithm is in the group of physical algorithms and is inspired by Newton's law of cooling. According to this law, the rate of heat loss of an object is proportional to the temperature difference between the object and its surroundings. Therefore, we consider each particle as a cooling or heating object, and by assuming another factor as the environment, heat transfer is formed between the two. After thermal exchange, the new temperature of the object determines its new position in the search space. In the original version of the algorithm, two phases of exploration and exploitation are considered to search for local and global optima, and each of these phases is associated with random distribution. The main purpose of random distributions is to create diversity in the search space, but it often causes premature convergence and stopping at local optima. Chaotic functions have suitable conditions to create jumps and get rid of the trap of local optima. By embedding chaos functions, it is possible to jump from the local optimum and converge to the global optimum. In this study, the Logistic and Gaussian chaos functions with three scenarios, embedding and significantly improving the results, have saved the algorithm from premature convergence. In these chaotic models, 6 different modes compete with each other and then the best answer is compared with the standard mode. For problems related to multi-frequency, size and layout constraints simultaneously, the standard mode of the algorithm faces extensive stagnation, which can be significantly improved by embedding chaos functions and chaotic jumps. With the formation of chaotic models, the weaknesses and strengths of algorithms in dealing with optimization problems are determined. Up to this point, there has been a competition between chaos functions, and the best chaos function has been identified for each meta-heuristic algorithm. In the following, we will compare the effectiveness of chaotic algorithms.

References

1. Kaveh, A.: Advances in Meta-heuristic Algorithms for Optimal Design of Structures, 2nd edn., 3rd edn. Springer, Cham, Switzerland (2021). https://doi.org/10.1007/978-3-319-05549-7
2. Peitgen, H.-O., Jürgens, H., Saupe, D.: Chaos and Fractals: New Frontiers of Science. Springer, Berlin (2006)
3. Kaveh, A., Sheikholeslami, R., Talatahari, S., Keshvari-Ilkhichi, M.: Chaotic swarming of particles: a new method for size optimization of truss structures. Adv. Eng. Softw. **67**, 136–147 (2014)
4. Kaveh, A., Yousefpoor, H.: Chaos-embedded meta-heuristic algorithms for optimal design of truss structures. Scientia Iranica A **29**(6), 2868–2885 (2022). https://doi.org/10.24200/sci.2022.59812.6441
5. Kaveh, A., Yosefpoor, H.: Chaotically enhanced meta-heuristic algorithms for optimal design of truss structures with frequency constraints. Periodica Polytechnica Civil Eng. **66**(3), 900–921 (2022). https://doi.org/10.3311/PPci.20220
6. Kaveh, A., Zarfam, P., Aziminejad, A., Yosefpoor, H.: Comparison of four chaotic meta-heuristic algorithms for optimal design of large-scale truss structures. Iran. J. Sci. Technol. Trans. Civil Eng. https://doi.org/10.1007/s40996-022-00908-8
7. Talatahari, S., Kaveh, A., Sheikholeslami, R.: Chaotic imperialist competitive algorithm for optimum design of truss structures. Struct. Multi. Optim. **46**(3), 355–367 (2012)
8. Kaveh, A., Javadi, S.M.: Chaos-based firefly algorithms for optimization of cyclically large-size braced steel domes with multiple frequency constraints. Comput. Struct. **214**, 28–39 (2019). https://doi.org/10.1016/j.compstruc.2019.01.006

Chapter 2
Chaotic Maps and Meta-heuristic Algorithms

2.1 Introduction

In meta-heuristic algorithms, two important stages of *exploration* and *exploitation* are considered in order to converge towards optimal answers. In the exploration phase, we settle on the points of the search space that have a privileged strategy, and in the exploitation phase, we carefully examine the area related to the neighborhood of the selected points. The Random Search algorithm (RS) is a fully exploration algorithm, while the Hill Climbing algorithm (HC) and Tabu Search algorithms (TS) are fully exploitation. These two steps are presented for a number of algorithms in Fig. 2.1.

The imbalance between these two stages will cause premature convergence and fall into the trap of local optima [1]. In order to improve the efficiency of these algorithms, the control of these two stages creates suitable conditions for improving their efficiency. To create a balance between these steps and increase the efficiency of meta-heuristic algorithms, chaos maps have good potential so that in different conditions they can improve the exploration step, the exploitation step or both steps simultaneously and prevent premature convergence. In this regard, some meta-heuristic algorithms need to be improved for the exploitation phase, and others need to be improved for the exploitation phase. Also, in a rare number of algorithms, both steps will need to be improved. In recent studies, chaos functions with three different scenarios have been performed on a large number of well-known meta-heuristic algorithms, and the optimization results have been significantly improved compared to the standard mode. Chaotic meta-heuristic algorithms, in addition to weighted optimization for optimization with discrete variables, optimization with multi-frequency constraints, and shape and geometry optimization, also bring significant improvement. In optimization with multi-frequency constraints, non-linear and non-convex penalty function and have implicit variables. Therefore, the gradient-based methods will not be justified, but by choosing different chaotic functions and expanding the statistical space, access to global optima will be accelerated.

A. Kaveh and H. Yousefpoor, *Chaotic Meta-heuristic Algorithms for Optimal Design of Structures*, Studies in Computational Intelligence 1129,
https://doi.org/10.1007/978-3-031-48918-1_2

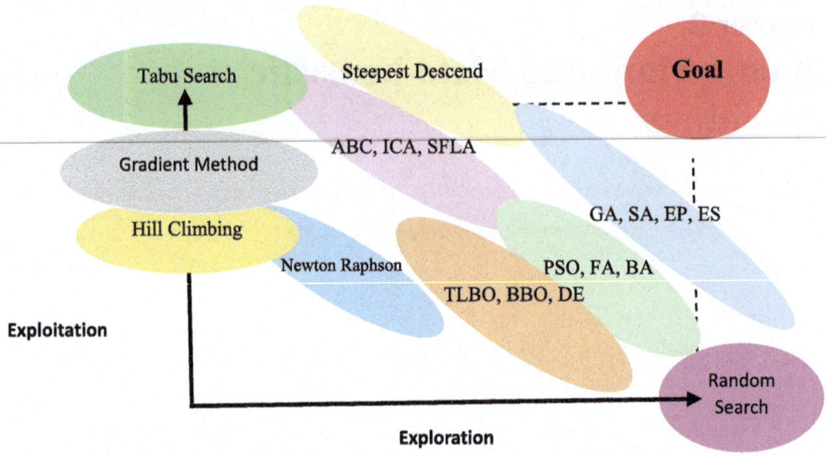

Fig. 2.1 The process of exploration and exploitation in a number of meta-heuristic algorithms

2.2 Chaotic Systems

Until a few decades ago, scientists considered the world to be a set of systems, which move according to the algebraic laws of nature in a very specific and predictable way, but with the progress of science, many natural events can no longer be explained by the previous deterministic views [2]. Attempts by scientists to describe such events led to chaos theory in mathematics. Chaos theory is called a branch of mathematics and physics that is related to systems whose dynamics show a very sensitive behavior in front of changing initial values; so that its future behavior is no longer predictable. These systems are called chaotic systems, which are non-linear dynamic systems, and the best examples for them are the butterfly effect, air currents, and the economic cycle. The main idea in chaos theory is that, in every disorder there is order. This means that order should not be sought only on an absolute scale. A phenomenon that seems completely random and unpredictable on a local scale, may be stable and predictable on a larger scale. There are similarities between chaos theory and statistics. Statistics seeks to discover order in chaos. The result of tossing a coin each time is random and unknown, because it has a local domain. But the expected consequences of this phenomenon when repeated in large numbers are stable and predictable. Many scientific predictions, including earthquake predictions, in short periods may seem completely random and disorderly, but on the scale of the 500-year and 2400-year return period, it will have a definite periodicity and some kind of order.

2.2.1 *Scientific History of Chaos Theory*

The study of this topic actually started with the studies of meteorological scientists. Some of the meteorologists were studying weather conditions and the impact of various things on the weather of the world and the region. For two years, they studied the weather of a certain region, with mild and completely moderate weather conditions, and recorded all the changes. In the autumn of the second year, the graph of these changes suddenly changed strangely, which indicated the occurrence of severe atmospheric changes, but what happened in practice did not show any change in itself. In order to investigate the lack of change, the studies were continued for another year, and the studies showed that there was a lake nearby where migratory birds used to visit, and the movement of their wings exerted pressure on the atmosphere, and without their presence A big storm formed in the area, and 12 ha of that area were destroyed.

With these studies, the *first slogan* of chaos theory was formed: "*A butterfly flutters in Africa and a tornado form in South America*". The similarity between butterfly effect and Lorenz attractor is shown in Fig. 2.2. In another study, genetic scientists were studying the chromosomes of frogs, to access genetic maps and control reproduction, they selected all frogs as males. After a year of study, they realized that five frogs have been added to their number! After studying, they found out that some of them had genetic mutation and their gender changed every six months of the year. This marked the *second slogan* of chaos theory: "*Life will find its way to survive*" The phrase "*butterfly effect*" was created after an article by "*Edward Norton Lorenz*". In the 139th Climate Summit, he presented an article with the title: "*Can the flapping of a butterfly's wings in Brazil cause a strong wind gust in Texas?*" The equation that *Lorenz* arrived at was composed of dimensionless equations in the following form, which he succeeded in solving with numerical methods. In these equations, the result of the last output of one day was entered as the initial conditions of the next day.

$$\frac{dx}{dt} = \sigma(y - x), \quad \frac{dy}{dt} = x(\rho - z) - y, \quad \frac{dz}{dt} = xy - \beta z \tag{2.1}$$

2.2.2 *Characteristics of Chaotic System*

Chaos theory has fascinated scientists and researchers for two reasons.

- Chaos theory provides useful practical and experimental tools for understanding and classifying complex behaviors. While these behaviors cannot be justified with other topics.
- Chaos theory includes most natural phenomena (they are universal). This theory is the theory of nerve cells, optical systems, mechanical oscillators, etc.... contains

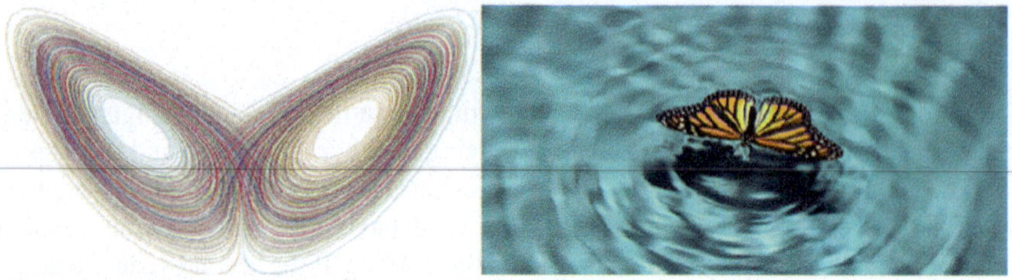

Fig. 2.2 Similarity of butterfly effect and Lorenz attractor

Fig. 2.3 Fractal structure for an equilateral triangle and a number of various shapes

Chaotic dynamic systems have the following characteristics:

- They are sensitive to initial conditions.
- Their alternating rotation is dense.
- They are combined topologically.

Sensitivity to initial conditions means that a small perturbation in the current path may lead to very different behavior in the future. Topological composition also means that over time, the system evolves in such a way that each region with an open set overlap with another certain region. Chaos theory needs a language to visualize its concepts and this forms the field that chaos geometry creates with fractals. In chaos geometry, we are dealing with different shapes that when we start drawing them, it is not clear what phenomenon we will end up with. For example, we can convert the equilateral triangle as shown in Fig. 2.3 into a map of a fractal structure.

2.2.3 *Bifurcation on Chaotic System*

Bifurcation is a qualitative change in dynamics that occurs due to a change in a system parameter. A bifurcation diagram shows the possible long-term values of a variable

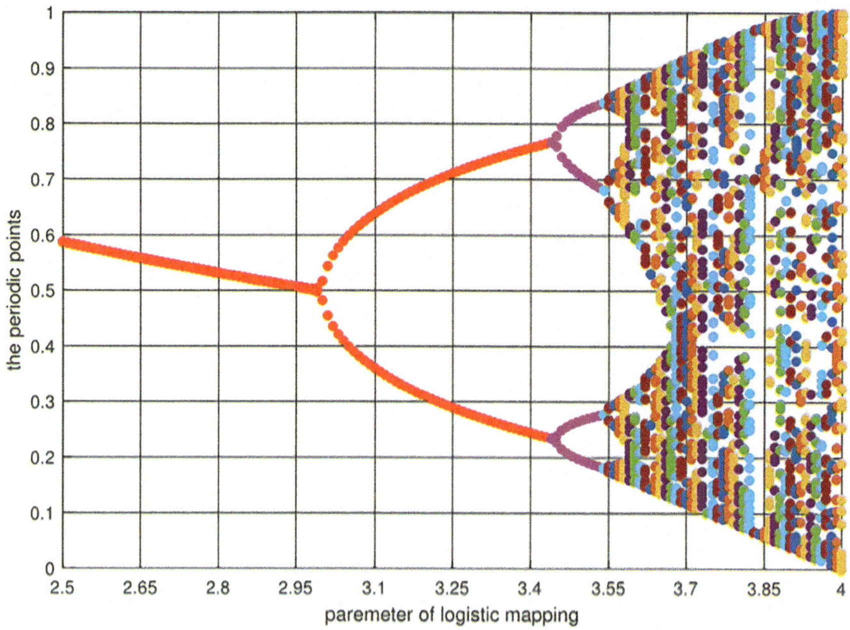

Fig. 2.4 Bifurcation of Logistic chaos map in the range of 3–3.57 for r parameter

of a system that can be obtained as a function of the system parameter. Figure 2.4 shows the Bifurcation of the Logistic chaos map.

2.2.4 Attractor on Chaotic System

The orbit of most chaotic functions is attracted to an orbit that is either periodic or chaotic through repeated iterations. The absorber is a set of points that, when the number of iterations increases, approaches the circuit, in other words, it will be a balanced solution to which the system converges. A strange absorber will contain an infinite number of bounded points in a given region of state space. Absorbers are the source of chaos. Lorenz was the first person to encounter a strange attractor and found that for certain ranges of a parameter, trajectories start from close initial conditions and move rapidly, leading to completely different future states. In fact, this issue expresses a strong and sensitive dependence on the initial conditions. Figure 2.5 shows an example of a chaotic attractor.

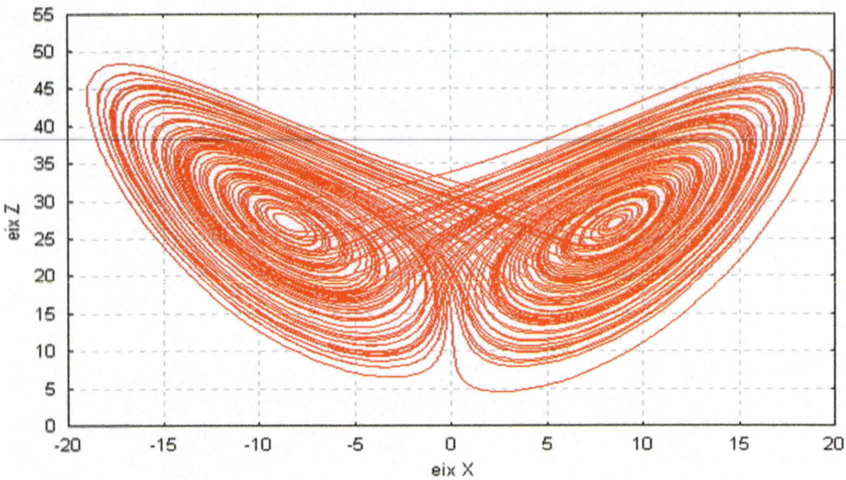

Fig. 2.5 Lorenz chaos attractor with bifurcation

2.3 Introduction Chaos Maps and Forming Chaos Series

In most meta-heuristic algorithms, in the standard mode the results are optimized slowly, and they stop when they reach the local optimal position. In such cases, premature convergence occurs. Chaotic series create suitable conditions to escape from the trap of local optima, which by creating disorder in the search space, it is possible to jump and settle in most vast and scattered positions of the search space [3]. Therefore, the local optimum will not have the opportunity to escape from their target. Chaos series consists of the arrangement of chaotic function sentences. These series have no traces of random behaviors, but they cause very disordered behaviors in the search space. One of the most important features of these series is sensitivity to initial conditions and non-periodic and ergodic behaviors, and the functions that make up the chaos series are very diverse and have no inverse. Some standard and well-known chaos maps are introduced in Table 2.1. For embedding chaos functions in meta-heuristic algorithms, some of them have good performance, which are: *Logistic, Tent, Gaussian, Liebovitch, Chebyshev, Sinusoidal and Piecewise*. In the chaos maps, Liebovitch converges with a very high probability from a local minimum to a general minimum. Therefore, this map is suitable for improving the exploration conditions of algorithms. The chaos Gaussian map is most likely in the local minimum range and is suitable for improving exploitation conditions. Finally, the chaotic map of Piecewise simultaneously improves both exploration and exploitation conditions. Therefore, by selecting these chaotic functions, the weakness of algorithms of any kind is improved. The numerical distribution of these chaos maps for 100 iterations is presented in Fig. 2.6. In the following, we will first introduce chaos functions and then chaos series to form chaos scenarios are presented.

Table 2.1 Introduction some standard and well-known chaos maps

1—Logistic map	17—Exponential map	33—Polynema type-A map	49—Ikeda map
2—Tent map	18—Finance system	34—Polynomic type-B map	50—Basin chaotic map
3—Gauss map	19—Folded-towel map	35—Polynomic type-C map	51—Hyper Newton-Leipnik
4—Liebovitch chaotic map	20—Fractal-dream system	36—Random rotate map	52—Hyper-Lu chaotic system
5—Chebyshev map	21 – Generalized Baker map	37—Sprott B chaotic system	53—Hyper-Rossler chaotic
6—Sinusoidal map	22—Grinch dragon fractal	38—Sprott-Linz A chaotic	54—Interval exchange map
7—Piecewise map	23—Henon map	39—Tangent map	55—Knot fractal map
8—Lorenz system	24—Hitzl-Zele map	40—Circular chaotic map	56—Pick over fractal map
9—Beta chaotic map	25—Horseshoe map	41—Tinkerbell map	57—Pulsed rotor
10—Busselton	26—Ikeda chaotic attractor	42—Tringle map	58—De Jong fractal map
11—Chen chaotic attractor	27—Lambic map	43—Ueda chaotic	59—Delayed-logistic system
12—Chen-Lee system	28—Linear map	44—Van der Pol oscillator	60—Hadley chaotic circulation
13—Chua circuit	29—Cells CNN system	45—Zaslofsky map	61—Hopa-Jong fractal
14—Complex cubic map	30—Lorenz map	46—Dyadic transformation	62—Hop along orbit fractal
15—Degenerate double rotor	31—Scroll-attractor	47—Complex squaring map	63—Lozi map
16—Duffing equation	32—Novel chaotic system	48—Duffing map	64—Lotka-Volterra system

2.3.1 Logistic Map

This map appears in nonlinear dynamic behaviors related to biological population. The statements of chaotic sequences in the Logistic map are obtained according to the following form:

$$X_{n+1} = a \cdot X_n \cdot (1 - X_n) \tag{2.2}$$

In this equation, X_n is the nth chaotic number, with n denoting the iteration number. Obviously, $X_n \in (0,1)$ under the conditions that initial $X_0 \in (0, 1)$ and that $X_0 \notin \{0.0, 0.25, 0.5, 0.75, 1\}$. In the experiments a = 4 is utilized.

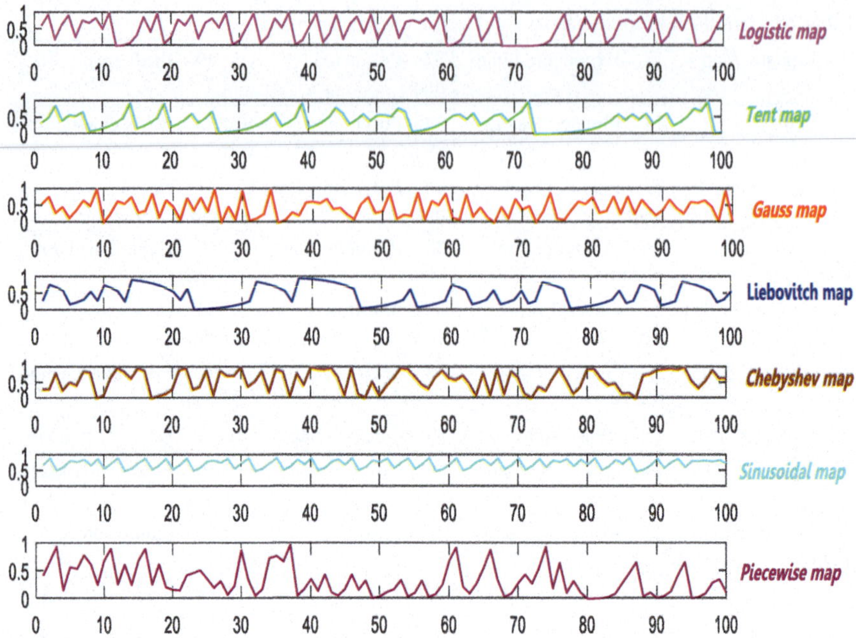

Fig. 2.6 Distribution of numerical values of chaos map in 100 iterations

2.3.2 Tent Map

The Tent map generates chaos signals in (0,1). This map is similar to Logistic in some respects and depicts very specific effects of chaotic behavior. The chaotic sequence sentences in this function are expressed by the following equation:

$$X_{n+1} = \begin{cases} \frac{X_n}{0.7} & if \quad X_n \prec 0.7 \\ \frac{10}{3 \cdot X_n}(1 - X_n) & otherwise \end{cases} \tag{2.3}$$

2.3.3 Gauss Map

Using this function in nonlinear dynamic behaviors has good results. The statements of chaotic sequences in the Gauss map are obtained according to the following form:

$$X_{n+1} = \begin{cases} 0 & X_n = 0 \\ \frac{1}{X_n} - \left[\frac{1}{X_n}\right] & X_n \neq 0 \end{cases} \tag{2.4}$$

2.3.4 Liebovitch Map

This function consists of three separate linear rules and there will be no common or repetitive sentences in these intervals. The chaotic sequence sentences in this function are expressed by Eq. (2.4):

$$X_{n+1} = \begin{cases} \alpha_1 X_n & 0 < X_n \le d_1 \\ \frac{d_2 - X_n}{d_2 - d_1}: & d_1 < X_n \le d_2 \\ 1 - \alpha_2 \times (1 - X_n) & d_2 < X_n \le 1 \end{cases} \tag{2.5}$$

Here the values $d_1 = 1/3$ and $d_2 = 2/3$ are selected. A_1 and α_2 are also calculated based on the Eq. (2.5):

$$\begin{aligned} \alpha_1 &= \frac{d_2}{d_1} \times (1 - (d_2 - d_1) \\ \alpha_2 &= \frac{1}{d_2 - 1}((d_2 - 1) - d_1(d_2 - d_1)) \end{aligned} \tag{2.6}$$

2.3.5 Chebyshev Map

The statements of chaotic sequences in the Chebyshev map are obtained according to the following form:

$$\begin{aligned} &X_{n+1} = \cos(n \cdot \cos^{-1} X_n) \\ &X_0 \in (0, 1) \\ &X_n \in \ nth\ chaotic\ number \end{aligned} \tag{2.7}$$

n denotes the iteration number. Sinusoidal.

2.3.6 Sinusoidal Map

This map is represented as follows:

$$X_{n+1} = a \cdot X_n^2 \cdot \sin(\pi \cdot X_n) \tag{2.8}$$

For a = 2.3 and $X_0 = 0.7$ it has the following simplified form:

$$X_{n+1} = \sin(\pi \cdot X_n) \tag{2.9}$$

It generated chaotic sequence in (0,1). Piecewise.

2.3.7 *Piecewise Map*

The piecewise map is the same as the Liebovitch map of three criteria, and the value of P is considered as the control parameter [2]. The range of changes of P is in the range of 0–0.5, which in this study we have considered 1/3, and the following category of equations indicates the relationship between the sentences of chaotic sequences in this function that expressed by Eq. (2.9):

$$X_{n+1} = \begin{cases} \frac{X_n}{p} & 0 \le X_n < p \\ \frac{X_n - p}{0.5 - p} & p \le X_n < \frac{1}{2} \\ \frac{1 - p - X_n}{0.5 - p} & \frac{1}{2} \le X_n < 1 - p \\ \frac{1 - X_n}{p} & 1 - p \le X_n < 1 \end{cases} \tag{2.10}$$

2.3.8 *Lorens Attractor System*

Lorenz attractor chaotic system is presented with differential equation according to Eq. 2.11.

$$\begin{aligned} \dot{x} &= \sigma \cdot (y - x) \\ \dot{y} &= x \cdot (\rho - z) - y \\ \dot{z} &= x \cdot y - \beta \cdot z \end{aligned} \tag{2.11}$$

We simulate this chaotic attractor system in MATLAB Simulink. This simulation includes three integrals, a number of Product and addition operators, constant values of sigma, rho and beta. The modeling structure in MATLAB Simulink is presented in Fig. 2.7. We consider the numerical values of these three parameters of the Lorenz attractor as follows:

$$Constant\ parameter \begin{cases} \sigma & \Rightarrow \quad sigma = 10 \\ \rho & \Rightarrow \quad rho = 28 \\ \beta & \Rightarrow \quad beta = \frac{8}{3} \end{cases} \tag{2.12}$$

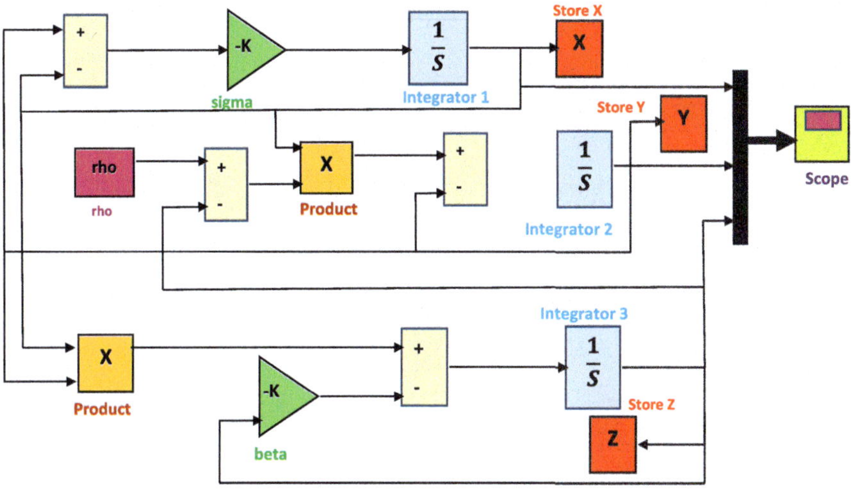

Fig. 2.7 Simulation of Lorenz attractor system in MATLAB Simulink

Lorenz's attractor chaotic function has a completely "*deterministic*" behavior and has two bifurcation that alternately takes over one of these bifurcations and is constantly going back and forth between the two. This system has the potential to investigate problems it is complex. And it has high efficiency to improve the optimization results. Among its other applications are cryptography, secure telecommunication control and image processing. The Lorenz attractor diagram is as shown in Fig. 2.8.

2.4 Chaos Series and Alternative Scenarios

In order to embed chaos functions in meta-heuristic algorithms, first, the number of sentences related to chaos series is determined in the exploration and exploitation parts, and then, based on the chosen chaos function and the number of sentences required, two groups of chaos series including **CHM1** and **CHM2** are formed [4–6]. For the first scenario, **CHM1** chaos series sentences replaces the random parameters related to the exploration phase, for the second scenario, **CHM2** chaos series sentences replaces the random parameters related to exploitation. Ultimately, for the third scenario, the simultaneous use of the **CHM1** and **CHM2** chaos series in both the exploration and exploitation stages are considered.

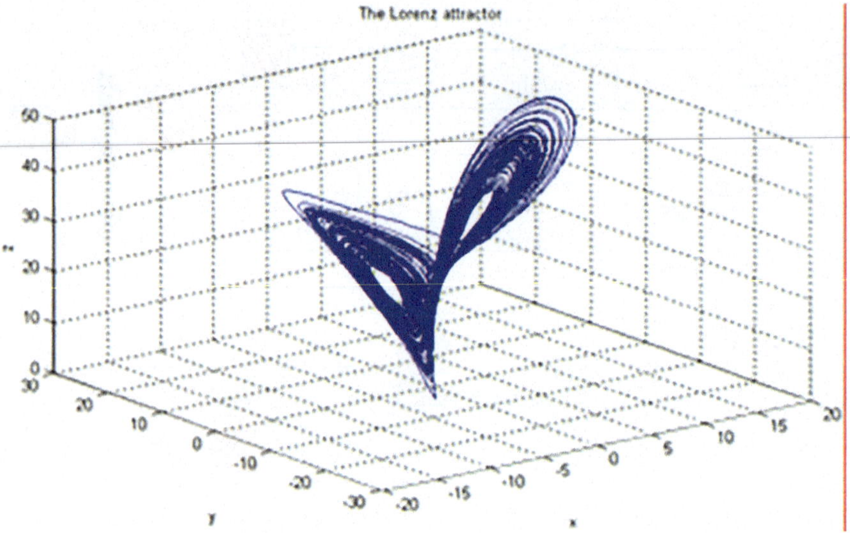

Fig. 2.8 The Lorenz chaotic attractor diagram

2.5 Meta-heuristic Algorithms and Chaos Map

Every meta-heuristic algorithm goes through two main stages of exploration and exploitation in the optimization stages to converge towards the optimal results. In other words, it first settles in scattered parts of the search space and then examines their neighborhoods. As an example, in the Genetic Algorithm, in the stage of mutation, the establishment takes place in scattered areas of the search space, and in the stage of crossover, in the neighborhood, the movement takes place towards better answers. Similarly, in the Imperialist Competitive Algorithm, settling in different parts of the search space with the action of revolution and moving in the neighborhood towards better answers is performed with the policy of assimilation. In both stages, random coefficients are predicted to produce diversity in the search space and their values are recommended based on a specific probabilistic distribution. In some meta-heuristic algorithms, uniform probabilistic distribution is uitalized, which naturally can be an unsuitable distribution, however, a Normal or Cauchy distribution, is associated with excellent results [7]. Especially the normal distribution having a very wide range, results in a good diversity. Since the random selection is always accompanied by doubt, the idea of using chaos maps that their values are definite and alternatives are good for random distributions. Scientific studies have published valuable results and considerable improvements for them. Chaos systems provide an inclination towards global responses and prevents them from falling into the trap of local optima [3, 8]. If these conditions are fulfilled, one will not observe premature convergence. Lorenz points out the following important points about series formed by chaos maps: Although series created by chaotic functions appear to be similar to sentences with random distributions, however, there are major differences. Some of these differences are: their values are deterministic, and have nonlinear behavior. In dynamic state, the sentences of the related series are non-repetitive, mostly have no inverse functions and finally non-convergent to a certain boundary [2]. Now, by replacing chaos series in meta-heuristic algorithms, the nonlinear and non-convex behavior of the objective function in truss weight optimization can easily be controlled and adjusted. In this book, in order to investigate the effects of chaos series in improving the optimization results of the algorithms, by selecting 11 well-known meta-heuristic algorithms in Chaps. 3–13, a wide range of chaos modes are formed and a challenging competition has been achieved. The selection of meta-heuristic algorithms is such that the standard mode of the algorithm has excellent efficiency in at least one of the exploration or exploitation cases to compensate for the weakness related to the other case by chaotic series.

2.6 Concluding Remarks

In meta-heuristic algorithms in standard mode, two important stages of exploration and exploitation are considered in order to converge towards optimal answers. The imbalance between these two steps causes us to improve the efficiency of the algorithm by embedding chaos functions instead of random numbers. Embedding chaos functions in exploration, exploitation, or both at the same time, with three scenarios. Because chaos systems play the main role in improving the performance of meta-heuristic algorithms, we prepare different types of them and embed them on algorithms, so that by creating a competition between them, the most suitable chaos function for the algorithm is selected in this competition. Accompanying chaotic functions causes sudden shocks and jumps from local optima. Therefore, the algorithm is saved from the trap of local optima and converges towards global optima. Therefore, premature maturity does not occur, and the algorithm is saved from the trap of local optima. In this book, in order to investigate the effects of chaos series in improving the optimization results of algorithms, by selecting 11 famous meta-heuristic algorithms in Chaps. 3–13, a wide range of chaos modes has been formed. Therefore, with a challenging competition between chaos functions, the most suitable ones will be introduced.

References

1. Kaveh, A.: Advances in Meta-heuristic Algorithms for Optimal Design of Structures, 2nd edn., 3rd edn. Springer, Cham, Switzerland (2021). https://doi.org/10.1007/978-3-319-05549-7
2. Peitgen, H.-O., Jürgens, H., Saupe, D.: Chaos and Fractals: New Frontiers of Science. Springer, Berlin (2006)

3. Kaveh, A., Sheikholeslami, R., Talatahari, S., Keshvari-Ilkhichi, M.: Chaotic swarming of particles: a new method for size optimization of truss structures. Adv. Eng. Softw. **67**, 136–47 (2014)
4. Kaveh, A., Yousefpoor, H.: Chaos-embedded meta-heuristic algorithms for optimal design of truss structures. Scientia Iranica A **29**(6), 2868–2885 (2022). https://doi.org/10.24200/sci.2022.59812.6441
5. Kaveh, A., Yosefpoor, H.: Chaotically enhanced meta-heuristic algorithms for optimal design of truss structures with frequency constraints. Periodica Polytechnica Civil Eng. **66**(3), 900–921 (2022). https://doi.org/10.3311/PPci.20220
6. Kaveh, A., Zarfam, P., Aziminejad, A., Yosefpoor, H.: Comparison of four chaotic meta-heuristic algorithms for optimal design of large-scale truss structures. Iran. J. Sci. Technol. Trans. Civil Eng. https://doi.org/10.1007/s40996-022-00908-8
7. Ott, E.: Chaos in Dynamical Systems. Cambridge University Press, Cambridge (2002)
8. Bucolo, M., Caponetto, R., Fortuna, L., Frasca, M., Rizzo, A.: Does chaos work better than noise? IEEE Circuit Syst. Mag. **2**(3), 4–19 (2002)

Chapter 3
Chaotic Cyclical Parthenogenesis Algorithm

3.1 Introduction

Meta-heuristic algorithms have gained more success in novel methods to access the optimization of complex problems in engineering sciences. These algorithms are usually inspired by natural phenomena and the reason is the unique characteristics of these phenomena. What is evident about nature, is the fact that if there is a point of empowerment, nature does it in the best way. The genetic evolution and reproduction of living organisms with the two operators of crossover and mutation, the swarm intelligence of animal and their nature in foraging, inspiration from physical laws and other inspirations such as political behaviors, modeling of education in the classroom, the migration process in wildlife habitats, each has brought merit results in the modeling of meta-heuristic algorithms [1]. Although meta-heuristic algorithms have various sources and inspirations, the common feature of all of them is that they have two basic steps of exploration and exploitation. In the exploration phase, meta-heuristic algorithms are established on the privileged points of the decision space and in the exploitation phase, they carefully examine the neighborhood of these points. Although significant progress is evident in meta-heuristic algorithms, these algorithms are always threatened by pests. Falling into the trap of local optima and stopping in them, premature convergence and slowing down of computational steps can be improved by applying sudden jumps related to chaos functions. The research shows that in most algorithms, the effort is to achieve a balance between the exploration and exploitation stages, but the suggested tricks of the algorithm are not enough in the standard mode, and by using alternative chaos series instead of the random parameters of the algorithm, a noticeable improvement in the optimization results and moving towards global optimality is achieved. Chaos functions can bring this improvement to meta-heuristic algorithms with different scenarios. In some cases, chaos functions replace the random parameters of the exploration stage, and in some other algorithms, we will need chaos functions in the exploitation stage. Even in some algorithms, we may use chaos functions simultaneously in both

A. Kaveh and H. Yousefpoor, *Chaotic Meta-heuristic Algorithms for Optimal Design of Structures*, Studies in Computational Intelligence 1129,
https://doi.org/10.1007/978-3-031-48918-1_3

stages. In this chapter, a new meta-heuristic algorithm is examined, which considers the social behavior of animal species such as aphids. Aphids are highly successful in reproduction from the point of view of animal science. With the mechanism of sexual and asexual reproduction and moving between the two in different environmental conditions, they have achieved greater compatibility with the surrounding environment and have won the competition with other organisms. The imbalance in the two stages of sexual and asexual reproduction is provided by using appropriate chaos functions and a significant improvement in the optimization results will be achieved. In order to find the chaos function with appropriate distribution, logistic, Tent and Gaussian chaos functions will be investigated with three scenarios. In the rest of this chapter, after introducing the Cyclical Parthenogenesis Algorithm (CPA), the basic steps of the algorithm will be presented in the standard mode, and then the Chaotic Cyclical Parthenogenesis Algorithm (CCPA), will be formed with three scenarios. In the end, the effectiveness of chaos functions will be challenged with well-known optimization examples of static constraints and multi-frequency constraints.

3.2 Standard Cyclical Parthenogenesis Algorithm (CPA)

This algorithm was presented by Kaveh et al. In this algorithm [2], the key aspects of reproduction in aphids have been discussed. Their ability to reproduce sexually and asexually provides the conditions to quickly grow a population of aphids and take advantage of favorable conditions. Research shows that in sexual reproduction, two different solutions share information, while in asexual reproduction, a new solution is produced solely using information from a female parent. Applying this inspiration in meta-heuristic algorithms creates the best conditions for the balance between exploration and exploitation and provides a high ability to escape from local optima and move towards global optima. In the stage of sexual reproduction, the vector of initial responses is located in the privileged areas of the decision space, and in the stage of asexual reproduction, it examines the neighborhood of privileged responses very carefully. In order to improve the results, a number of chaos functions replace the random selection of these dual stages and the results of the chaotic modes are compared with the standard mode. Considering that the standard mode competes with 9 chaotic modes, therefore a wide range of response space is obtained and the global optimality cannot be far from the target of chaotic algorithms.

3.3 Basic Steps of Cyclical Parthenogenesis Algorithm

Step 1. Creating an initial population of aphids: Within the scope of the search space, a population of aphids is formed as the initial population. The selection of this population is done randomly. In the following relation, the method of selecting the initial population is presented from Eq. 3.1:

$$x_{ij}^{0} = x_{j,\min} + rand(x_{j,\max} - x_{j,\min}) \quad j = 1, 2, ..., nA \tag{3.1}$$

where X_{ij}^{0} denotes the jth variable of the ith population of aphids, $X_{j,max}$ and $X_{j,min}$ respectively expresses the upper and lower bounds of the variable j. The total population of aphids is nA, which are located in colonies with nC number and each with nM population. It is clear that nM = nA/nC and nM is constant during the optimization operation.

Step 2. Formation of the population of children: To form the population of offspring in each colony, the number of Fr × nM offspring without mating is formed. The parents of these children are female and their selection is done randomly and from the best answers. In MATLAB coding, this selection and formation of the children population is as Eqs. (3.2) and (3.3).

$$rf_i = round(1 + (Fr \times nM - 1). \times \mathbf{rand}) \tag{3.2}$$

$$x_{ij}^{k+1} = F_j^k + \alpha_1 \times \frac{randn}{NITs} \times (x_{j,\max} - x_{j,\min}) \quad j = 1, 2, ..., n\text{var} \tag{3.3}$$

In relation to Eq. (3.2), the index related to the female parent is determined and in relation to Eq. (3.3) new children related to the state without mating, it is placed in the new cell array. Now it is time to form offspring by mating. The number of these offspring is (1 − Fr) × nM in which each male parent M randomly selects a female parent F and is placed in a new cell array according to the relationship of Eq. (3.4) new children related to mating.

$$x_{ij}^{k+1} = M_j^k + \alpha_2 \times \mathbf{rand} \times (F_j^k - M_j^k) \quad j = 1, 2, ..., n\text{var} \tag{3.4}$$

Step 3. Fly the best aphid and die the worst aphid: After the formation of a new generation of offspring, the target function is evaluated and with a probability of Pf, one of the best winged aphids is selected from colony 1 and by reproducing it, it replaces the worst aphid in colony 2. To keep the colony population stable, removing the worst colony 2 aphid is compared to the death of the aphid and replacing the best aphid with flying. The probability of this step is based on Eq. (3.5):

$$pf = \frac{NITs - 1}{\max\ NITs - 1} \tag{3.5}$$

Step 4. Replacing the best aphid: In each colony, the population of parents is compared with that of children, and the number of nM from the best is selected to form the next generation.

Step 5. Check the termination conditions and repeat the operation from Step 2 if necessary.

3.4 Chaos Enhanced Cyclical Parthenogenesis Algorithm (CCPA)

In this algorithm, two important modes are selected to form the population of offspring, including reproduction with and without mating [3–5]. These two steps have the role of exploration and exploitation of the algorithm. By replacing chaos maps instead of random selection, we will have a significant improvement in optimization results. This replacement is done with the following proposed scenarios:

Scenario 1. Apply the chaos map in reproduction stage without mating: In this case, the first chaos map ***CHM1*** is applied in Eq. (3.2) and replaces the random selection of the algorithm that the result will be Eq. (3.6):

$$rf_i = round(1 + (Fr \times nM - 1) \times \mathbf{CHM1}) \tag{3.6}$$

Scenario 2. Apply the chaos map in reproduction stage with mating: In this case, the second chaos map ***CHM2*** is applied in Eq. (3.4) and replace the random selection of the algorithm the result will be Eq. (3.7):

$$x_{ij}^{k+1} = M_j^k + \alpha_2 \times \mathbf{CHM2} \times (F_j^k - M_j^k) \quad j = 1, 2, \ldots, n\text{var} \tag{3.7}$$

Scenario 3. Apply the chaos maps in both steps simultaneously: In this case, both chaos maps are applied simultaneously in Eqs. (3.2) and (3.4) and replace the random selection of the algorithm.

3.5 Constructive Role of Chaos Functions in Sensitivity Analysis

In the algorithm of Cyclical parthenogenesis Algorithm, the Diversity Index is very important. This index is performed at each stage of calculation and sensitivity analysis regarding the number of aphids, the number of colonies and other parameters used in the algorithm. In chaotic algorithms, we will no longer need sensitivity analysis. The only thing that chaotic algorithms are sensitive to is the starting number of the chaotic series. In this regard, before the main optimization, the pre-processing operation is performed to determine the appropriate value of the starting sentence of the chaos series.

3.6 Truss Weight Optimization with Static Constraints

In order to form chaotic algorithms, the number of three chaotic functions is selected and embedded in the Cyclical parthenogenesis Algorithm. This function includes logistic, Tent and Gaussian chaos function. Also, each of these functions will be accompanied by three scenarios. In these scenarios, chaos functions respectively replace the distributions related to exploration, exploitation or both stages simultaneously. 9 chaotic modes and standard mode will compete with each other to determine the best efficiency. In order to increase the diversity in the statistical space, each model is run independently 30 times and the statistical population corresponding to the best value, the best average and the coefficient of variation is extracted. By examining the effectiveness of chaotic models and triple scenarios, a significant improvement in the optimization results is achieved, and we can determine the chaos function compatible with the algorithm that creates the balance between the exploration and exploitation stages with the best efficiency. We can also learn about the weakness of the algorithm in the exploration or exploitation stages.

3.6.1 Formulation of the Structural Optimization Problems

Optimization problems for truss structures include objective function, design constraints, and bound of decision variables. In optimization relationships, the weight of the structure is selected as the objective function and all design constraints must be satisfied with the least possible weight. The general form related to these problems is defined as Eq. (3.8):

$$\begin{aligned}
&Find && A = \{A_1, A_2, A_3, \ldots, A_n\} \\
&to\ Minimize && W(A) = \sum_{i=1}^{n} \gamma_i \times A_i \times L_i \\
&Subjected\ to && g_j(A) \leq 0; \quad j = 1, 2, 3, \ldots, m \\
& && h_k(A) \leq 0; \quad k = 1, 2, 3, \ldots, p \\
& && \{A_L\} \leq \{A\} \leq \{A_U\}
\end{aligned} \tag{3.8}$$

In this category of Eq. (3.8), A is the cross-section vector of the members, W is the total weight of the structure, n is the number of members of the structure, g_i and h_i are design constraints that can include stress, slender constraints and node displacement. Also, A_i and A_u are the lower and upper bounds of the decision variables. Despite the selected algorithms are used to optimize unbounded problems. In modeling, the penalty function method is used to convert the bound function to the unbound function. In this method, if there is no violation, the amount of the fine will be zero. Otherwise, and if there is a violation, the value of the penalty function is obtained from Eqs. (3.9) to (3.13):

$$\sigma_i \leq \sigma^{\max} \Rightarrow V_i = \max(0, \frac{\sigma_i}{\sigma^{\max}} - 1); \quad i = 1, 2, 3, \ldots, n \tag{3.9}$$

$$\delta_j \leq \delta^{\max} \Rightarrow V_j = \max(\frac{\delta_j}{\delta^{\max}} - 1); \quad j = 1, 2, 3, \ldots, n \tag{3.10}$$

$$\lambda_k \leq \lambda^{\max} \Rightarrow V_k = \max(0, \frac{\lambda_k}{\lambda^{\max}} - 1); \quad k = 1, 2, 3, \ldots, p \tag{3.11}$$

$$F_{penalty}(A) = 1 + \gamma_p \times \left\{ \sum_{i=1}^{m} V_i + \sum_{j=1}^{n} V_j + \sum_{k=1}^{p} V_k \right\} \tag{3.12}$$

$$to\ Minimize\ Mer(A) = W(A) \times F_{penalty}(A) \tag{3.13}$$

Equations (3.9)–(3.11) are related to stress, displacement and slenderness, respectively. The penalty function is presented in Eq. (3.12) and the objective function is formed after the penalty (merit function) in Eq. (3.13).

3.6.2 *Introduction of Selected Chaos Map*

Chaotic map shown no signs of random behavior, yet they cause very erratic behaviors in the environment. The most important features of these map are sensitivity to initial conditions and non-periodic and ergodic behavior can be introduced [4]. The use of chaos sequences to evolve variables has significant advantages over other methods. In deterministic searches, compared to random search, more speed and convergence towards the general answer is achieved. In this research, the Logistic, Tenet and Gauss maps have been selected as shown in Fig. 3.1. In a Logistic map it is very probable to converge from a local to a global minimum. Therefore, this map is suitable for improving the exploration specifications of algorithm. The Gaussian map is most probable in the local minimum range and is suitable for improving exploitation attribute.

Finally, Tenet map simultaneously improves both conditions. Therefore, by selecting these maps, the weakness of algorithms of any kind is improved. How to apply then to meta-heuristic algorithms is shown in the flowchart of Fig. 3.2.

3.6.2.1 Logistics Map

This map appears in nonlinear dynamic behaviors related to biological populations [6]. The sentences of chaotic sequences in the logistic map are obtained according to the following equations:

$$CHM_{k+1} = \mathrm{a} \times CHM_k(1 - CHM_k) \tag{3.14}$$

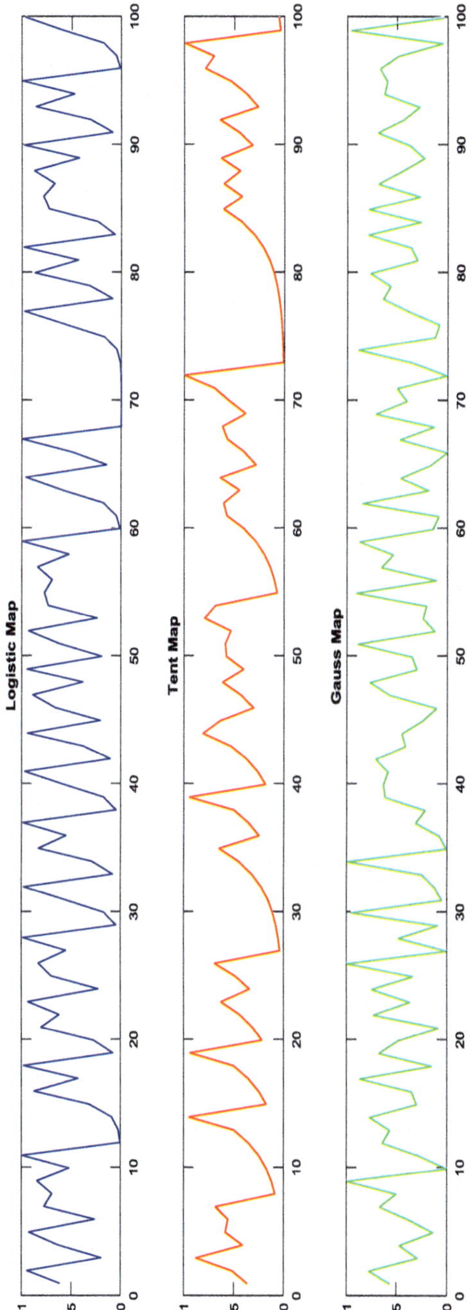

Fig. 3.1 The Chaotic value distribution during 100 iterations

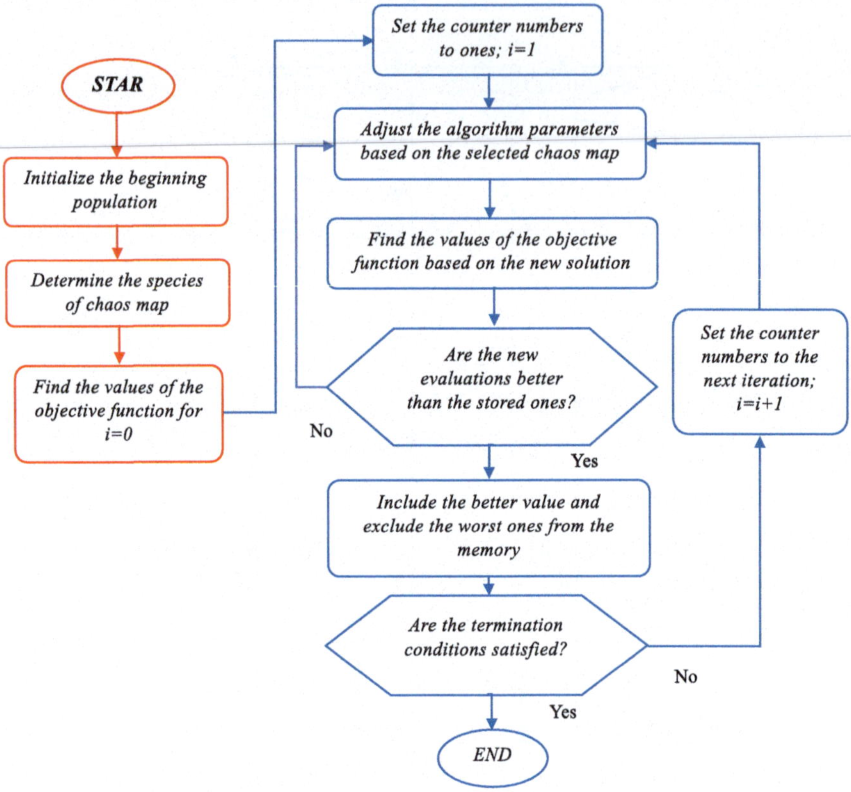

Fig. 3.2 The flowchart for the chaos algorithm

$$CHM_{k+1} \in (0, 1); \quad CHM_0 \in (0, 1); \quad CHM_0 \notin (0, 0.25, 0.5, 0.75, 1) \tag{3.15}$$

In the performed studies, a = 4 has been considered.

3.6.2.2 Tent Map

This map is similar to logistics in some respects and depicts very specific effects of chaotic behavior. The chaotic sequence sentences in this function are expressed by the following equation:

$$CHM_{k+1} = \begin{cases} \frac{CHM_k}{0.7} & CHM_k < 0.7 \\ \frac{10}{3} \times CHM_k(1 - CHM_k) & CHM_k \geq 0.7 \end{cases} \tag{3.16}$$

3.6.2.3 Gauss Map

Using this function in nonlinear dynamic behaviors has good results [7]. The statements of chaotic sequences in the Gaussian function are obtained according to Eq. (3.17).

$$CHM_{k+1} = \begin{cases} 0 & CHM_k = 0 \\ \frac{1}{CHM_k} - \left[\frac{1}{CHM_k}\right] & CHM_k \neq 0 \end{cases} \quad (3.17)$$

3.6.3 Numerical Examples of Optimal Truss Design

In this study, the purpose of the optimal design of trusses is to find the optimal value for the cross-section of members that have less weight and in addition to the limitations related to allowable stress, allowable deformation of joints and slenderness of members according to regulations to be. Famous optimization examples have been selected in this regard and the efficiency of algorithms, chaos map and proposed scenarios have been investigated. These examples contain:

- The 52-bar planar truss
- The 72-bar spatial truss
- The 120-bar dome shaped truss

3.6.3.1 A 52-Bar Planar Truss

The topology and nodal numbering of the 52-bar planer truss are shown in Fig. 3.3. The material density is 7860 kg/m^3 and the modulus of elasticity is 207 GPa. The members are subjected to stress limitation 180 MPa in both tension and compression. For loading, the intensity of the loads in the X and Y directions is 100 and 200 kN, respectively, and both loads are applied simultaneously. Also, truss members are classified into 12 groups according to geometric symmetry. The cross-section areas of the members were chosen from the 64 discrete values of the AISC code. In order to ensure the performance of chaos map and algorithms, as well as to increase the accuracy and sensitivity of calculations, each of the modes has been performed independently 30 times and the results related to the best response and the average value of responses are presented in Table 3.1. The graph of these results is also shown in Fig. 3.4. The coefficient of variation of the responses, which indicates the stability and robustness of the responses, has been calculated and another criterion for the efficiency of chaos map and algorithms has been obtained. After reviewing the optimization results for different combinations of algorithms with chaos map and comparing it with the initial and standard mode, a significant improvement in reducing the weight of the 52-bar planar truss results. The enhanced results for chaotic

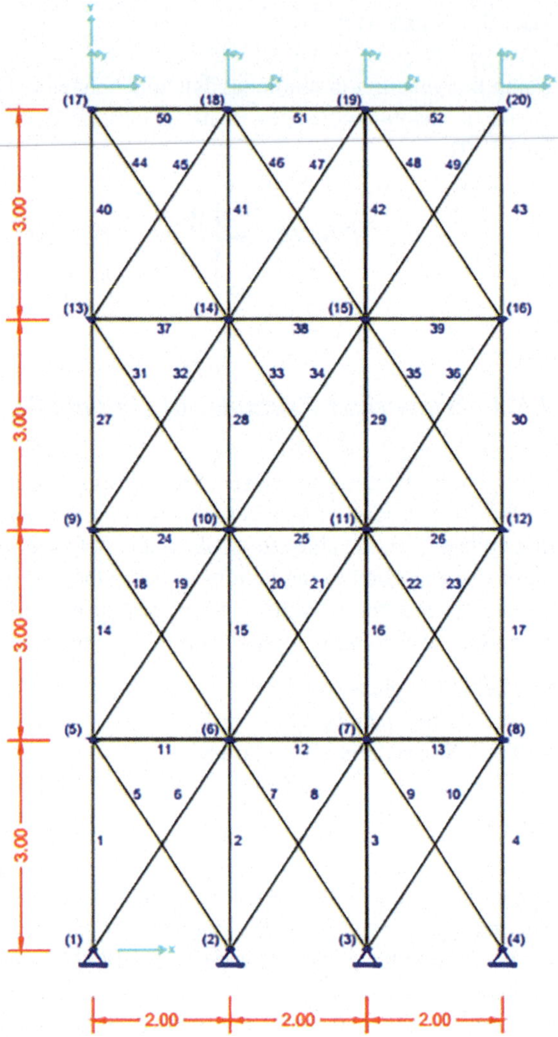

Fig. 3.3 Schematic of a 52-bar planar truss

cyclical parthenogenesis algorithm by logistic chaos map and scenario 2, a weight of 1902.605 kg has the optimal response.

3.6.3.2 A 72-Bar Spatial Truss

The topology and nodal numbering of the 72-bar spatial truss are shown in Fig. 3.5. The material density is 0.1 lb/in^3 and the modulus of elasticity is 10,000 ksi. Permissible members' stress of 25 ksi is assumed. The maximum displacement for each of the upper nodes is limited at 0.25 in in all directions. The two loading conditions

Table 3.1 Optimal design comparison for the 52-bar planar truss

Number group	Element group	Standard	Logistic map			Tent map			Gauss map		
		CPA	CCPA 21	CCPA 22	CCPA 23	CCPA 31	CCPA 32	CCPA 33	CCPA 41	CCPA 42	CCPA 43
1	A_1–A_4	4658.055	4658.055	4658.055	4658.055	4658.055	4658.055	4658.055	4658.055	4658.055	4658.055
2	A_5–A_{10}	1283.868	1161.288	1161.288	1161.288	1161.288	1283.868	1161.288	1161.288	1161.288	1283.868
3	A_{11}–A_{13}	494.193	494.193	494.193	494.193	494.193	494.193	388.386	506.451	494.193	494.193
4	A_{14}–A_{17}	3303.219	3303.219	3303.219	3303.219	3303.219	3303.219	3303.219	3303.219	3303.219	3303.219
5	A_{18}–A_{23}	1008.385	939.998	939.998	939.998	939.998	939.998	1008.385	1045.159	939.998	939.998
6	A_{24}–A_{26}	494.193	494.193	494.193	506.451	494.193	494.193	285.161	494.193	506.451	494.193
7	A_{27}–A_{30}	2290.318	2290.318	2238.705	2238.705	2290.318	2238.705	2290.318	2238.705	2238.705	2290.318
8	A_{31}–A_{36}	1008.385	1008.385	1008.385	1008.385	1008.385	1008.385	1008.385	1008.385	1008.385	1008.385
9	A_{37}–A_{39}	494.193	506.451	494.193	494.193	494.193	494.193	494.193	494.193	494.193	494.193
10	A_{40}–A_{43}	1374.191	1374.191	1283.868	1374.191	1374.191	1283.868	1283.868	1374.191	1283.868	1374.191
11	A_{44}–A_{49}	1161.288	1161.288	1161.288	1161.288	1161.288	1161.288	1161.288	1161.288	1161.288	1161.288
12	A_{50}–A_{52}	494.193	494.193	494.193	494.193	506.451	494.193	494.193	494.193	494.193	494.193
Best weight		1948.464	1916.571	**1902.605**	1911.702	1916.571	1923.450	1910.218	1929.584	1903.184	1936.836
Mean weight		1951.439	1920.324	**1907.341**	1924.241	1927.314	1932.381	1931.152	1938.381	1911.823	1939.856
Coefficient variation		**0.14103**	0.16771	0.16904	0.3538	0.28820	0.27764	0.54033	0.27933	0.25465	0.15886

The bolded items in the table indicate the optimal values for the best weight, the best mean and the best coefficient of variation

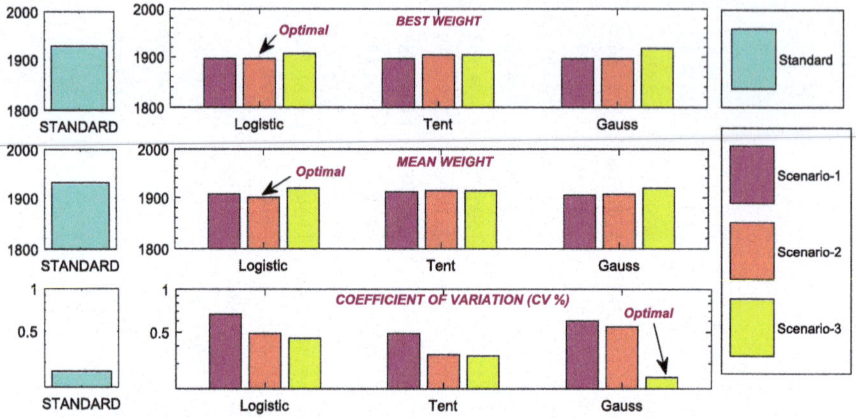

Fig. 3.4 Optimization results in standard and chaos map for the 52-bar planar truss

are applied according to Table 3.2. Also, truss members are classified into 16 groups according to considering the geometric symmetry. The cross section of the members, which are the same as the decision variables, is considered in the range of 0.1–4 in^2.

In order to ensure the performance of chaos map and algorithms, as well as to increase the accuracy and sensitivity of calculations, each of the modes has been performed independently 30 times and the results related to the best response and the average value of responses are presented in Table 3.3. A graph of these results is also shown in Fig. 3.6. The coefficient of variation of the responses, which indicates the stability and robustness of the responses, has been calculated and another criterion for the efficiency of chaos map and algorithms has been obtained. Examining the optimization results for different combinations of algorithms with chaos map and comparing it with the initial and standard modes, shows a significant improvement in reducing the weight of the 72-bar spatial truss. The enhanced result for cyclical parthenogenesis algorithm by the logistic chaos map and scenario 2 weighing 379,072 lb has the optimal response.

3.6.3.3 A 120-Bar Dome Shaped Truss

The topology and nodal numbering of the 120-bar dome shaped truss shown in Fig. 3.7. The material density is 0.288 lb/in^3 and the modulus of elasticity is 30,450 lb/in^2. Permissible members' tensile and compressive stresses are proposed in accordance with AISC ASD code. The maximum deformation for each node in all directions is set at 0.1969 in. All non-abutment nodes are affected by vertical loading, with intensities of − 13.49 kips in node one and − 6.744 kips in spheres 2–14 and 2.248 kips in the other nodes, respectively. Also, truss members are classified into seven groups according to their shape and geometric symmetry. The minimum cross-sectional area of the members, which is the lower limit of the decision variables, is

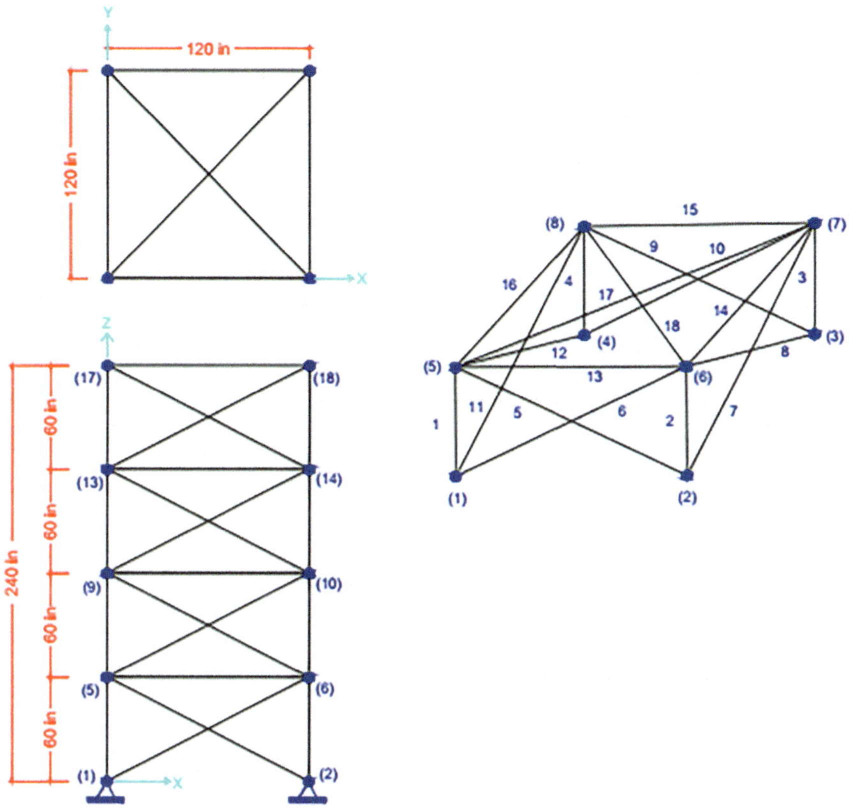

Fig. 3.5 Schematic of a 72-bar spatial truss and typical story

Table 3.2 Loading conditions for the 72-bar spatial truss

Node	Load case 1			Load case 2		
	F_x (kips)	F_y (kips)	F_z (kips)	F_x (kips)	F_y (kips)	F_z (kips)
17	5	5	− 5	0	0	− 5
18	0	0	0	0	0	− 5
19	0	0	0	0	0	− 5
20	0	0	0	0	0	− 5

0.775 in^2, and the maximum cross-sectional area is 20 in^2. The allowable stress in tension and pressure is determined based on the following equations.

$$\sigma_i = \begin{cases} 0.6F_y & for \quad \sigma_i \geq 0 \\ \sigma_i^- & for \quad \sigma_i < 0 \end{cases} \tag{3.18}$$

Table 3.3 Optimal design comparison for the 72-bar spatial truss

Number group	Element group	Standard	Logistic map			Tent map			Gauss map		
		CPA	CCPA 21	CCPA 22	CCPA 23	CCPA 31	CCPA 32	CCPA 33	CCPA 41	CCPA 42	CCPA 43
1	A_1–A_4	1.9005	1.8900	1.8937	1.8686	1.9127	1.8576	1.8306	1.8730	1.8986	1.8797
2	A_5–A_{12}	0.5146	0.51291	0.51255	0.50821	0.5130	0.51334	0.5166	0.51137	0.51167	0.51478
3	A_{13}–A_{16}	0.1000	0.1000	0.1000	0.10009	0.10001	0.10021	0.10018	0.1000	0.1000	0.10004
4	A_{17}–A_{18}	0.1000	0.1000	0.1000	0.1000	0.1000	0.10049	0.10043	0.1000	0.10001	0.10001
5	A_{19}–A_{22}	1.2782	1.2777	1.2696	1.2712	1.2806	1.2811	1.3224	1.2751	1.2536	1.2822
6	A_{23}–A_{30}	0.4964	0.50998	0.50575	0.51243	0.5071	0.52035	0.51631	0.51453	0.51344	0.51102
7	A_{31}–A_{34}	0.10109	0.1000	0.10002	0.10000	0.10002	0.1001	0.10003	0.10001	0.10011	0.1000
8	A_{35}–A_{36}	0.1000	0.10006	0.1000	0.10006	0.1000	0.10056	0.10007	0.1000	0.10007	0.1000
9	A_{37}–A_{40}	0.59545	0.52693	0.50509	0.51357	0.52216	0.53018	0.5163	0.52007	0.51038	0.51739
10	A_{41}–A_{48}	0.51926	0.51427	0.51736	0.52432	0.51413	0.51741	0.50553	0.51587	0.52235	0.5183
11	A_{49}–A_{52}	0.10005	0.1000	0.10006	0.1000	0.1000	0.10001	0.10008	0.1000	0.10001	0.1001
12	A_{53}–A_{54}	0.10343	0.10011	0.10006	0.1000	0.10023	0.10400	0.10002	0.10003	0.10003	0.1000
13	A_{55}–A_{58}	0.15573	0.15666	0.15661	0.15665	0.15665	0.15657	0.15645	0.15655	0.15698	0.15678
14	A_{59}–A_{66}	0.53895	0.53965	0.54881	0.54464	0.54221	0.53766	0.55442	0.5460	0.53874	0.54031
15	A_{67}–A_{70}	0.41679	0.41098	0.41713	0.4059	0.40799	0.39967	0.38818	0.40286	0.41189	0.40736
16	A_{71}–A_{72}	0.55317	0.57226	0.55779	0.57257	0.56904	0.56709	0.58259	0.5709	0.57064	0.5733
Best weight		379.933	379.075	**379.072**	379.223	379.286	379.200	379.468	379.177	379.312	379.390
Mean weight		380.387	379.400	**379.246**	379.422	380.229	380.229	383.170	379.473	379.512	379.548
Coefficient variation		0.10519	0.05553	0.04491	0.04799	0.03593	0.2837	0.3258	0.04989	0.02974	**0.02458**

The bolded items in the table indicate the optimal values for the best weight, the best mean and the best coefficient of variation

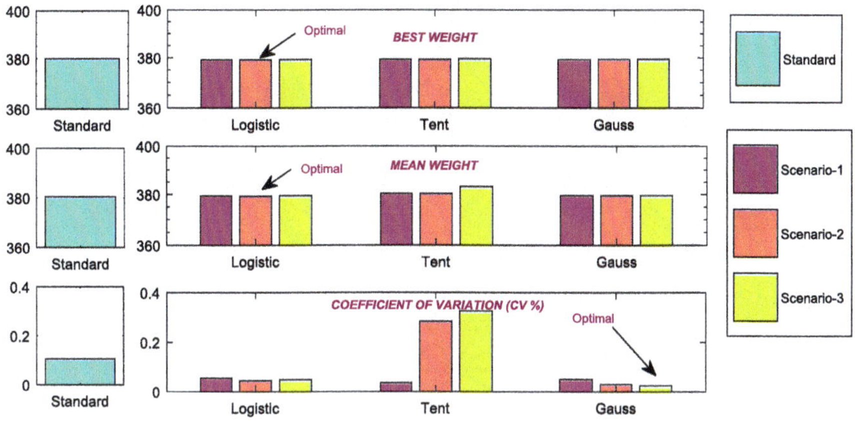

Fig. 3.6 Optimization results in standard mode and chaos map for the 72-bar spatial truss

For compressive stresses, we have the following equations:

$$\sigma_i^- = \begin{cases} (1 - \frac{\lambda^2}{2C^2})F_y/FS & for \quad \lambda < C \\ \frac{12\pi^2 E}{23\lambda^2} & for \quad \lambda \geq C \end{cases} \tag{3.19}$$

In this equation we have:

$$FS = (\frac{5}{3} + \frac{3\lambda}{8C} - \frac{\lambda^3}{8C^3}); \quad C = \sqrt{(\frac{2\pi^2 E}{F_y})};$$
$$\lambda = \frac{kl}{r}; \quad r = aA^b; \quad a = 0.4993; \quad b = 0.6777 \tag{3.20}$$

The components used include E modulus of elasticity, Fy the yield stress of steel, C the ratio of the limit slenderness which separates the elastic and inelastic buckling region in comparison with the existing slandering λ. Also, k is the effective length factor of the l and r is the radius of rotation of the limb. In order to ensure the performance of chaos map and algorithms, as well as to increase the accuracy and sensitivity of calculations, each of the modes has been performed independently 30 times and the results related to the best response and the average value of responses are presented in Table 3.4. A bar graph of these results is shown in Fig. 3.8.

Also, the coefficient of variation of the responses, which indicates the stability and robustness of the responses, has been calculated and another criterion for the efficiency of chaos map and algorithms has been obtained. The minimum and maximum cross-sectional area of all members is 0.775 in^2 and 20 in^2 respectively. By examining the optimization results for different combinations of algorithms chaos map and comparing it with standard mode, a significant improvement in reducing the weight of the 120-bar dome shaped truss is achieved. In the enhanced result for

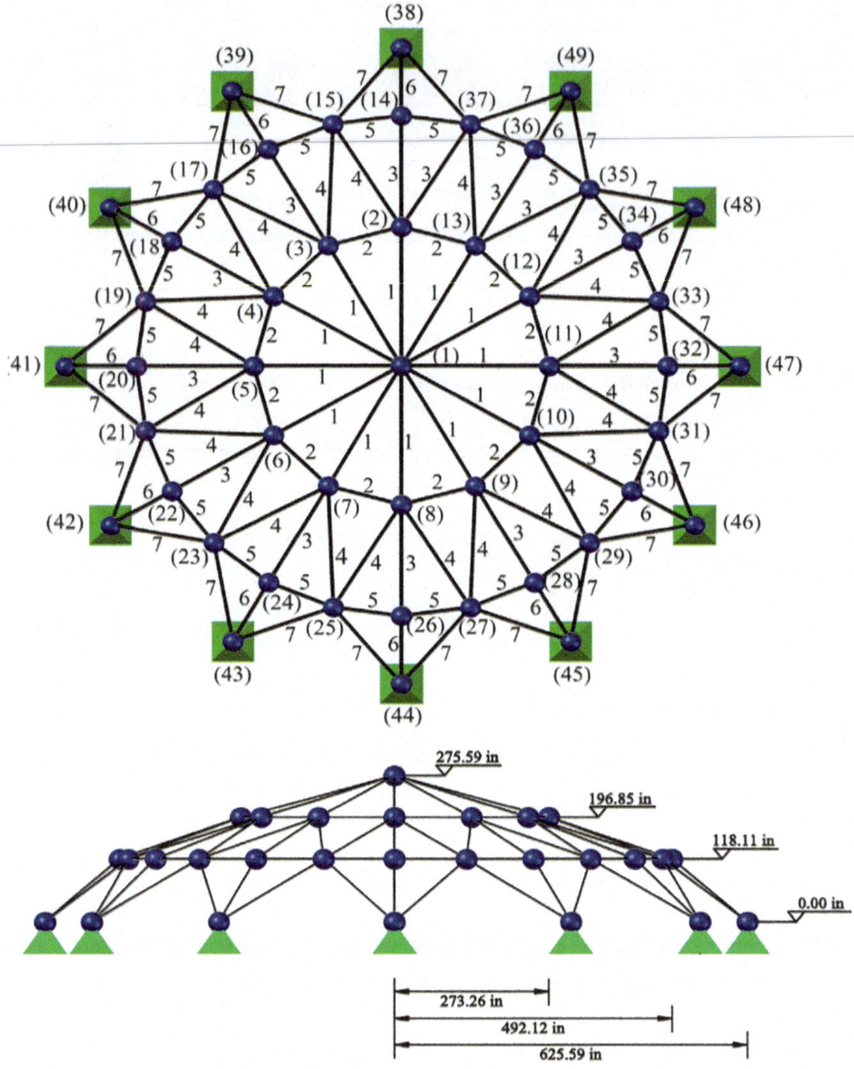

Fig. 3.7 Schematic of a 120-bar dome shaped truss

cyclical parthenogenesis algorithm by the logistic chaos map and scenario 3 weighing 33,242.4819 lb has the optimal response.

Table 3.4 Optimal design comparison for the 120-bar dome shaped truss

Number group	Standard	Logistic map			Tent map			Gauss map		
	CPA	CCPA 21	CCPA 22	CCPA 23	CCPA 31	CCPA 32	CCPA 33	CCPA 41	CCPA 42	CCPA 43
1	3.02512	3.02487	3.02443	3.02412	3.02464	3.0254	3.02552	3.02397	3.02505	3.0241
2	14.591	14.7517	14.7443	14.7935	14.7698	15.6058	14.9393	14.6240	14.671	14.8425
3	5.17932	5.16096	5.09438	5.06922	5.12044	5.0610	5.05949	5.07114	5.18775	5.05328
4	3.13214	3.13142	3.13111	3.12896	3.13491	3.06655	3.08893	3.13701	3.12999	3.13542
5	8.47334	8.4308	8.48537	8.48944	8.4492	8.1537	8.47033	8.56723	8.44546	8.47236
6	3.32423	3.28349	3.29537	3.28278	3.28788	3.41232	3.48379	3.28691	3.29267	3.28271
7	2.49491	2.49552	2.49626	2.49676	2.49587	2.49528	2.49447	2.49727	2.49511	2.49677
Best weight	33,246.78	33,247.21	33,246.02	**33,242.48**	33,248.89	33,305.02	33,340.02	33,247.39	33,244.96	33,249.62
Mean weight	33,250.01	33,250.87	33,248.23	**33,247.61**	33,250.22	33,386.16	33,419.68	33,250.26	33,249.16	33,252.51
Coefficient variation	0.00668	0.00697	0.005244	0.010151	**0.0031352**	0.10643	0.1123	0.007138	0.00867	0.010215

The bold items in the table indicate the optimal values for the best weight, the best mean and the best coefficient of variation

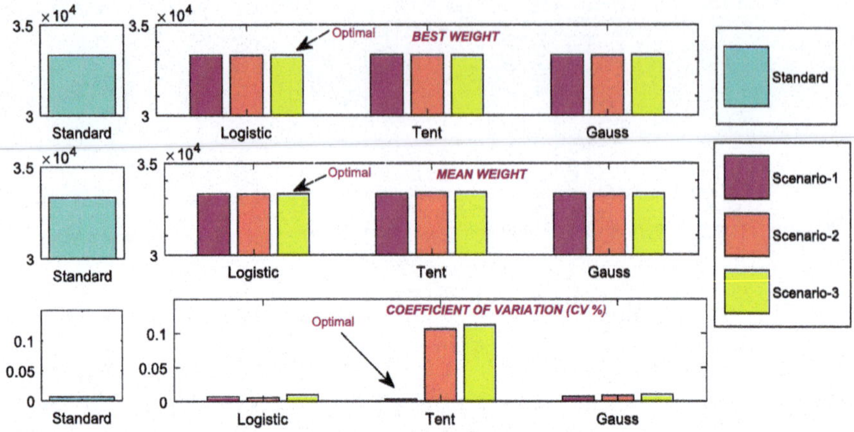

Fig. 3.8 Optimization results in standard mode and chaos map for the 120-bar dome shaped truss

3.7 Truss Weight Optimization with Multiple Frequency Constraints

The natural frequencies of the structures have useful information about the dynamic behavior of the structures and we can easily control the dynamic behavior of a structural system by accessing them quickly. One of these cases is preventing the destructive and dangerous phenomenon of resonance. Considering that the frequency limit relations are highly non-linear and non-convex and have complex implicit variables, the application of gradient-based methods will be very difficult and time-consuming, and in most cases the calculation process will stop at local optima. In this regard, the use of meta-Heuristic algorithms has brought good results, and by combining chaos functions with these algorithms, we can escape from the trap of local optima and move the results towards global optima. By creating a balance between the two stages of exploration and exploitation, these functions prevent premature convergence of the algorithm and accelerate and improve the convergence towards global optimality. *Gaussian*, *Liebovitch* and *Piecewise* chaos functions provide suitable conditions for improving the optimization results. By embedded these functions in the stages of exploration, exploitation or both parts at the same time, the scenarios of Chaotic Cyclical parthenogenesis Algorithm (CPA) are formed. Also, in order to increase diversity in the statistical space, each model is run 20 times independently, and the statistical population corresponding to the best value of cross-sections, the best average and coefficient of variation is determined. By examining the effectiveness of chaotic models, we will have a significant improvement in the optimization results and we can determine the chaos function compatible with the algorithm that creates the balance between the exploration and exploitation stages with the best efficiency. We can also detect the weakness of the algorithm in the exploration or exploitation stages.

3.7.1 Formulation of the Structural Optimization with Frequency Constraints

In optimization problems for the cross section and geometric shape of trusses that are associated with frequency constraints, the goal is to minimize the weight of the structure so that the constraints for a number of natural frequencies for vibration modes are satisfied. The cross section of the members along with the coordinates of some nodes are introduced as decision variables. These variables are selected continuously. The upper and lower bounds are also specified for variables in some cases. These optimization problems are defined in mathematical form according Eq. (3.21):

$$\begin{aligned} &Find && X=\{A,S\},\ A=\{A_1,A_2,\dots,A_{na}\},\ S=\{S_1,S_2,\dots,S_{ns}\} \\ &to\ Minimize: && W(A,S)=\sum_{i=1}^{nm}\rho_i\times L_i(S)\times A_i \\ &subjected\ to: && \begin{cases} \tilde{\omega}_i^L\le\omega_i\le\tilde{\omega}_i^U & i=1,2,\dots,n\omega \\ A_j^L\le A_j\le A_j^U & j=1,2,\dots,na \\ S_k^L\le S_k\le S_k^U & k=1,2,\dots,ns \end{cases} \end{aligned} \tag{3.21}$$

In this relation, X is the vector of decision variables, A is the variable of the cross-section of the members, na is the symbol to represent the number of variables of the cross-section, A_i is the value of the cross-section of the ith variable, S is the variable of shape and arrangement, ns is the number of shape variables with the same coordinates Nodes are, S_i is the numerical value of the ith variable, W expresses the weight of the truss, nm indicates the total number of members, ρ_i the specific gravity of the material belonging to the ith member of the truss, L_i the length corresponding to the ith member which can be expressed through the variables Determine, ω_i expression of the ith natural frequency of the truss, ω_i^L and ω_i^u, respectively, represent the lower limit and the upper limit of the fixed frequency of the base, $n\omega$ indicates the total number of frequency restrictions, Aj^L and Aj^u, respectively, express the lower limit and the upper limit of the A_j cross section, and S_k^L and S_k^U to Arrange the lower limit and limit of kth variable S_k. Given that meta-heuristic algorithms are used for unbounded problems; we use the penalty function in modeling to convert the bound to unbound state. In this method, if no violation has been committed, the amount of the fine will be zero, otherwise, if there is a violation, the amount of the penalty function will be obtained from Eqs. (3.22) to (3.25):

$$V_i=\begin{cases} 0 & if\ \ \tilde{\omega}_i^L\le\omega_i\le\tilde{\omega}_i^U \\ 1-\frac{\omega_i}{\tilde{\omega}_i^L} & if\ \ \tilde{\omega}_i^L>\omega_i \\ 1-\frac{\tilde{\omega}_i^U}{\omega_i} & if\ \ \omega_i>\tilde{\omega}_i^U \end{cases} \tag{3.22}$$

$$\gamma=\sum_{i=1}^{n\omega}V_i \tag{3.23}$$

$$F_{penalty} = (1 + \varepsilon_1 \times \gamma)^{\varepsilon_2} \quad (3.24)$$

$$to\ Minimize\quad Mer(A, S) = W(A, S) \times F_{penalty} \quad (3.25)$$

In these relations γ represents the set of violations and ε_1 and ε_2 are selected based on the search and extraction ratio. In this study, ε_1 units and ε_2 are selected with incremental linear changes in the range of 1.3–3 at the end of the iteration. Finally, "*Mer*" is the merit function or the objective function after the penalty is imposed.

3.7.2 *Introduction of Selected Chaos Map*

In most meta-heuristic algorithms, the optimization results stagnate and stop when they reach the local optimal position. In such cases, premature convergence occurs. In order to escape the trap of local optimization, chaos functions create suitable conditions that by creating disorder in the search space, the possibility of jumping to most positions of the search space is done. Therefore, global optimum will not have the opportunity to escape the turmoil of chaotic maps. These functions have no traces of random behaviors, but cause very disordered behaviors in the search space. One of the most important features of these functions is sensitivity to initial conditions and non-periodic and ergodic behaviors and these functions do not have inverse. The use of chaos series to evolve variables in COA algorithms has significant advantages over other methods. In definite searches, compared to random search, more speed and convergence towards the general answer is achieved [8]. In this study, the Gaussian, Liebovitch, and Piecewise chaos maps have been selected. In the chaos maps, Liebovitch converges with a very high probability from a local minimum to a general minimum. Therefore, this function is suitable for improving the exploration conditions of algorithms. The chaos Gaussian map is most likely in the local minimum range and is suitable for improving exploitation conditions. Finally, the chaotic map of Piecewise simultaneously improves both exploration and exploitation conditions. Therefore, by selecting these chaotic functions, the weakness of algorithms of any kind is improved. The numerical distribution of these chaos maps for 100 iterations is presented in Fig. 3.9.

3.7.2.1 Gaussian Map

Using this function in nonlinear dynamic behaviors has good results. The statements of chaotic sequences in the Gaussian function are obtained according to Eq. (3.26):

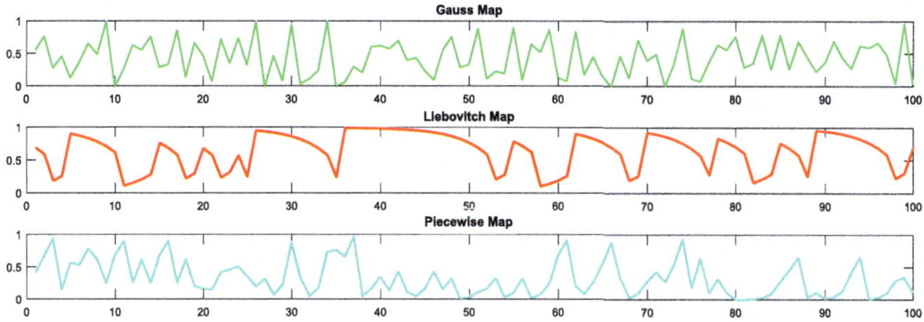

Fig. 3.9 Distribution of numerical values of chaos functions in 100 repetitions

$$X_{n+1} = \begin{cases} 0 & X_n = 0 \\ \frac{1}{X_n} - \left[\frac{1}{X_n}\right] & X_n \neq 0 \end{cases} \tag{3.26}$$

3.7.2.2 Liebovitch Map

This function consists of three separate linear rules and there will be no common or repetitive sentences in these intervals [9]. The chaotic sequence sentences in this function are expressed by Eq. (3.27):

$$X_{n+1} = \begin{cases} \alpha_1 X_n & 0 < X_n \leq d_1 \\ \frac{d_2 - X_n}{d_2 - d_1}: & d_1 < X_n \leq d_2 \\ 1 - \alpha_2 \times (1 - X_n) & d_2 < X_n \leq 1 \end{cases} \tag{3.27}$$

Here the values $d_1 = 1/3$ and $d_2 = 2/3$ are selected. A_1 and α_2 are also calculated based on Eqs. (3.28) and (3.29):

$$\alpha_1 = \frac{d_2}{d_1} \times (1 - (d_2 - d_1)) \tag{3.28}$$

$$\alpha_2 = \frac{1}{d_2 - 1}((d_2 - 1) - d_1(d_2 - d_1)) \tag{3.29}$$

3.7.2.3 Piecewise Map

The piecewise map is the same as the Liebovitch map of three criteria, and the value of P is considered as the control parameter [10]. The range of changes of P is in the range of 0–0.5, which in this study we have considered 1/3, and the following category of equations indicates the relationship between the sentences of chaotic sequences in this function that expressed by Eq. (3.30):

$$X_{n+1} = \begin{cases} \frac{X_n}{p} & 0 \le X_n < p \\ \frac{X_n - p}{0.5 - p} & p \le X_n < \frac{1}{2} \\ \frac{1 - p - X_n}{0.5 - p} & \frac{1}{2} \le X_n < 1 - p \\ \frac{1 - X_n}{p} & 1 - p \le X_n < 1 \end{cases} \tag{3.30}$$

3.7.3 Numerical Examples of Optimal Truss Design

In this research, the purpose of optimal truss design is to find the optimal value for the cross-sectional area of members and the shape of structures, which has a lower weight and at the same time provides limitations related to the oscillation frequency in different vibration modes to avoid the destructive phenomenon of resonance and high vibration. Famous examples in this regard are reviewed and research results are compared to determine the efficiency of algorithms, chaos map and scenarios. In order to form a statistical community and interpret the results, each of the examples has been independently examined 20 times. For each of the studied optimal, the results related to the best weight, average weight and coefficient of variation are reported. These examples contain:

- The planar 10-bar truss
- The simply supported 37-bar planar truss
- The 200-bar planar truss structure

3.7.3.1 A Planar 10-Bar Truss

The planar 10-bar truss as depicted in Fig. 3.10 is a well-known benchmark problem in the field of weight optimization of the structures with frequency constraints. This truss only considers "optimizing the size of the sections" and the geometric shape of the structure is constant during the optimization process. The non-structural concentrated mass of 454 kg affected the free nodes. The mechanical properties of the structure are: The density of materials is 2770 kg/m^3, the modulus of elasticity is 69,800 MPa, the frequency limits of the structure in the first, second and third modes are greater than 7, 15 and 20 Hz, respectively. The cross-sectional area of each member is selected as an independent decision variable. Therefore, ten design variables with a lower limit of 0.645 cm^2 and an upper limit of 50 cm^2 are considered. In order to ensure the performance of chaotic functions and algorithms, as well as to increase the accuracy and sensitivity of calculations, each of the modes has been performed independently 20 times and the results related to the best response and the average value of responses are presented in Statistical Table 3.5. Also, the coefficient of variation of responses,

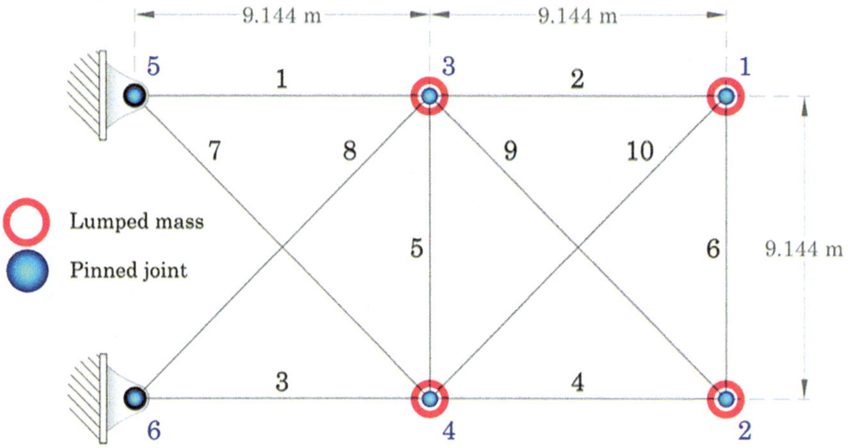

Fig. 3.10 Schematic of a planar 10-bar truss

which is a measure of the robustness of responses, has been calculated and used to compare the efficiency of chaos maps and standard algorithms. For quick access to optimization information, a bar chart of each component is shown in Fig. 3.11.

Examining the optimization results for different combinations of algorithms with chaos maps and comparing it with the standard mode, shows a significant improvement in reducing the weight of the planar 10-bar truss. The enhanced results for chaotic cyclical parthenogenesis algorithm by Gaussian map and scenario 2, with weighing 523.7471 kg has the optimal response.

3.7.3.2 A Simply Supported 37-Bar Planar Truss

A 37-member planar truss with a simple abutment as shown in Fig. 3.12 is a well-known benchmark problem for optimizing the weight and shape of a truss with a frequency limit. This truss considers both optimizations of the section size and geometric coordinates of the nodes and the geometric shape of the structure is determined during the optimization process. The lower edge member is considered to be a rod member with a rectangular cross-section and an area of 40 cm^2. The other members are modeled as rod members. These decision variables are grouped according to symmetry in the geometric shape. The y-axis coordinates form the shape variables in the upper year, taking into account the symmetry. In the lower axis, a non-structural concentrated mass of 10 kg affects all free nodes. The mechanical properties of the structure are: The density of the materials is 7800 kg/m^3, the modulus of elasticity is 21000 MPa, the frequency limits of the structure in the first, second and third modes are greater than 20, 40 and 60 Hz, respectively. For the cross section of the members, the lower limit is 1 cm^2 and the upper limit is 10 cm^2. The total size and shape decision variables are 19 variables, of which 14 variables

Table 3.5 Optimal design comparison for the planar 10-bar truss

Number group	Standard	Gauss map			Liebovitch map			Piecewise map		
	CPA	CCPA 21	CCPA 22	CCPA 23	CCPA 31	CCPA 32	CCPA 33	CCPA 41	CCPA 42	CCPA 43
A_1	35.5891	35.398	34.8838	35.1535	34.8371	34.9961	36.8726	34.7869	35.3876	34.674
A_2	13.9729	14.0658	14.376	14.5802	15.4727	14.5258	9.83077	15.0615	14.6476	14.9595
A_3	34.7018	34.8383	34.5673	34.9879	34.9925	34.8679	37.5801	35.3603	34.8944	35.2571
A_4	15.348	15.5912	14.7694	15.2398	13.7915	14.2378	18.3652	14.1526	14.7858	14.5811
A_5	0.645	0.645198	0.645005	0.645003	0.645	0.645	4.82988	0.645	0.645	0.645
A_6	4.56985	4.56282	4.58685	4.54341	4.58641	4.59696	4.02891	4.57458	4.55853	4.5671
A_7	22.2432	22.0455	23.4511	22.795	25.6229	24.3865	20.6511	24.8133	23.5452	24.0311
A_8	25.3207	25.1993	24.1436	24.2458	21.9375	23.6268	22.0557	22.8008	23.8563	23.2528
A_9	13.049	12.7533	12.3608	12.7576	11.9393	12.068	13.5259	12.0689	12.5227	12.3917
A_{10}	11.7617	12.0677	12.8414	12.2893	13.1142	12.77	11.2213	12.7838	12.3009	12.6616
Best weight (kg)	524.763	524.352	**523.747**	524.554	524.351	524.046	524.059	524.471	524.462	524.267
Mean weight (kg)	524.903	524.791	524.499	524.947	524.599	525.618	524.805	524.561	525.169	**524.484**
Coefficient variation	0.02587	0.06398	0.0834	0.08902	0.04116	0.4806	0.19133	**0.015002**	0.18952	0.02372
ω_1 (HZ)	7.01234	6.99999	6.99538	7.00001	6.99125	6.99673	6.94397	7.0001	7	6.99959
ω_2 (HZ)	16.1799	16.1568	16.1486	16.184	16.1264	16.1624	17.591	16.2	16.192	16.1863
ω_3 (HZ)	20.0032	20	19.9998	20	20	20.0002	20	20.0001	20	20.0003

The bolded items in the table indicate the optimal values for the best weight, the best mean and the best coefficient of variation

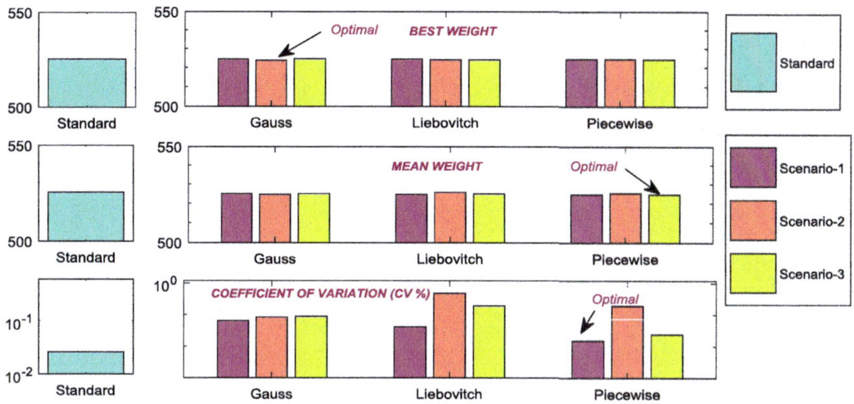

Fig. 3.11 Optimization results in standard mode and selection of chaos map for the planar 10-bar truss

are related to the size and cross-sectional area of the members and 5 variables are related to the shape of the structure and the coordinates of the nodes. There are also three frequency limits. In order to ensure the performance of turbulent functions and algorithms, as well as to increase the accuracy and sensitivity of calculations, each of the independent modes has been performed 20 times and the results related to the best response and the average value of responses. Is presented in statistical Table 3.6. Also, the coefficient of variation of responses, which is a measure of the robustness and robustness of responses, has been calculated and used to compare the efficiency of turbulence functions and algorithms. For quick access to optimization information, a bar chart of each component is shown in Fig. 3.13.

Examining the optimization results for different combinations of algorithms with turbulence functions and comparing it with the standard mode, shows a significant improvement in reducing the weight of A simply supported 37-bar planar truss. The results for chaotic cyclical parthenogenesis algorithm, the Gaussian chaos map with scenario 3 weighing 358.9183 kg has the optimal response.

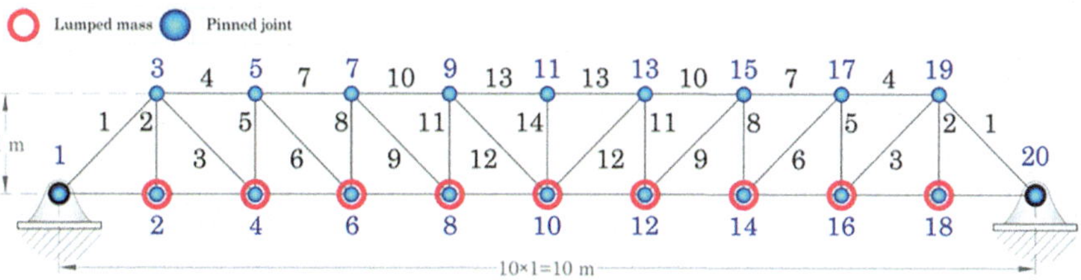

Fig. 3.12 Schematic of A simply supported 37-bar planer truss

Table 3.6 Optimal design comparison for the simply supported 37-bar truss

Decision variable	Standard	Gauss map			Liebovitch map			Piecewise map		
	CPA	CCPA 21	CCPA 22	CCPA 23	CCPA 31	CCPA 32	CCPA 33	CCPA 41	CCPA 42	CCPA 43
Y_3, Y_{19} (m)	0.93836	0.92145	0.95911	0.86834	0.96829	0.96351	0.94164	0.94217	0.95068	0.90753
Y_5, Y_{17} (m)	1.3264	1.2907	1.3476	1.2232	1.3444	1.3419	1.3206	1.3217	1.3321	1.3013
Y_7, Y_{15} (m)	1.5147	1.4714	1.534	1.3962	1.5354	1.5189	1.4983	1.504	1.5075	1.4931
Y_9, Y_{13} (m)	1.6415	1.6072	1.6636	1.5249	1.6694	1.6485	1.6095	1.6331	1.6277	1.632
Y_{11} (m)	1.7077	1.684	1.7379	1.5931	1.7413	1.71	1.6943	1.7031	1.6985	1.705
A_1, A_{27}	2.952	3.0046	2.9661	3.0497	2.9481	3.0552	2.9296	3.043	2.946	3.0984
A_2, A_{26}	1	1	1.0001	1.0006	1.0008	1	1	1.0026	1.0234	1
A_3, A_{24}	1	1	1	1	1	1	1	1.0004	1.0114	1.0005
A_4, A_{25}	2.6558	2.7173	2.4569	2.8707	2.5256	2.4587	2.6741	2.6168	2.5312	2.6673
A_5, A_{23}	1.1833	1.1281	1.1154	1.058	1.1505	1.1415	1.1745	1.1614	1.1606	1.186
A_6, A_{21}	1.2824	1.2024	1.2853	1.2344	1.284	1.2296	1.1786	1.2326	1.2037	1.2172
A_7, A_{22}	2.5081	2.726	2.4081	2.7286	2.5469	2.5054	2.7002	2.6342	2.5366	2.5837
A_8, A_{20}	1.3462	1.4048	1.4612	1.2952	1.358	1.3558	1.4374	1.3058	1.3508	1.3743
A_9, A_{18}	1.5071	1.4516	1.4781	1.5182	1.4932	1.5088	1.4881	1.5355	1.5618	1.5007
A_{10}, A_{19}	2.4757	2.6848	2.5265	2.6786	2.4103	2.5414	2.5786	2.4695	2.6146	2.4847
A_{11}, A_{17}	1.1674	1.2495	1.2634	1.17	1.2155	1.1727	1.098	1.1532	1.1054	1.2399
A_{12}, A_{15}	1.3689	1.2816	1.3007	1.3334	1.3527	1.3365	1.3291	1.3501	1.3662	1.3271
A_{13}, A_{16}	2.5183	2.5989	2.4642	2.648	2.3968	2.4895	2.5685	2.5447	2.6147	2.5105
A_{14}	1.0002	1	1.0007	1.0014	1.001	1	1.0012	1	1	1
Best weight (kg)	359.8199	359.8841	359.8344	**358.918**	359.803	359.709	359.7128	359.8181	359.8039	359.848

(continued)

Table 3.6 (continued)

Decision variable	Standard	Gauss map			Liebovitch map			Piecewise map		
	CPA	CCPA 21	CCPA 22	CCPA 23	CCPA 31	CCPA 32	CCPA 33	CCPA 41	CCPA 42	CCPA 43
Mean weight (kg)	359.9175	359.9379	359.9763	**359.451**	359.973	359.795	359.8275	359.8892	359.9564	359.941
Coefficient variation	0.023882	0.022051	0.039902	0.14332	0.03100	**0.01488**	0.026985	0.019185	0.046966	0.02077
ω_1 (HZ)	19.9926	20.0065	20	19.3394	19.9996	20.233	20.0516	19.9398	20.0008	19.996
ω_2 (HZ)	40.0131	39.9738	39.9998	39.9999	39.9996	39.9596	40.0224	39.8939	39.997	39.9998
ω_3 (HZ)	60.0446	59.9862	60.0008	59.9399	59.9996	59.5863	59.786	59.887	60.0171	60.003

The bolded items in the table indicate the optimal values for the best weight, the best mean and the best coefficient of variation

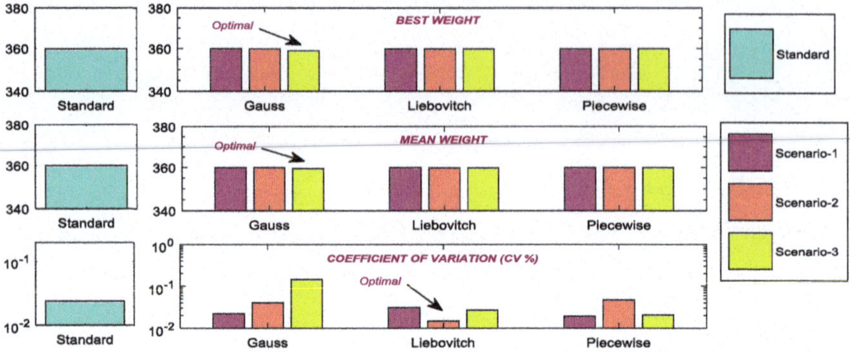

Fig. 3.13 Optimization results in standard mode and selection of chaos map for the simply supported 37-bar truss

3.7.3.3 A 200-Bar Planar Truss Structure

The 200-bar planar truss structure as shown in Fig. 3.14 is a well-known benchmark problem for weight optimization with frequency limits. This truss only considers "optimizing the size of the sections" and the geometric shape of the structure is constant during the optimization process. The decision variables related to the sections size of the members are classified into 29 groups according Table 3.7. Non-structural concentrated mass in nodes 1–5 and in the amount of 100 kg affect the structure. The mechanical characteristics of the structure are: material density 7860 kg/m^3, modulus of elasticity 210,000 MPa, frequency limits of the structure in the first, second and third modes are greater than 5, 10 and 15 Hz, respectively. For the cross section of the members, the lower limit range is 0.1 cm^2 and the upper limit is 25 cm^2. In order to ensure the performance of turbulent functions and algorithms, as well as to increase the accuracy and sensitivity of calculations, each of the modes has been performed independently 20 times and the results related to the best response and the average value of responses are presented in Statistical Table 3.8.

Also, the coefficient of change of responses, which is a measure of the robustness and robustness of responses, has been calculated and used to compare the efficiency of turbulence functions and algorithms. For quick access to optimization information, a bar chart of each component is shown in Fig. 3.15. Examining the optimization results for different combinations of algorithms with turbulence functions and comparing it with the standard mode, shows a significant and significant improvement in reducing the weight of the 200-bar planar truss structure. The results for chaotic cyclic parthenogenesis algorithm by Liebovitch chaos map and scenario 2 with a weight of 2157.2685 kg has the optimal response.

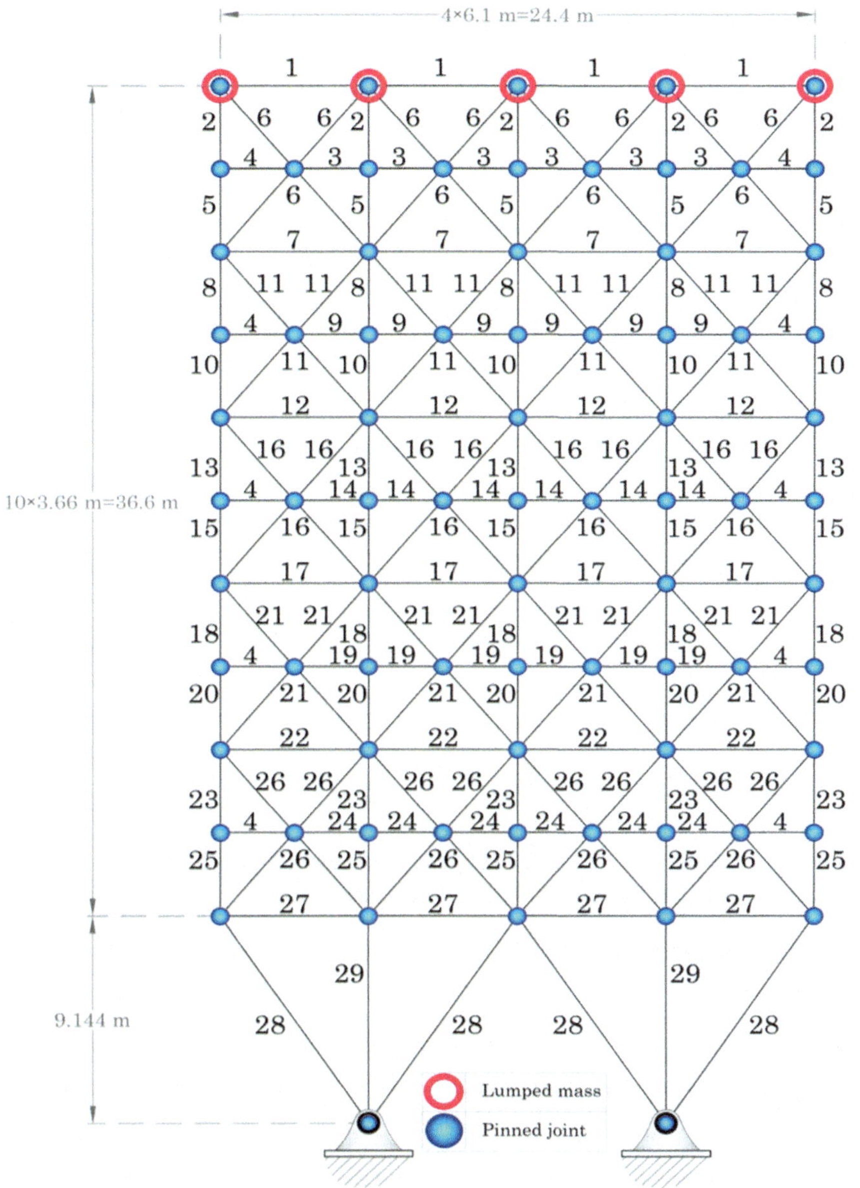

Fig. 3.14 Schematic of a 200-bar planar truss structure

Table 3.7 Classification of members in the 200-bar planar truss structure

Group	Element group	Group	Element group
A_1	1, 2, 3, 4	A_{16}	82, 83, 85, 86, 88, 89, 91, 92, 103, 104, 106, 107, 109, 110, 112, 113
A_2	5, 8, 11, 14, 17	A_{17}	115, 116, 117, 118
A_3	19, 20, 21, 22, 23, 24	A_{18}	119, 122, 125, 128, 131
A_4	18, 25, 56, 63, 94, 101, 132, 139, 170, 177	A_{19}	133, 134, 135, 136, 137, 138
A_5	26, 29, 32, 35, 38	A_{20}	140, 143, 146, 149, 152
A_6	6, 7, 9, 10, 12, 13, 15, 16, 27, 28, 30, 31, 33, 34, 36, 37	A_{21}	120, 121, 123, 124, 126, 127, 129, 130, 141, 142, 144, 145, 147, 148, 150, 151
A_7	39, 40, 41, 42	A_{22}	153, 154, 155, 156
A_8	43, 46, 49, 52, 55	A_{23}	157, 160, 163, 166, 169
A_9	57, 58, 59, 60, 61, 62	A_{24}	171, 172, 173, 174, 175, 176
A_{10}	64, 67, 70, 73, 76	A_{25}	178, 181, 184, 187, 190
A_{11}	44, 45, 47, 48, 50, 51, 53, 54, 65, 66, 68, 69, 71, 72, 74, 75	A_{26}	158, 159, 161, 162, 164, 165, 167, 168, 179, 180, 182, 183, 185, 186, 188, 189
A_{12}	77, 78, 79, 80	A_{27}	191, 192, 193, 194
A_{13}	81, 84, 87, 90, 93	A_{28}	195, 197, 198, 200
A_{14}	95, 96, 97, 98, 99, 100	A_{29}	196, 199
A_{15}	102, 105, 108, 111, 114		

3.8 Concluding Remarks

This chapter is dedicated to introducing a new population-based meta-heuristic optimization method, called Cyclical Parthenogenesis Algorithm (CPA). This algorithm has imitated the social behavior and reproduction of some animal species such as aphids. Mating and without Mating form the two main parts of the exploration and exploitation stage. The statistical distributions used in these two important parts can be replaced by chaos functions. By embedded the chaos maps in Cyclical Parthenogenesis Algorithm (CCPA), the local optima mutated to global optima. Therefore, the algorithm is saved from the trap of local optima and the problem of premature convergence is solved. Logistic, Tenet, and Gaussian chaos maps are investigated for problems with static state, and each one consists of three scenarios. These functions can improve the parts related to search, extraction, or both simultaneously, and a significant improvement in the optimization results can be achieved. For multi-frequency constraints, where the problem is strongly non-linear and non-convex, Gaussian, Liebovitch and Piecewise chaos maps have significant results.

Table 3.8 Optimal design comparison for the 200-bar planar truss structure

Number group	Standard	Gauss map			Liebovitch map			Piecewise map		
	CPA	CCPA 21	CCPA 22	CCPA 23	CCPA 31	CCPA 32	CCPA 33	CCPA 41	CCPA 42	CCPA 43
A_1	0.289608	0.293019	0.296588	0.279431	0.321632	0.290973	0.302725	0.305881	0.301036	0.314605
A_2	0.478876	0.481785	0.45132	0.467379	0.45274	0.47605	0.459191	0.437918	0.475204	0.453254
A_3	0.1	0.1	0.102453	0.100015	0.101953	0.1	0.1	0.100277	0.100028	0.100167
A_4	0.1	0.10002	0.1	0.100553	0.101086	0.1	0.1	0.100201	0.100003	0.100889
A_5	0.534043	0.511026	0.526107	0.57492	0.495129	0.514842	0.526126	0.511712	0.542686	0.514067
A_6	0.855027	0.802494	0.821307	0.810291	0.828989	0.816734	0.828703	0.836873	0.825909	0.818738
A_7	0.106872	0.1	0.104966	0.1	0.1	0.1	0.1	0.10006	0.102934	0.100103
A_8	1.45968	1.43353	1.42315	1.45716	1.3927	1.44981	1.39917	1.42379	1.42626	1.44631
A_9	0.1	0.100097	0.100883	0.1	0.100025	0.1	0.1	0.1	0.100014	0.10004
A_{10}	1.67063	1.5509	1.55082	1.60393	1.6465	1.58069	1.63389	1.54839	1.58925	1.64161
A_{11}	1.15734	1.14041	1.16683	1.17245	1.14493	1.17757	1.14078	1.13178	1.15376	1.12905
A_{12}	0.1	0.12283	0.120206	0.135671	0.171711	0.1	0.1	0.150499	0.157137	0.155883
A_{13}	2.6988	2.98258	2.98067	2.94719	2.86799	3.00066	2.96865	2.92751	2.90215	2.92327
A_{14}	0.107052	0.102668	0.101199	0.100655	0.1	0.100747	0.1	0.1	0.100461	0.100215
A_{15}	3.02663	3.35048	3.4592	3.19098	3.15603	3.30353	3.19757	3.24423	3.14083	3.15229
A_{16}	1.62581	1.59531	1.54881	1.64317	1.64943	1.5436	1.58055	1.5782	1.5897	1.58294
A_{17}	0.115866	0.179198	0.193128	0.259028	0.255962	0.219206	0.273799	0.231779	0.338909	0.269646
A_{18}	5.12565	5.27223	5.11597	5.03086	4.99662	5.12183	4.95754	5.19051	5.12288	5.12214
A_{19}	0.15735	0.109914	0.103126	0.1	0.116359	0.1	0.1	0.162592	0.103628	0.100344
A_{20}	5.35157	5.50947	5.3785	5.29687	5.46964	5.62978	5.38271	5.40244	5.63183	5.48943
A_{21}	1.98312	2.09841	2.0438	2.0835	2.05333	2.12834	2.093	2.04858	2.06948	2.16342

(continued)

Table 3.8 (continued)

Number group	Standard	Gauss map			Liebovitch map			Piecewise map		
	CPA	CCPA 21	CCPA 22	CCPA 23	CCPA 31	CCPA 32	CCPA 33	CCPA 41	CCPA 42	CCPA 43
A_{22}	0.810999	0.500568	0.577766	0.70096	0.674592	0.597895	0.580611	0.691715	0.603365	0.61213
A_{23}	7.8834	7.88482	8.08354	7.59061	7.59844	7.62196	7.51811	7.68869	8.14687	7.65708
A_{24}	0.1	0.1	0.2377	0.164721	0.1	0.1	0.133458	0.110313	0.177151	0.224254
A_{25}	8.1585	7.97413	8.29101	7.74267	8.06875	8.02438	7.89413	7.85961	8.37709	7.50853
A_{26}	2.92169	2.78865	2.85578	2.82315	2.91003	2.77154	2.64498	2.82733	2.75363	2.78613
A_{27}	11.466	10.6676	10.4392	10.2975	10.3786	10.6055	10.6285	10.4381	10.0403	9.92378
A_{28}	20.9808	21.2479	21.1001	21.2661	21.3168	21.0961	21.8031	21.4201	20.974	21.7303
A_{29}	10.1727	10.497	10.574	11.1136	10.4244	10.9458	10.5525	10.649	11.2971	10.7432
Best weigh	2163.26	2157.791	2158.299	2157.67	2158.062	**2157.269**	2157.622	2157.407	2158.534	2158.601
Mean weight	2170.262	2160.858	2160.178	2159.245	2159.725	2159.246	2158.804	**2158.735**	2164.521	2162.947
Coefficient variation	0.25081	0.23417	0.085413	0.10542	0.084546	0.14705	**0.049231**	0.052349	0.37636	0.30742
ω_1 (HZ)	5.00185	5.004	4.99976	5.00008	5.00096	4.99841	5.00886	4.99903	4.99977	5.00002
ω_2 (HZ)	12.1946	12.164	12.2009	12.0157	12.4239	12.0408	12.2493	12.2233	12.2837	12.3311
ω_3 (HZ)	15.09	15.0466	15.0535	15.0621	15.0852	14.9765	15.1111	15.0341	15.0767	15.111

The bolded items in the table indicate the optimal values for the best weight, the best mean and the best coefficient of variation

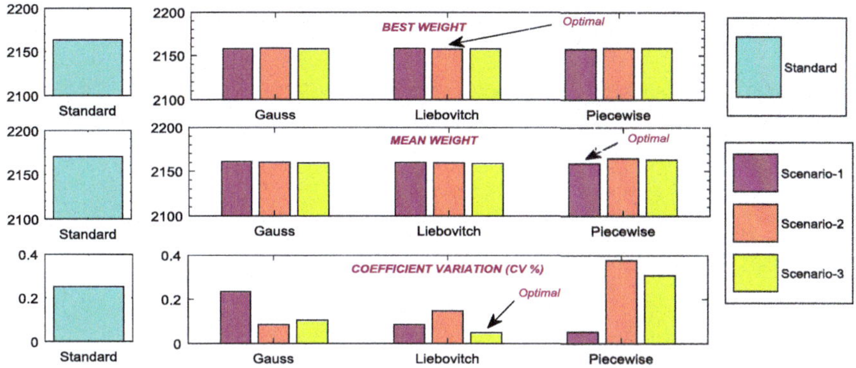

Fig. 3.15 Optimization results in standard mode and selection of chaos map for the 200-bar planar truss structure

References

1. Kaveh, A.: Advances in Meta-heuristic Algorithms for Optimal Design of Structures, 3rd edn. Springer, Cham, Switzerland 2021. https://doi.org/10.1007/978-3-319-05549-7
2. Kaveh, A., Zolghadr, A.: Cyclical parthenogenesis algorithm: a new meta-heuristic algorithm. Adv. Eng. Softw. (in press) (2016)
3. Kaveh, A., Yousefpoor, H.: Chaos-embedded meta-heuristic algorithms for optimal design of truss structures Scientia Iranica A **29**(6), 2868–2885 (2022). https://doi.org/10.24200/sci.2022.59812.6441
4. Kaveh, A., Yosefpoor, H.: Chaotically enhanced meta-heuristic algorithms for optimal design of truss structures with frequency constraints. Periodica Polytechnica Civil Eng. **66**(3), 900–921 (2022). https://doi.org/10.3311/PPci.20220
5. Kaveh, A., Zarfam, P., Aziminejad, A., Yosefpoor, H.: Comparison of four chaotic meta-heuristic algorithms for optimal design of large-scale truss structures. Iran. J. Sci. Technol. Trans. Civil Eng. https://doi.org/10.1007/s40996-022-00908-8
6. Ott, E.: Chaos in Dynamical Systems. Cambridge University Press, Cambridge (2002)
7. Bucolo, M., Caponetto, R., Fortuna, L., Frasca, M., Rizzo, A.: Does chaos work better than noise? IEEE Circuit Syst. Mag. **2**(3), 4–19 (2002)
8. Kaveh, A., Sheikholeslami, R., Talatahari, S., Keshvari-Ilkhichi, M.: Chaotic swarming of particles: a new method for size optimization of truss structures. Adv. Eng. Softw. **67**, 136–47 (2014)
9. Erramilli, A., Singh, R., Pruthi, P.: Modeling Packet Traffic with Chaotic Maps. KTH, Stockholm (1994)
10. Peitgen, H.-O., Jürgens, H., Saupe, D.: Chaos and Fractals: New Frontiers of Science. Springer, Berlin (2006)

Chapter 4
Chaotic Teaching Learning Based Optimization

4.1 Introduction

Considering the challenges in solving complex optimization problems in the real world, classical methods have faced many problems and gradually meta-heuristic methods with different inspirations and features have replaced them. The main goal of these new methods is to efficiently explore the space of decision variables and find global or near-global solutions to it [1]. In this regard, two important stages of exploration and exploitation are considered in meta-heuristic algorithms to converge to optimal answers. Based on these steps, the strategy of each algorithm requires that it is first established in the points of the search space where the objective function has privileged values, and then the neighborhood of these privileged points is examined with as much accuracy as possible. This strict search causes the tendency towards local optima. From the comparison between these local optima, the global pseudo-optimum corresponding to each step is selected. These two steps are known as exploration and exploitation steps respectively. The configuration of each meta-heuristic algorithm is formed using the interaction between these two features. The exploration phase maintains diversity in the algorithm, and always seeks to discover new points. This step in meta-heuristic algorithms takes place in the initial iterations and in which a general experience of the search space is concluded. The exploration phase is associated with random movements and therefore has high risk. One of the characteristics of this stage is jumps, mutation and sudden movements. Therefore, in the first stage, an extensive search is performed to ensure that good and excellent answers are selected with a suitable distribution which subsequently starts extracting and exploiting the answers. This step is also called local search. In this step, we move towards the optimal points. The above step preserves the convergence of the algorithm and has low risk. This step has a local look at the search and may get stuck in the trap of local answers. Now, if only exploration is considered in the algorithm and it always looks for new discoveries, without investigation and exploitation, that algorithm becomes a random search. In such a situation, the algorithm often moves

A. Kaveh and H. Yousefpoor, *Chaotic Meta-heuristic Algorithms for Optimal Design of Structures*, Studies in Computational Intelligence 1129,
https://doi.org/10.1007/978-3-031-48918-1_4

to the nearest solution, but because its next move is random and not purposeful, it moves away from the solution again. In cases where the algorithm is only looking for extraction and exploitation, in these cases it will be highly dependent on the starting point. If the selected point is appropriate, the global optimal solution is obtained. Therefore, each of these alone does not provide the necessary efficiency for the algorithm, and the capability of the algorithm lies in the joint interaction of exploration and exploitation. Research shows that this balance is disturbed at some stages of the optimization process and the processing operation stops at local optima. Therefore, this premature convergence, disturbs the optimization results and the replacement of chaos functions instead of the random distributions of these two stages provides a suitable acceleration to improve the optimization operation again. The source of inspiration in the recent algorithm is the teaching and learning process in the classroom. Two important strategies in this algorithm include the strategy of the teacher's effect on the students and the mutual influence of the students on each other in the learning process. These two phases are called teacher phase and student phase respectively. If the interaction and balance between these two steps is not achieved, the operation process will stop in local optima and chaos accelerators can lead the algorithm from the trap of local optima to global optima. In the rest of this chapter, after introducing the Teaching Learning Based Optimization (TLBO), the basic steps of the algorithm will be presented in the standard mode, and then the Chaotic Teaching–Learning Based Optimization (CTLBO), will be formed with three scenarios. Finally, with well-known optimization examples of static constraints and multi-frequency constraints, the efficiency of several chaos functions will be challenged.

4.2 Standard Teaching–Learning-Based Optimization (TLBO)

The TLBO was introduced by Rao et al. in 2011. In this algorithm [2], the teaching and learning process in the classroom is considered. The algorithm consists of two main stages, which includes the effect of the teacher in the learning process and the mutual influence of students on each other. This algorithm, like other population-based meta-heuristic algorithms, starts with a series of random initial solutions, which are the students, and the best and most intelligent of them in each cycle of repetition is known as the teacher. The academic level of students improves in two phases in each repetition. The first phase is based on knowledge transferred from the teacher to the students and is known as the teacher's phase. In this phase, the average academic position of students improves based on a normal distribution. In the second phase, students learn from each other and students exchange scientific information with each other. In the second phase, students learn from each other and students exchange scientific information with each other, as a result of which weak students move towards stronger students with a random distribution. At each step, the new values are evaluated with the objective function and if there is an improvement over the previous values, they are replaced. In the following, the basic steps of this algorithm are presented.

4.3 Basic Steps in Standard Teaching–Learning-Based Optimization

Step 1. Forming the initial parameters of the algorithm: these parameters include the student population nL, the number of decision variables nV, the maximum number of iterations of NFEs, which is also known as the stopping criterion.
Step 2. Formation of the initial population of students and evaluate it: considering the lower and upper bounded of the search space, this population is formed randomly. Then, by applying to the objective function, the answers are estimated and sorted them.
Step 3. Among the sorted population, the best one is introduced as T teacher. Then the average position of the students is calculated and based on the teacher phase, the improved academic level of the students is determined based on Eqs. (4.1) and (4.2):

$$stepsize_i = T - TF_i \times MeanL \tag{4.1}$$

$$\begin{aligned} newL &= L + \mathbf{rand_{i,j}} \times stepsize \\ i &= 1, 2, ..., nL \quad and \quad j = 1, 2, ..., nV \end{aligned} \tag{4.2}$$

In these relationships, the *TF* is teaching factor that randomly selected as 1 or 2, and it indicates the success of the teacher in increasing the average level of students. At the end of the step, if the evaluated values are better than the previous values, the results will replace them.
Step 4. In the student phase, the interactions of the students with each other are checked. In this phase, each student exchanges information with another randomly selected student (except himself/herself). It is possible to improve the information when the performance of the selected student is better, in which case their position changes according Eqs. (4.3) and (4.4):

$$\begin{aligned} if \quad PFit_i < PFit_{rp} \Rightarrow stepsize_i = L_i - L_{rp} \\ if \quad PFit_i > PFit_{rp} \Rightarrow stepsize_i = L_{rp} - L_i \end{aligned} \tag{4.3}$$

$$\begin{aligned} newL &= L + \mathbf{rand_{i,j}} \times stepsize \\ i &= 1, 2, ..., nL \quad and \quad j = 1, 2, ..., nV \end{aligned} \tag{4.4}$$

The resulting values are evaluated and replaced if they are better than the previous values. Among the populations, the best ones are introduced in each stage. With this choice, the elitism of the algorithm is provided.

Step 5. The termination conditions are checked and if necessary, the operation from Step 2 is repeated.

4.4 Chaos Enhanced Teaching–Learning-Based Optimization (CTLBO)

The process of teaching and learning in the classroom forms the inspiration for this algorithm, which is mainly investigated in two phases. These phases include the effect of the teacher in the learning process and the mutual influence of the students on each other, which play a significant role in the exploration and exploitation phases, respectively [3–5]. In each of these phases, random distributions can be replaced by series related to chaos functions. These chaos functions are proposed to improve exploration, exploitation or both simultaneously. Therefore, the suggested scenarios for this replacement are as follows:

Scenario 1. Replacing the chaos function in the phase of the teacher's effect in the learning process: In this case, the first chaos function **CHM1** is replaced in Eq. (4.2) and the result will be Eq. (4.5):

$$newL = L + \mathbf{CHM1}_{\mathrm{i,j}} \times stepsizeT \tag{4.5}$$

Scenario 2. Replacing the chaos function in the stage of interaction influence of students on each other: In this case, the second chaos function **CHM2** is replaced in Eq. (4.4) and the result will be Eq. (4.6):

$$newL = L + \mathbf{CHM2}_{\mathrm{i,j}} \times stepsizeS \tag{4.6}$$

Scenario 3. Placing the chaos function in both stages simultaneously: In this case, two chaos functions are replaced simultaneously in Eqs. (4.2) and (4.4).

4.5 Truss Weight Optimization with Static Constraints

In order to check the effectiveness of chaos functions in Teaching–Learning-based Optimization (TLBO), Logistic, Tenet and Gossin chaos functions have been chosen. By embedded chaos maps and formation of chaotic algorithms, the competition between chaotic functions is formed to access the best optimal weight. To determine the best performance of the chaotic algorithm, three scenarios have been formed. In these scenarios, the chaos functions replace the distributions related to exploration, exploitation or both phases simultaneously. In each example, 9 chaotic models are formed and compared with the standard state. To form the statistical space and increase diversity in the research, each model is independently evaluated 30 times. Also, to analyze the statistical results, the best weight, the best average and the coefficient of variation are provided. In a number of studies, the application of chaos functions has increased the coefficient of variation. This performance indicates sudden jumps and increasing the difference between the best and worst answers. Due to the wide scope of investigations, there is a strong competition between chaos functions and as a result, the possibility of approaching the global optimum increases. This optimum cannot be far from the crosshairs of chaotic mutations. In each investigation, the chaos function compatible with the algorithm and the scenario corresponding to the best improvement are selected. We can also find out about the weakness of the algorithm in the exploration or exploitation stages.

4.5.1 Formation of the Objective Function and Constraint Conditions

For the optimal design in the structures, the objective function is generally expressed as follows:

$$\text{minimize} \quad W(X) = \sum_{i=1}^{n} \gamma_i \cdot x_i \cdot L_i \tag{4.7}$$

In this regard, W(X) expresses the weight of the structure, n is the number of members of the structure, γ_i demonstrate the material density that make up the structure for member i, L_i is the length of member i. X is the cross-section vector of the members, which is selected based on the upper bound and lower bound of the decision variables space. This optimization should satisfy the design constraints by forming a penalty function. For structures, the constraints are as follows:

$$\begin{aligned} &\delta_{\min} \le \delta_i \le \delta_{\max} && i = 1, 2, \ldots, m \\ &\sigma_{\min} \le \sigma_j \le \sigma_{\max} && j = 1, 2, \ldots, n \\ &\sigma_k^b \le \sigma_k \le 0 && k = 1, 2, \ldots, n_c \end{aligned} \tag{4.8}$$

In which m is the number of non-support nodes; n_c represents the number of compression elements; σ_i and δ_i are the stress and nodal displacement, respectively; σ_i^b denotes allowable buckling stress in member i When working under compression.

4.5.2 *Introduction of Selected Chaos Map*

Chaos dynamic series are formed using chaos functions. These series can provide convergence towards global responses and create a suitable potential to avoid falling into the trap of local optima. By embedded chaos functions, the non-linear and non-convex behavior of the objective function in the optimization of truss structures can be easily controlled and adjusted. Among the advantages of the series formed by the mediation of chaos functions are non-repetition and non-convergence towards a certain limit. The meta-heuristic Teaching–Learning-based Optimization (TLBO), in dealing with non-linear and non-convex problems with one of the problems related to the phase of the teacher or the phase of the student will have premature convergence. The *Logistic chaos function* [6] converges from a local minimum to the global value and will be effective for modifying the exploration conditions of the algorithm. But the *Gaussian chaos function* [7] is formed in the range of local optima and will improve the exploitation conditions of the algorithm. Also, the *tent chaos function* will be suitable for improving both conditions at the same time. Therefore, by choosing these functions, the weakness of algorithms of any kind is improved. The relationships related to these chaos functions have been introduced in the previous chapter.

4.5.3 *Numerical Examples of Optimal Truss Design*

In this section, common truss optimization examples as benchmark problems are used for optimization using some chaotic meta-heuristic algorithm. Chaotic modes are used for both discrete and continuous variable problems. By examining the competition between chaos functions and triple scenarios, the improved results are compared with the standard case. In fact, 10 independent answers compete with each other. Due to the variety of reviews, the optimal responses will be largely close to the global optimality. The examples provided are as follows:

- The 25-bar spatial truss
- The 47-bar power transmission tower
- The 200-bar planar truss structure

4.5.3.1 A 25-Bar Spatial Truss

The topology and nodal numbering of the 25-bar spatial truss is as shown in Fig. 4.1. The material density is 0.1 lb/in^3 and the modulus of elasticity of the members is 10,000 ksi. The structure includes 25 members, which are divided into eight groups according to Table 4.1. Also, allowable compressive and tensile stresses have been determined for each type.

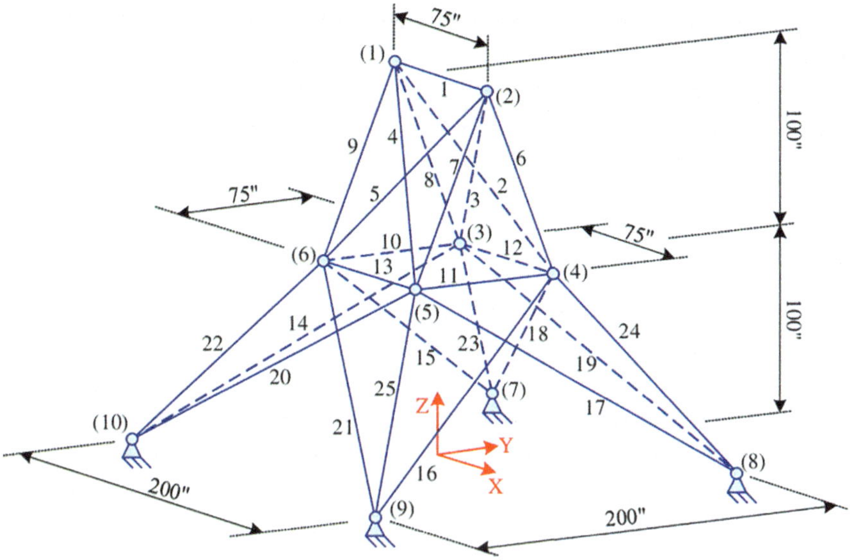

Fig. 4.1 Schematic of a twenty-five-bar spatial truss

Table 4.1 Grouping of members and stresses limitation for the 25-member spatial truss

Number group	Element group	Compressive stress limitation (ksi)	Tensile stress limitation (ksi)
1	A_1	35.092	40
2	A_2–A_5	11.590	40
3	A_6–A_9	17.305	40
4	A_{10}–A_{11}	35.092	40
5	A_{12}–A_{13}	35.092	40
6	A_{14}–A_{17}	6.759	40
7	A_{18}–A_{21}	6.959	40
8	A_{22}–A_{25}	11.082	40

This spatial truss is subjected to two Loading conditions as shown in Table 4.2. Also, the maximum displacement limitations for the nodes in all directions of the axes is $\pm$ 0.35 in. The range of choice of members' cross-section or range of decision variables is in the range of 0.01–3.4 in^2. In order to ensure the performance of chaotic maps and algorithms, as well as to increase the accuracy and sensitivity of calculations, each of the modes has been performed independently 30 times and the results related to the best response and the average value of responses are presented in Table 4.3. The graph of these results is also shown in Fig. 4.2. The coefficient of variation of the responses, which indicates the stability and robustness of the responses, has been calculated and another criterion for the efficiency of chaos functions and

Table 4.2 Loading condition for the 25-member spatial truss

Node	Load case 1			Load case 2		
Point	F_x (kips)	F_y (kips)	F_z (kips)	F_x (kips)	F_y (kips)	F_z (kips)
1	0	20	− 5	1	10	− 5
2	0	− 20	− 5	0	10	− 5
3	0	0	0	0.5	0	0
4	0	0	0	0.5	0	0

algorithms has been obtained. After reviewing the optimization results for different combinations of algorithms with chaos map and comparing it with the initial and standard mode, a significant improvement in reducing the weight of the 25-bar spatial truss results. The enhanced results for Chaotic Teaching–Learning-based Optimization (CTLBO), by Logistic chaos map and scenario 3, a weight of 544.2504 kg has the optimal response.

4.5.3.2 A 47-Bar Power Transmission Tower

The topology and nodal numbering of the power transmission tower that consists of 47 members and 22 nodes, as shown in Fig. 4.3. The density of material is 0.3 lb/in^3 and the modulus of elasticity of the members is 30,000 ksi. Both stress and buckling limits must be satisfied for all member. The allowable stress of tensile members is 20 ksi and for compressive members 15 ksi is recommended. Euler buckling compressive stress for members is also calculated from the following equation.

$$\sigma_i^e = \frac{-KEA_i}{L_i^2}; \quad i = 1, 2, ..., 47 \tag{4.9}$$

where K is a constant coefficient that must be selected according to the geometric type of the section, E is the modulus of elasticity of the material, A_i is cross-sectional area of member and L_i is the length of member. Here we consider k to be 3.96. Structural nodes are affected by a combination of triple loading modes. The first group includes loads of 6 kip in the positive direction of the X-axis and 14 kip in the negative direction of the Y-axis, which are acting in nodes 17 and 22. The second group consists of loads of 6 kg in the positive direction of the X-axis and 14 kg in the negative direction of the Y-axis, which acting only at node 17.

The third group includes loads of 6 kP in the positive direction of the X-axis and 14 kP in the negative direction of the Y-axis, which acting only in node 22. The first group represents the oblique loads of both power transmission lines and the second and third groups represent the state where one of the two lines breaks. Also, truss members are classified into 27 groups according to geometric symmetry. The cross-section areas of the members, were chosen from the 64 discrete values of the AISC code. In order to ensure the performance of chaos map and algorithms, as well as

Table 4.3 Optimal design comparison for the 25-bar spatial truss

Number group	Element group	Standard	Logistic map			Tent map			Gauss map		
		TLBO	CTLBO 21	CTLBO 22	CTLBO 23	CTLBO 31	CTLBO 32	CTLBO 33	CTLBO 41	CTLBO 42	CTLBO 43
1	A_1	0.010018	0.01	0.01	0.01	0.010193	0.01	0.011958	0.010772	0.01	0.01
2	A_2–A_5	1.975	1.9858	1.9917	1.9829	1.9709	1.9635	1.9921	1.9492	1.9684	1.9444
3	A_6–A_9	2.997	2.9835	2.9711	2.9815	3.0087	3.0191	2.9711	3.0304	3.0107	3.0295
4	A_{10}–A_{11}	0.010293	0.01	0.01	0.01	0.010053	0.01	0.010058	0.010168	0.010066	0.010004
5	A_{12}–A_{13}	0.010029	0.010387	0.01	0.010011	0.01	0.01	0.010494	0.011256	0.01	0.01
6	A_{14}–A_{17}	0.69063	0.68331	0.68369	0.68319	0.66966	0.68781	0.68367	0.68051	0.6808	0.68788
7	A_{18}–A_{21}	1.6816	1.6769	1.677	1.6783	1.6793	1.6808	1.676	1.6874	1.6797	1.6909
8	A_{22}–A_{25}	2.648	2.6598	2.6614	2.6579	2.6706	2.6453	2.6615	2.6482	2.6554	2.6376
Best weight		544.7586	544.5067	544.4059	**544.2504**	544.5631	544.6927	544.3725	544.5591	544.5368	544.4636
Mean weight		544.995	544.9874	544.8229	**544.7604**	544.879	544.9063	544.8246	544.8052	544.7831	544.9125
CV		0.034468	0.052891	0.05324	0.067707	0.044183	0.030633	0.075565	0.040728	**0.010066**	0.057458

The bold items in the table indicate the optimal values for the best weight, the best mean and the best coefficient of variation

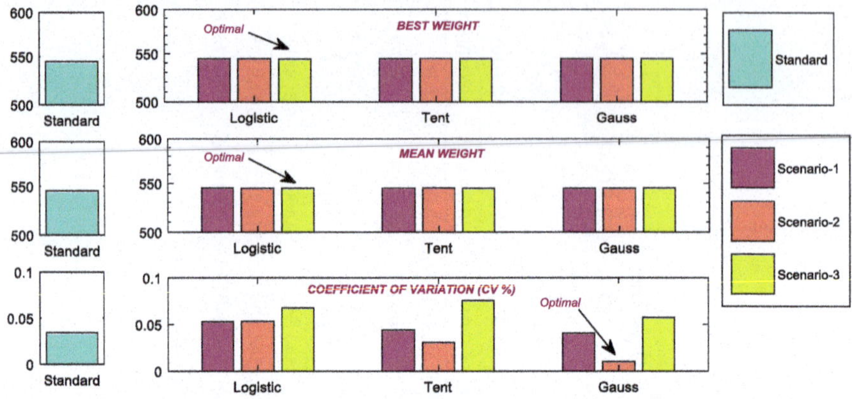

Fig. 4.2 Optimization results in standard mode and selection of chaos map for the 25-member spatial truss

to increase the accuracy and sensitivity of calculations, each of the modes has been performed independently 30 times and the results related to the best response and the average value of responses are presented in Table 4.4. The graph of these results is also shown in Fig. 4.4. The coefficient of variation of the responses, which indicates the stability and robustness of the responses, has been calculated and another criterion for the efficiency of chaos maps and algorithms has been obtained. After reviewing the optimization results for different combinations of algorithms with chaos map and comparing it with the initial and standard mode, a significant improvement in reducing the weight of the 47-bar power transmission tower results. The enhanced results for Chaotic Teaching–Learning-based Optimization (CTLBO), by Gauss chaos map and scenario 2, a weight of 2363.293 kg has the optimal response.

4.5.3.3 A 200-Bar Planar Truss Structure

The topology and nodal numbering of the 200-bar planar truss structure are shown in Fig. 4.5. The material density is 0.283 lb/in^3 and the module of elasticity is 30,000 ksi. Allowable stress of members 10 klb has been assumed. No deformation restrictions are intended for optimization. Truss loading is done in three independent groups. The first group is the lateral load of the structure, which includes a load of 1 kip along the positive axis of Xs and is applied in nodes 1, 6, 15, 20, 29, 34, 43, 48, 57, 62 and 71.

The second group covers the gravity load of the structure, which includes a load of 10 kip in the negative direction of the axis of the Ys and in nodes 1, 2, 3, 4, 5, 6, 8, 10, 12, 14, 15, 16, 17, 18, 19 and... 71, 72, 73, 74 and 75 apply. The third group applies both loading groups together. Truss members are also classified into 29 groups according Table 4.5. The minimum cross section of the members, which is the lower limit of the decision variables, is considered 0.1 in^2 and the maximum cross section

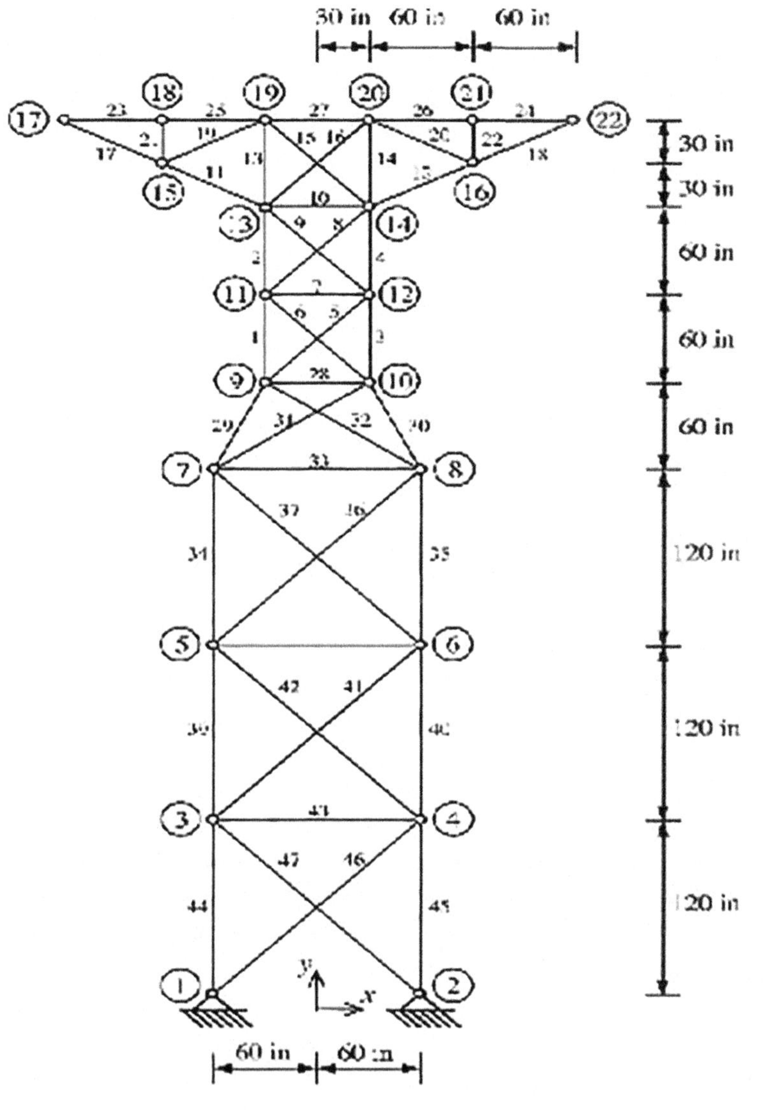

Fig. 4.3 Schematic of a 47-bar power transmission tower

is 16 in^2. In order to ensure the performance of chaotic maps and algorithms, as well as increasing the accuracy and sensitivity of calculations, each state is independently executed 30 times and the results of the best response and the average amount of responses are presented in statistical Table 4.6, and shown in Fig. 4.6. Also, the coefficient of changes in responses, which indicates the stability and robustness of the responses, has been calculated and another criterion for the efficiency of chaotic functions and algorithms has been achieved. After reviewing the optimization results

Table 4.4 Optimal design comparison for the 47-bar power transmission tower

Number group	Element group	Standard	Logistic map				Tent map			Gauss map	
		TLBO	CTLBO 21	CTLBO 22	CTLBO 23	CTLBO 31	CTLBO 32	CTLBO 33	CTLBO 41	CTLBO 42	CTLBO 43
1	A_1–A_3	3.84	3.84	3.84	3.84	3.84	3.84	3.84	3.84	3.84	3.84
2	A_2–A_4	3.38	3.38	3.38	3.38	3.38	3.38	3.38	3.38	3.38	3.38
3	A_5–A_6	0.994	0.785	0.785	0.785	0.766	0.785	0.994	0.766	0.766	0.994
4	A_7	0.766	0.196	0.111	0.25	0.111	0.111	1.228	0.111	0.111	0.307
5	A_8–A_9	0.994	0.994	0.785	0.994	0.994	0.785	0.994	0.785	0.785	0.994
6	A_{10}	1.8	1.99	1.99	1.99	1.8	2.38	1.8	1.99	1.85	1.99
7	A_{11}–A_{12}	2.13	2.13	2.13	2.13	2.13	2.13	2.13	2.13	2.13	2.13
8	A_{13}–A_{14}	1.228	1.228	1.228	1.228	1.228	1.266	1.266	1.228	1.228	1.228
9	A_{15}–A_{16}	1.563	1.563	1.563	1.563	1.62	1.563	1.563	1.563	1.563	1.563
10	A_{17}–A_{18}	2.13	2.13	2.13	2.13	2.13	2.13	2.13	2.13	2.13	2.13
11	A_{19}–A_{20}	0.111	0.111	0.111	0.111	0.111	0.111	0.111	0.111	0.111	0.141
12	A_{21}–A_{22}	0.111	0.111	0.111	0.111	0.111	0.111	0.141	0.111	0.111	0.111
13	A_{23}–A_{24}	1.8	1.8	1.8	1.8	1.8	1.8	1.8	1.8	1.8	1.8
14	A_{25}–A_{26}	1.8	1.8	1.8	1.8	1.8	1.8	1.8	1.8	1.8	1.8
15	A_{27}	1.563	1.457	1.457	1.457	1.8	1.457	1.457	1.457	1.457	1.457
16	A_{28}	0.442	0.442	0.391	0.391	0.442	0.563	0.391	0.563	0.543	0.391
17	A_{29}–A_{30}	3.63	3.63	3.63	3.63	3.63	3.63	3.63	3.63	3.63	3.63
18	A_{31}–A_{32}	1.563	1.563	1.563	1.563	1.563	1.563	1.563	1.563	1.457	1.563
19	A_{33}	0.391	0.563	0.563	0.563	0.442	0.563	0.563	0.442	0.25	0.563
20	A_{34}–A_{35}	3.09	3.09	3.10	3.09	3.09	3.09	2.98	3.09	3.09	3.09
21	A_{36}–A_{37}	1.457	1.457	1.457	1.457	1.457	1.457	1.457	1.457	1.228	1.457

(continued)

Table 4.4 (continued)

Number group	Element group	Standard	Logistic map				Tent map			Gauss map	
		TLBO	CTLBO 21	CTLBO 22	CTLBO 23	CTLBO 31	CTLBO 32	CTLBO 33	CTLBO 41	CTLBO 42	CTLBO 43
22	A_{38}	0.25	0.196	0.111	0.141	0.25	0.111	0.196	0.196	0.371	0.111
23	A_{39}–A_{40}	3.73	3.70	3.70	3.71	3.73	3.71	3.70	3.69	3.74	3.70
24	A_{41}–A_{42}	1.563	1.563	1.563	1.563	1.563	1.563	1.62	1.563	1.563	1.563
25	A_{43}	0.141	0.111	0.141	0.111	0.111	0.196	0.141	0.196	0.111	0.111
26	A_{44}–A_{45}	4.84	4.53	4.53	4.53	4.55	4.52	4.526	4.532	4.59	4.53
27	A_{46}–A_{47}	1.457	1.457	1.563	1.457	1.457	1.62	1.563	1.563	1.457	1.457
Best weight		2445.799	2405.098	2401.543	2403.892	2409.445	2420.091	2441.354	2403.060	**2363.293**	2414.966
Mean weight		2448.287	2410.264	2408.298	2407.587	2415.326	2422.019	2450.84	2412.73	**2375.124**	2423.183
CV		0.0682	0.10542	0.0824	0.0759	0.11556	**0.05951**	0.19433	0.2681	0.32143	0.2143

The bold items in the table indicate the optimal values for the best weight, the best mean and the best coefficient of variation

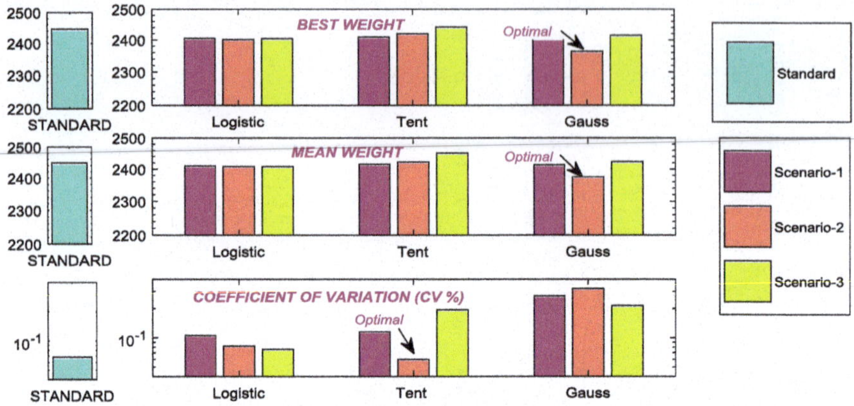

Fig. 4.4 Optimization results in standard mode and selection of chaos map for the 47-bar power transmission tower

for different combinations of algorithms with chaos map and comparing it with the initial and standard mode, a significant improvement in reducing the weight of the 200-bar planar truss results. The enhanced results for Chaotic Teaching–Learning-based Optimization (CTLBO), by Gauss chaos map and scenario 1, a weight of 25,514.061 kg has the optimal response.

4.6 Truss Weight Optimization with Multiple Frequency Constraints

In cases where the structures are affected by dynamic loads such as earthquakes and storms, in order to prevent the phenomenon of resonance, their natural frequency should be limited in a certain range [4]. In order to apply this type of restrictions, the natural frequencies of the structures contain all the required information about the dynamic behavior of the structures. In low frequency vibration problems, the response of the structure depends on the basic frequencies and the mode shape, and by applying the frequency limit, the dynamic behavior of the structure can be easily controlled. In a number of optimization problems, the effects of some modes can be reduced by using these relationships. For example, in the design of airplane wings, the effort is to reduce bending and twisting modes. In addition to being non-linear and non-convex, frequency limit relationships have complex implicit variables that most meta-heuristic algorithms for optimization are caught in the trap of local optima. Therefore, the embedding of chaos functions provides suitable conditions for leaping and mutating towards the global optimum. These functions have good conditions for problems with multi-frequency constraints and significantly improve the results by creating a balance between the exploration and exploitation steps. *Gaussian,*

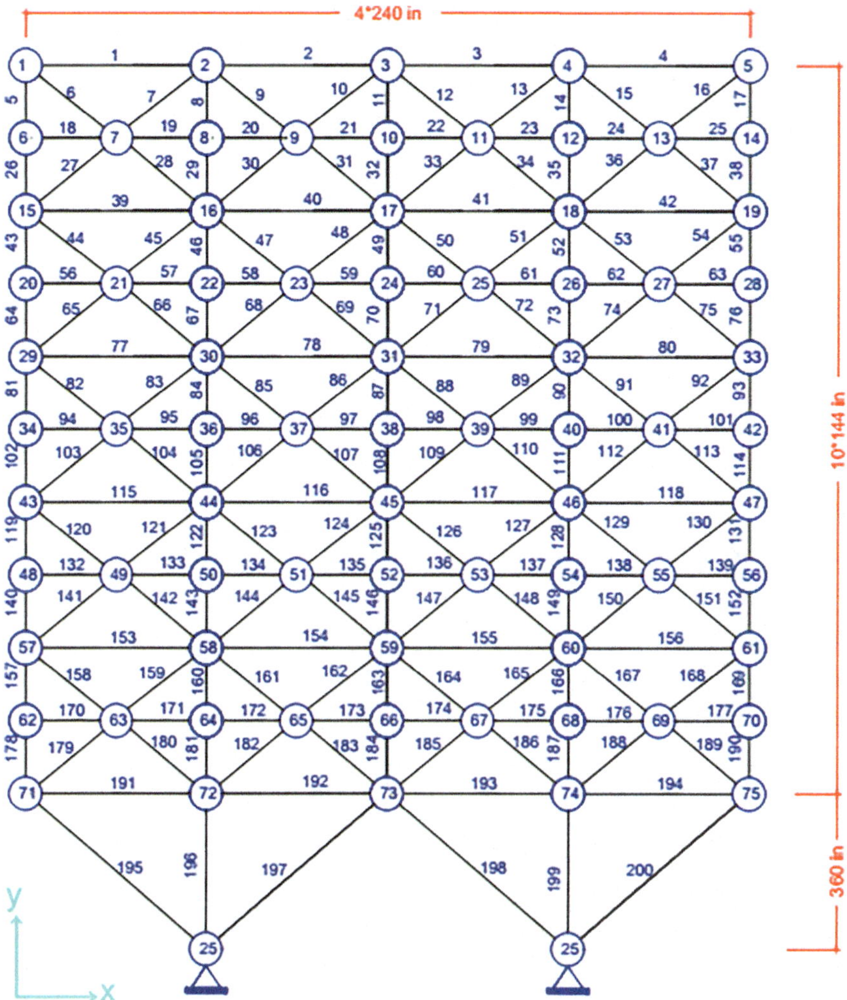

Fig. 4.5 Schematic of a 200-bar planar truss structure

Liebovitch and *Piecewise* chaos functions provide suitable conditions for the meta-heuristic algorithm of Teaching–Learning-based Optimization (TLBO), to improve the steps related to the teacher phase and the student phase in three scenarios. Also, in order to increase the diversity in the statistical space, each model was implemented 20 times and the best weight, the best average and the coefficient of variation were presented. By choosing the best answer among the improved designs, the appropriate chaos function is selected and we learn about the weakness of the algorithm in the exploration or exploitation mode.

Table 4.5 Classification of members in the 200-bar planar truss structure

Group	Element group	Group	Element group
A_1	1, 2, 3, 4	A_{16}	82, 83, 85, 86, 88, 89, 91, 92, 103, 104, 106, 107, 109, 110, 112, 113
A_2	5, 8, 11, 14, 17	A_{17}	115, 116, 117, 118
A_3	19, 20, 21, 22, 23, 24	A_{18}	119, 122, 125, 128, 131
A_4	18, 25, 56, 63, 94, 101, 132, 139, 170, 177	A_{19}	133, 134, 135, 136, 137, 138
A_5	26, 29, 32, 35, 38	A_{20}	140, 143, 146, 149, 152
A_6	6, 7, 9, 10, 12, 13, 15, 16, 27, 28, 30, 31, 33, 34, 36, 37	A_{21}	120, 121, 123, 124, 126, 127, 129, 130, 141, 142, 144, 145, 147, 148, 150, 151
A_7	39, 40, 41, 42	A_{22}	153, 154, 155, 156
A_8	43, 46, 49, 52, 55	A_{23}	157, 160, 163, 166, 169
A_9	57, 58, 59, 60, 61, 62	A_{24}	171, 172, 173, 174, 175, 176
A_{10}	64, 67, 70, 73, 76	A_{25}	178, 181, 184, 187, 190
A_{11}	44, 45, 47, 48, 50, 51, 53, 54, 65, 66, 68, 69, 71, 72, 74, 75	A_{26}	158, 159, 161, 162, 164, 165, 167, 168, 179, 180, 182, 183, 185, 186, 188, 189
A_{12}	77, 78, 79, 80	A_{27}	191, 192, 193, 194
A_{13}	81, 84, 87, 90, 93	A_{28}	195, 197, 198, 200
A_{14}	95, 96, 97, 98, 99, 100	A_{29}	196, 199
A_{15}	102, 105, 108, 111, 114		

4.6.1 *Truss Size and Layout Optimization with Multi Frequency Constraints*

In truss design with multi-frequency constraints and optimization of size and shape, the goal is to minimize the weight of the structure while satisfying some limiting on natural frequencies of the structure. In this type of optimization, the decision variables include the cross-section of the members or the geometric coordinates of some nodes. There will be no changes in the topology of the structure, and the connection conditions remain unchanged during the optimization process. The decision variables should be chosen within the corresponding range of the upper bound and the lower bound of their search space. Optimization problems can be represented by the mathematical relationship according Eq. (4.10):

$$\begin{aligned}
&Find && X = [x_1, x_2, \ldots, x_n] \\
&to\ minimize && P(X) = f(X) \times f_{penalty}(X) \\
&subjected\ to && \omega_j \le \omega_j^* && for\ some\ natural\ frequencies\ j \\
& && \omega_k \ge \omega_k^* && for\ some\ natural\ frequencies\ k \\
& && x_{i,\min} \le x_i \le x_{i,\max}
\end{aligned} \tag{4.10}$$

Table 4.6 Optimal design comparison for 200– bar planar truss structure

Number group	Standard	Logistic map			Tent map			Gauss map		
	TLBO	CTLBO 21	CTLBO 22	CTLBO 23	CTLBO 31	CTLBO 32	CTLBO 33	CTLBO 41	CTLBO 42	CTLBO 43
A_1	0.114526	0.103266	0.10299	0.11724	0.143901	0.101351	0.129418	0.148	0.1611	0.135897
A_2	1.03985	0.956579	1.22393	0.93484	0.9553	0.941239	0.966012	0.946	0.9553	0.998903
A_3	0.117996	0.108242	0.110766	0.104948	0.108631	0.100277	0.121147	0.101	0.1	0.10168
A_4	0.1	0.1	0.100236	0.106225	0.102363	0.1	0.10049	0.101	0.1004	0.100115
A_5	1.95852	1.93042	1.99314	1.93127	1.94154	1.93341	1.94053	1.9461	1.9662	2.0065
A_6	0.276925	0.275805	0.264903	0.283578	0.311233	0.258503	0.300406	0.2979	0.3055	0.293183
A_7	0.1	0.113258	0.106881	0.122309	0.1	0.100527	0.109354	0.101	0.1	0.102691
A_8	3.06448	3.01761	3.05175	3.13427	3.24185	3.13134	3.09105	3.1072	3.1618	3.08104
A_9	0.12921	0.1078	0.158345	0.10008	0.133385	0.866071	0.109461	0.101	0.1152	0.122218
A_{10}	4.05135	4.03048	4.04926	4.06089	4.37226	3.85129	4.08181	4.1062	4.2405	4.07415
A_{11}	0.404649	0.403106	0.430658	0.389666	0.457593	0.516709	0.48483	0.4049	0.4046	0.416374
A_{12}	0.113312	0.231064	0.156036	0.257661	0.177943	0.482961	0.11871	0.1844	0.1	0.103959
A_{13}	5.20321	5.23735	5.39107	5.27569	5.55548	5.05041	5.17357	5.4299	5.4132	5.33165
A_{14}	0.215375	0.565002	0.105332	0.305508	0.106538	0.359028	0.214968	0.101	0.1545	0.101
A_{15}	6.28051	6.23436	6.38324	6.25731	7.07889	6.14677	6.17556	6.4299	6.3976	6.25704
A_{16}	0.564362	0.732084	0.558962	0.658518	0.619647	0.764901	0.602926	0.5755	0.5555	0.525132
A_{17}	0.513539	0.103304	0.131234	0.170068	0.128856	0.198731	0.592307	0.1349	0.4405	0.192131
A_{18}	7.62574	7.8308	7.79296	7.79576	8.04232	7.64721	7.7218	7.9747	8.0928	7.52387
A_{19}	1.22555	0.16712	0.726828	0.100111	0.163398	0.411206	0.113518	0.101	0.1004	0.695692
A_{20}	8.60563	8.85312	8.77319	8.63541	9.08694	9.62978	8.77562	8.9747	8.9918	8.79145
A_{21}	1.41028	0.76966	1.02058	0.738673	0.745399	0.746353	0.835812	0.70648	0.8925	0.811709

(continued)

Table 4.6 (continued)

Number group	Standard	Logistic map			Tent map			Gauss map		
	TLBO	CTLBO 21	CTLBO 22	CTLBO 23	CTLBO 31	CTLBO 32	CTLBO 33	CTLBO 41	CTLBO 42	CTLBO 43
A_{22}	0.19623	1.23804	0.163675	1.81057	0.151994	0.905774	0.591056	0.4225	0.2544	1.43745
A_{23}	11.4351	10.6177	10.9458	11.0266	10.8481	10.5831	10.9938	10.8685	11.1214	10.1496
A_{24}	0.224448	0.104885	0.905955	0.140675	0.113596	0.103505	0.797625	0.101	0.1	0.740378
A_{25}	11.9284	11.9519	11.9072	11.8422	11.8114	11.3227	11.9659	11.8684	12.3304	11.1536
A_{26}	1.28688	1.53247	1.49744	1.79926	0.919436	1.12549	1.70706	1.0359	1.011	1.8649
A_{27}	6.97565	7.433	6.8991	6.5646	7.35603	7.61553	6.6081	6.6859	6.4103	7.41621
A_{28}	12.66797	12.1696	12.98569	12.47078	11.1591	12.9978	12.35856	10.911	10.5814	11.58264
A_{29}	13.6741	13.7156	13.7172	14.004	13.6227	13.3185	13.9579	13.8464	14.1288	14.1
Best weigh	27,406.80	27,020.71	27,387.59	27,299.42	26,007.82	27,213.68	27,164.03	**25,514.06**	25,699.75	26,905.89
Mean weight	28,428.45	27,304.66	27,515.02	27,446.97	26,086.63	27,576.09	28,011.64	26,224.27	**26,067.12**	27,094.79
Coefficient variation	3.555	1.2437	0.47102	0.78233	1.6189	1.6578	2.1787	2.4661	1.4919	**0.46121**

The bold items in the table indicate the optimal values for the best weight, the best mean and the best coefficient of variation

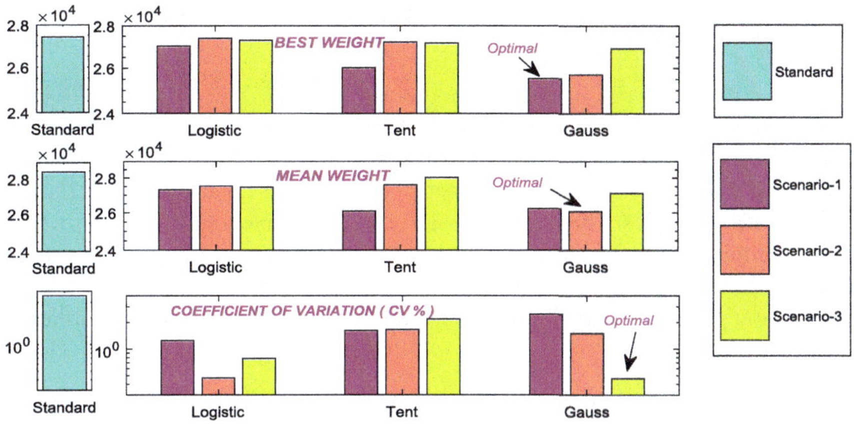

Fig. 4.6 Optimization results in standard and chaos map for the 200-bar planar truss structure

In this equation, X is the vector of decision variables, including the geometric coordinates of the nodes and the cross-sectional area of the members, n is the number of variables or the dimension of the search space, which is introduced based on the grouping of members and according to the symmetry of the structure. P(X) is the penalized objective function or merit function that must reach a minimum value. f(X) is the cost function that is considered as the weight in the optimization of structures. $f_{penalty}(X)$ is the penalty function that is used to apply constraints. In cases where some constraints related to the problem are violated, this function enters the calculation process with a suitable weight. w_j expresses the jth natural frequency of the structure and w_j^* is its upper bound. W_k expresses the kth natural frequency of the structure and w_k^* is its lower bound. X_{imin} and X_{imax} are the lower and upper bound of the decision variable x_i, respectively. The cost function is expressed by the mathematical relationship according to Eq. (4.11):

$$f(X) = \sum_{i=1}^{nm} \gamma_i \cdot L_i \cdot A_i \tag{4.11}$$

In this equation, γ_i is the specific material density of member i, L_i is the length of member I, and A_i is the cross-sectional area of member i. The penalty function is defined by the mathematical relation according to Eq. (4.12):

$$f_{penalty}(X) = (1 + \varepsilon_1 \cdot \nu)^{\varepsilon_2}, \quad \nu = \sum_{i=1}^{q} \nu_i \tag{4.12}$$

where q is the number of frequency constraints and ν_i expresses the violation value which is defined from the following equation:

$$\nu_i = \begin{cases} 0 & if\ the\ ith\ constrain\ is\ satisfied \\ \left|1 - \frac{\omega_i}{\omega_i^*}\right| & else \end{cases} \quad (4.13)$$

The parameters ε_1 and ε_2 are chosen based on creating a balance between exploration and exploitation in the decision variables space. In these investigations, the value of ε_1 is set as unity, and the parameter ε_2 starts from 1.5 and increases linearly to 6 in all test examples. These values penalize the unfeasible solutions more severely as the optimization process proceeds. As a result, in the early stages, the agents are free to explore the search space easily, but in the final iterations, the conditions of exploitation prevail and among them, the best local optimum is selected as the sub global optimum.

4.6.2 Introduction of Selected Chaos Map

The main reasons for the success of using chaos functions in meta-heuristic algorithms is to increase the speed of optimization in non-linear and non-convex problems. One of these cases is multi-frequency limits to prevent the destructive phenomenon of resonance. Also, optimizing the layout and size simultaneously increases the complexity of the objective function and the balance between the exploration and exploitation steps is not achieved. In such cases, chaos functions have good potential to improve optimization results. Choosing the best chaos function and the best scenario will require challenges and competition between different types of chaos functions. Functions that can mutate from local optima to global optima, and solve the premature convergence problem for the meta-heuristic algorithm, are selected by forming a wide range of chaotic states [8]. In this study, in order to diversify and form a wide statistical space, three chaos functions, with three scenarios, embedded in the Teaching–Learning-based Optimization (TLBO). These functions include: *Gossin map*, *Liebovitch map* [9] and *piecewise map* [10]. Investigations show that Liebovitch chaos map converges local optima to global optima with a very high probability, and this function will be suitable for improving exploitation conditions. The Gaussian chaos map is likely to be in the range of local optima, and it is suitable for improving the exploitation conditions. Finally, the piecewise chaos map simultaneously improves both conditions. Therefore, by choosing these functions, the weakness of algorithms of any kind is improved. The relationships of these chaos functions have been introduced in the previous chapter.

4.6.3 Numerical Examples of Optimal Truss Design

Three numerical examples on layout and size optimization are presented in order to check the capability of chaos functions and compare it with the standard case.

By creating a competition between chaotic modes and triple scenarios, a significant improvement in results is achieved. In fact, 10 independent answers compete with each other. Due to the variety of checks, the optimal responses will be very close to the global optimal. The examples provided are as follows:

- The 52-bar dome-like truss
- The 72-bar spatial truss
- The 120-bar spatial dome

4.6.3.1 A 52-Bar Dome-Like Truss

The 52-bar dome-like truss as shown in Fig. 4.7 is a well-known benchmark problem for optimizing the weight and shape of trusses with frequency constraints. This truss considers both optimizations of the section size and geometric coordinates of the nodes and the geometric shape of the structure is determined during the optimization process. The decision variables related to the section size are classified into 8 groups according to the symmetry in the geometric shape. Geometric coordinates of all symmetric free nodes can be changed by 2 m from the initial position along the coordinate axes. In this case, the number of decision variables related to the shape of the structure and the geometric coordinates of the nodes is limited to 5 variables, and the sum of the variables, including shape and size, will be 13 variables. In all free nodes, a non-structural concentrated mass of 50 kg has affected all free nodes. The mechanical characteristics of the structure are: density of materials 7800 kg/m^3, modulus of elasticity 210,000 MPa, frequency limitations of the structure in the first mode are less than 15.916 Hz and in the second mode are greater than 28.648 Hz. For the cross section of the members, the lower limit is 1 cm^2 and the upper limit is 10 cm^2.

In order to ensure the performance of turbulent functions and algorithms, as well as to increase the accuracy and sensitivity of calculations, each of the modes has been performed independently 30 times and the results related to the best response and the average value of responses are presented in Statistical Table 4.7. Also, the coefficient of variation of responses, which is a measure of the robustness and robustness of responses, has been calculated and used to compare the efficiency of turbulence functions and algorithms. For quick access to optimization information, a bar chart of each component is shown in Fig. 4.8. After reviewing the optimization results for different combinations of algorithms with chaos map and comparing it with the initial and standard mode, a significant improvement in reducing the weight of the 52-bar dome-like truss results. The enhanced results for Chaotic Teaching–Learning-based Optimization (CTLBO), by Liebovitch chaos map and scenario 2, a weight of 192.5833 kg has the optimal response.

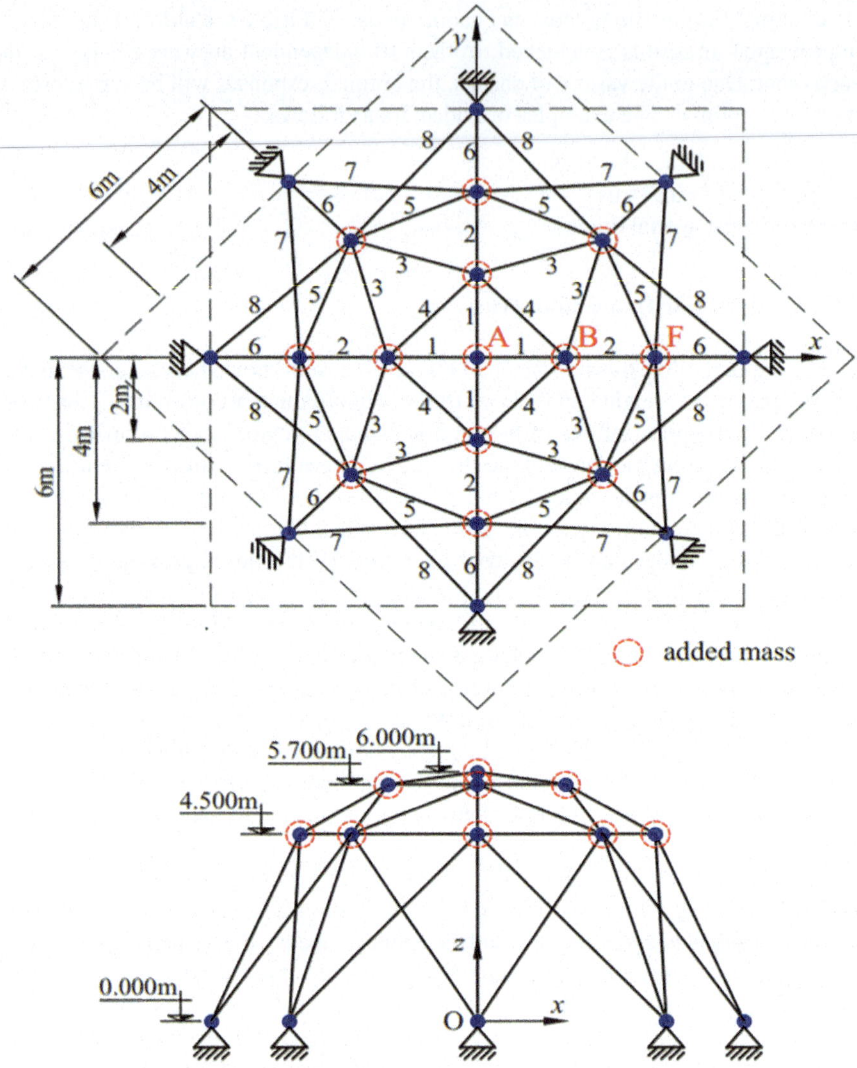

Fig. 4.7 Schematic of A 52-bar dome-like truss

4.6.3.2 A 72-Bar Spatial Truss

The 72-bar spatial truss as shown in Fig. 4.9. is a well-known benchmark problem with weight-limit optimization. This truss only considers "optimizing the size of the sections" and the geometric shape of the structure is constant during the optimization process. The decision variables related to the section size are classified into 16 groups according to the symmetry in the geometric shape. In nodes 1–4, the concentrated non-structural mass of 2270 kg has affected the truss. The mechanical characteristics

Table 4.7 Optimal design comparison for the 52-bar dome-like truss

Decision variable	Standard	Gauss map			Liebovitch map			Piecewise map		
	TLBO	CTLBO 21	CTLBO 22	CTLBO 23	CTLBO 31	CTLBO 32	CTLBO 33	CTLBO 41	CTLBO 42	CTLBO 43
Z_A (m)	5.8504	5.9702	5.9613	5.9833	5.9953	6.06	5.977	5.9488	5.8975	5.9944
X_B (m)	2.3586	2.2842	2.2527	2.2027	2.2877	2.3427	2.2401	2.2376	2.2176	2.2456
Z_B (m)	3.7007	3.7224	3.7294	3.7554	3.737	3.7563	3.7398	3.7287	3.7133	3.7459
X_F (m)	3.9676	3.9749	3.9642	3.9566	3.9871	4.0319	3.9645	3.9556	3.9335	3.9731
Z_F (m)	2.5108	2.5	2.5	2.5	2.5	2.5002	2.5002	2.5	2.501	2.5002
A_1 (cm^2)	1	1	1.0001	1	1	1.0003	1.0002	1.0002	1	1.0013
A_2 (cm^2)	1	1.096	1.1173	1.1565	1.0856	1.0275	1.1252	1.1365	1.1552	1.1247
A_3 (cm^2)	1.1427	1.1961	1.2054	1.2139	1.1875	1.1561	1.2097	1.2161	1.2195	1.2212
A_4 (cm^2)	1.5894	1.4579	1.4675	1.4652	1.4422	1.4522	1.4542	1.4464	1.4686	1.4202
A_5 (cm^2)	1.3817	1.398	1.4023	1.403	1.3975	1.4181	1.4027	1.393	1.3822	1.4047
A_6 (cm^2)	1.0105	1	1	1	1	1	1	1	1	1
A_7 (cm^2)	1.6416	1.5865	1.5912	1.5776	1.5905	1.5555	1.6013	1.5941	1.6039	1.5942
A_8 (cm^2)	1.3936	1.379	1.3667	1.3721	1.3817	1.3914	1.3599	1.3741	1.3676	1.3595
Best weight (kg)	194.7714	192.9113	192.921	192.9099	192.9511	**192.5833**	193.0481	193.0055	192.8472	192.9188
Mean weight (kg)	199.4808	196.5568	193.0511	193.0708	194.8092	**193.0411**	194.8546	198.3462	196.9014	193.1454
Coefficient variation	1.7479	2.444	**0.04594**	0.048783	2.0155	0.1338	2.0017	2.4248	2.3152	0.068056
ω_1 (HZ)	12.4307	11.5879	11.4211	11.1687	11.5951	11.7997	11.3474	11.3811	11.3026	11.3781
ω_2 (HZ)	28.6704	28.6311	28.6319	28.6258	28.6322	28.5888	28.6401	28.6371	28.6202	28.6284

The bold items in the table indicate the optimal values for the best weight, the best mean and the best coefficient of variation

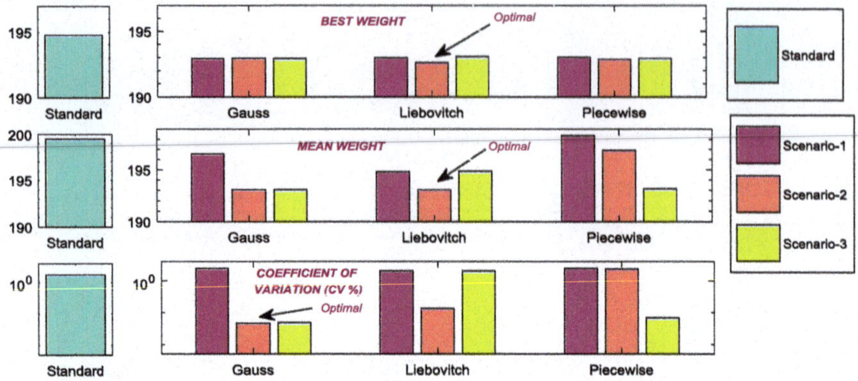

Fig. 4.8 Optimization results in standard mode and selection of chaos map for the 52-bar dome-like truss

of the structure are: material density 2770 kg/m^3, modulus of elasticity 69,800 MPa, frequency limits of the structure in the first mode are equal to 4 Hz and in the second mode are greater than 6 Hz. For the cross section of the members, the lower limit is 0.645 cm^2 and the upper limit is 20 cm^2. In order to ensure the performance of chaotic functions and algorithms, as well as to increase the accuracy and sensitivity of calculations, each of the modes has been performed independently 30 times and the results related to the best response and the average value of responses are presented in Table 4.8. Also, the coefficient of variation of responses, which is a measure of the robustness and robustness of responses, has been calculated and used to compare the efficiency of turbulence functions and algorithms. For quick access to optimization information, a bar chart of each component is shown in Fig. 4.10.

After reviewing the optimization results for different combinations of algorithms with chaos map and comparing it with the initial and standard mode, a significant improvement in reducing the weight of the 72-bar spatial truss results. The enhanced results for Chaotic Teaching–Learning-based Optimization (CTLBO), by Piecewise chaos map and scenario 2, a weight of 323.0594 kg has the optimal response.

4.6.3.3 A 120-Bar Spatial Dome

The 120-bar spatial dome as shown in Fig. 4.11 is a well-known benchmark problem with weight-limit optimization. This truss only considers "optimizing the size of the sections" and the geometric shape of the structure is constant during the optimization process. The decision variables related to the size of the members' sections and according to the symmetry in the geometric shape of the dome along the X and Y axes, are classified into 7 groups. Non-structural concentrated mass in all free nodes affects the structure. Their values are 3 kg in node 1500 kg in nodes 2–13 and 100 kg in other nodes. The mechanical characteristics of the structure are: material density

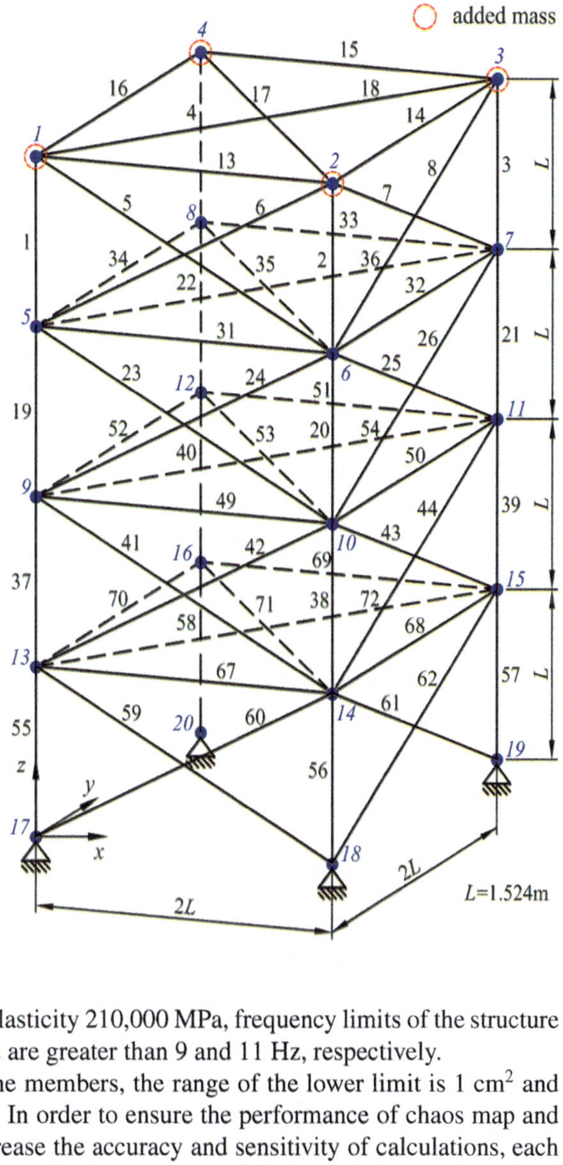

Fig. 4.9 Schematic of a 72-bar spatial truss

7971.81 kg/m^3, modulus of elasticity 210,000 MPa, frequency limits of the structure in the first and second modes are greater than 9 and 11 Hz, respectively.

For the cross section of the members, the range of the lower limit is 1 cm^2 and the upper limit is 129.3 cm^2. In order to ensure the performance of chaos map and algorithms, as well as to increase the accuracy and sensitivity of calculations, each of the modes has been performed independently 30 times and the results related to the best response and the average value of responses are presented in Statistical Table 4.9. Also, the coefficient of change of responses, which is a measure of the robustness and robustness of responses, has been calculated and used to compare the efficiency of turbulence functions and algorithms. For quick access to optimization information, the bar chart of each component is shown in Fig. 4.12. After reviewing the optimization results for different combinations of algorithms with chaos map and comparing it with the initial and standard mode, a significant improvement in

Table 4.8 Optimal design comparison for the 72-bar spatial truss

Element group	Standard	Gauss map			Liebovitch map			Piecewise map		
	TLBO	CTLBO 21	CTLBO 22	CTLBO 23	CTLBO 31	CTLBO 32	CTLBO 33	CTLBO 41	CTLBO 42	CTLBO 43
A_1–A_4	3.47194	3.70101	3.48713	3.60054	3.78312	3.50561	3.57221	3.60945	3.68127	3.78023
A_5–A_{12}	8.02775	7.85482	7.78138	7.85636	7.84179	7.8695	7.99954	7.97798	7.81484	7.8969
A_{13}–A_{16}	0.509056	0.500378	0.5	0.511278	0.500497	0.500407	0.50303	0.500046	0.500394	0.503224
A_{17}–A_{18}	0.598847	0.566195	0.571144	0.571429	0.565448	0.584176	0.580296	0.573312	0.567089	0.576285
A_{19}–A_{22}	8.3062	8.0383	8.28622	8.32761	8.09924	8.2668	8.09963	7.98236	8.7517	8.29416
A_{23}–A_{30}	8.12682	7.92444	7.92989	7.98546	7.86479	7.92666	7.94487	7.82793	7.84567	7.86395
A_{31}–A_{34}	0.503693	0.502951	0.5	0.500124	0.500772	0.50326	0.50011	0.501665	0.501636	0.5
A_{35}–A_{36}	0.502658	0.501478	0.5	0.50783	0.5	0.5	0.501379	0.502086	0.5	0.5
A_{37}–A_{40}	12.3254	13.2527	13.2483	13.3603	13.1866	12.8194	12.4773	12.9704	12.5582	13.2434
A_{41}–A_{48}	7.98206	7.96439	7.96708	7.91841	8.03097	8.1572	7.99108	7.90408	7.9767	8.06857
A_{49}–A_{52}	0.501951	0.502879	0.500539	0.500052	0.501598	0.506425	0.5	0.504073	0.50073	0.5
A_{53}–A_{54}	0.500063	0.501054	0.500914	0.5	0.501734	0.503869	0.5	0.500087	0.501422	0.508477
A_{55}–A_{58}	17.9989	17.6807	17.6022	17.244	17.6549	17.8499	18.4503	18.0276	17.782	17.2539
A_{59}–A_{66}	8.23607	7.98237	8.08609	8.09535	7.94086	8.00325	7.91977	8.11181	8.02018	8.0053
A_{67}–A_{70}	0.5	0.502659	0.5	0.51334	0.500366	0.50164	0.5	0.500468	0.512838	0.500976
A_{71}–A_{72}	0.500324	0.502391	0.5	0.5	0.501763	0.5	0.512085	0.511041	0.509138	0.5
Best weight (kg)	327.3835	323.3697	323.5514	324.1844	323.0722	324.771	324.2566	323.9809	**323.0594**	324.0367
Mean weight (kg)	327.6981	325.1703	325.6066	**325.1481**	325.4119	325.8839	325.9446	326.0193	325.8049	325.8593

(continued)

Table 4.8 (continued)

Element group	Standard	Gauss map			Liebovitch map			Piecewise map		
	TLBO	CTLBO 21	CTLBO 22	CTLBO 23	CTLBO 31	CTLBO 32	CTLBO 33	CTLBO 41	CTLBO 42	CTLBO 43
Coefficient variation	**0.077136**	0.41409	0.39655	0.21657	0.45831	0.33982	0.33552	0.35584	0.51174	0.33478
ω_1 (HZ)	4.0001	4.0003	4.0001	3.9997	4.0003	4	4	4.0001	4	4
ω_3 (HZ)	6.0001	5.9985	5.9969	5.9923	5.9859	5.9616	5.9923	5.979	5.9939	5.9603

The bold items in the table indicate the optimal values for the best weight, the best mean and the best coefficient of variation

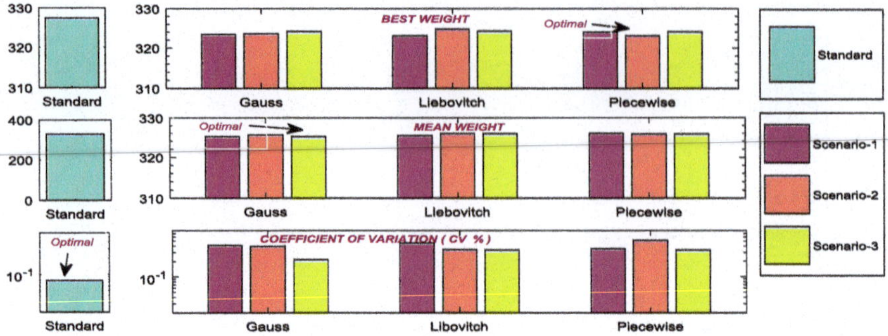

Fig. 4.10 Optimization results in standard mode and selection of chaos map for the 72-bar spatial truss

reducing the weight of the 120-bar spatial dome results. The enhanced results for Chaotic Teaching–Learning-based Optimization (CTLBO), by Liebovitch chaos map and scenario 3, a weight of 8708.51 kg has the optimal response.

4.7 Concluding Remarks

The Teaching–Learning-based Optimization (TLBO), is inspired by the learning process in the classroom. The algorithm consists of two main steps. These steps include the effect of the teacher in improving the average academic level of the students and the mutual influence of the students on each other. Like other population-based algorithms, it starts with a series of initial and random solutions, these solutions are the students' academic level, and the smartest of them is selected as the teacher in each cycle of repetition. First, the average academic level of the students is improved by the teacher, then the students learn from each other. These two phases are known as the teacher's phase and the student's phase and constitute the exploitation and exploration phases. The random distributions used in these two important parts can be replaced with chaos functions and a significant improvement can be concluded. By embedding Chaos functions in the Teaching–Learning-based Optimization (CTLBO), the algorithm is saved from the trap of local optima, and with chaotic jumps, local optima jump to the global optima. As a result, the problem of premature convergence of the algorithm is solved and the balance between the exploration and exploitation steps is achieved. Logistic, Tent and Gaussian chaos functions for problems with static conditions are embedded in the meta-heuristic Teaching–Learning-based Optimization (CTLBO). Also, each chaos function is associated with three scenarios, and these scenarios can improve the parts related to exploration, exploitation, or both at the same time. As a result, the nine chaotic modes compete with each other and the best optimal response is compared with the standard mode. In problems with multi-frequency constraints that are strongly non-linear and non-convex, chaotic models are formed by embedding Gaussian, Liebovitch and

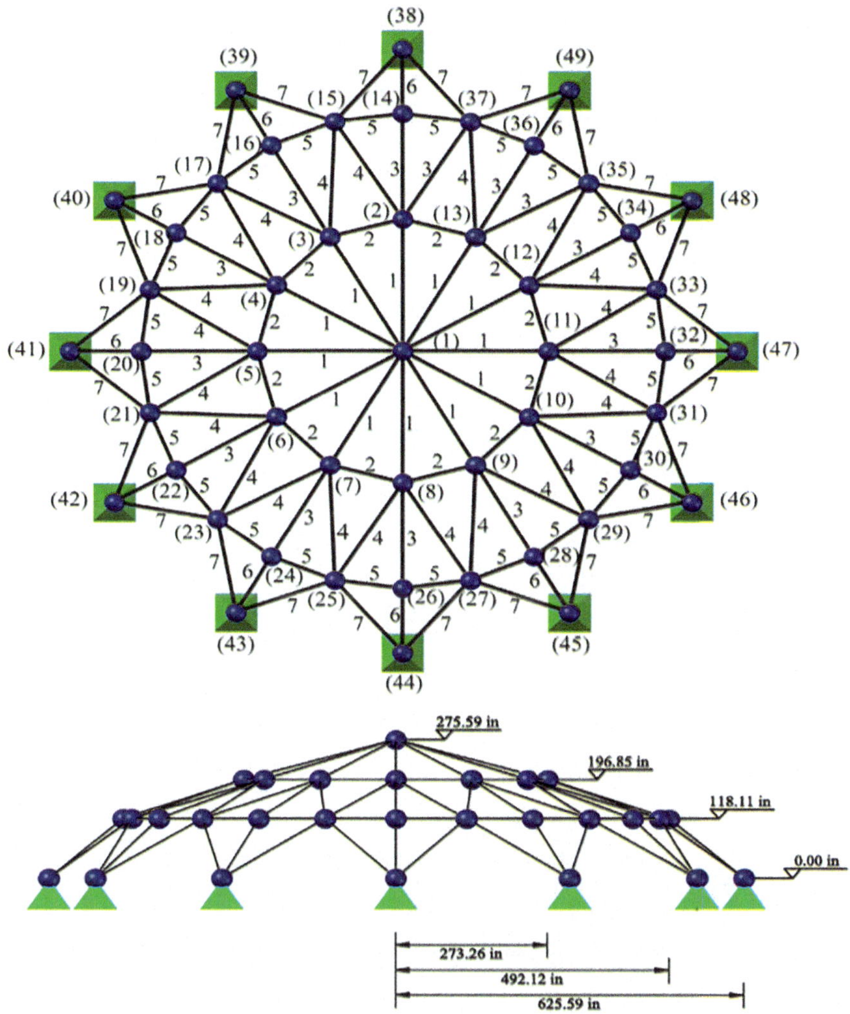

Fig. 4.11 Schematic of a 120-bar spatial dome

Piecewise functions in the Teaching–Learning-based Optimization (CTLBO). As before, each chaos function is associated with three scenarios. Finally, nine chaotic modes compete with each other and the best answer is compared with the standard mode. In both groups of optimization problems, chaotic modes have resulted in significant improvement.

Table 4.9 Optimal design comparison for the 120-bar spatial dome

Decision variable	Standard	Gauss map			Liebovitch map			Piecewise map		
	TLBO	CTLBO 21	CTLBO 22	CTLBO 23	CTLBO 31	CTLBO 32	CTLBO 33	CTLBO 41	CTLBO 42	CTLBO 43
A_1	19.5193	19.5775	19.5988	19.4465	19.4293	19.5141	19.514	19.4526	19.4736	19.5915
A_2	40.3491	40.0623	40.1193	40.5495	40.5881	40.3268	40.325	40.6899	40.5	40.295
A_3	10.4362	10.6327	10.5794	10.6002	10.623	10.5489	10.625	10.5863	10.6038	10.6208
A_4	21.3053	21.1446	21.1129	21.136	21.1032	21.1431	21.0889	21.015	21.0951	21.1436
A_5	9.86125	9.88773	9.82804	9.87759	9.88817	9.88427	9.86931	9.80155	9.85378	9.71522
A_6	12.1834	11.773	11.8165	11.8632	11.8256	11.9115	11.7806	11.8154	11.8265	11.7692
A_7	14.5943	14.8306	14.8729	14.7574	14.7872	14.7778	14.8565	14.9035	14.8222	14.855
Best weight (kg)	8713.148	8708.661	8708.61	8708.831	8708.672	8708.756	**8708.51**	8708.85	8708.616	8709.105
Mean weight (kg)	8716.157	8709.57	8709.616	8709.813	8708.951	8709.645	8709.369	8709.258	**8709.245**	8709.853
Coefficient variation	0.026596	0.010765	0.009924	0.015615	**0.001935**	0.013158	0.009658	0.005437	0.005675	0.007743
ω_1 (HZ)	9.0001	9	9.00006	9.00004	9	9	9	9.00002	9.00002	9.0001
ω_2 (HZ)	11.0007	11	10.9999	11.0001	11	11	11	11	11	11

The bold items in the table indicate the optimal values for the best weight, the best mean and the best coefficient of variation

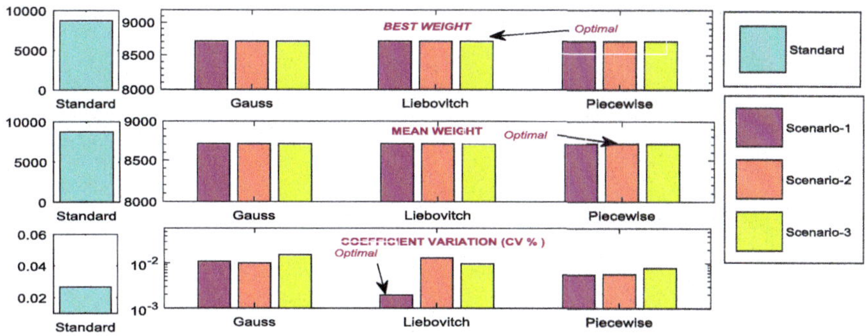

Fig. 4.12 Optimization results in standard mode and selection of chaos map for the 120-bar spatial dome

References

1. Kaveh, A.: Advances in Meta-heuristic Algorithms for Optimal Design of Structures, 2nd edn., 3rd edn. Springer, Cham, Switzerland, (2021). https://doi.org/10.1007/978-3-319-05549-7
2. Rao, R.V., Savsani, V.J., Vakharia, D.P.: Teaching–learning-based optimization: a novel method for constrained mechanical design optimization problems. Compute. Aided Des. **43**(3), 303–315 (2011)
3. Kaveh, A., Yousefpoor, H.: Chaos-embedded meta-heuristic algorithms for optimal design of truss structures. Scientia Iranica A **29**(6), 2868–2885 (2022). https://doi.org/10.24200/sci.2022.59812.6441
4. Kaveh, A., Yosefpoor, H.: Chaotically enhanced meta-heuristic algorithms for optimal design of truss structures with frequency constraints. Periodica Polytechnica Civil Eng **66**(3), 900–921 (2022). https://doi.org/10.3311/PPci.20220
5. Kaveh, A., Zarfam, P., Aziminejad, A., Yosefpoor, H.: Comparison of four chaotic meta-heuristic algorithms for optimal design of large-scale truss structures. Iran. J. Sci. Technol. Trans. Civil Eng. https://doi.org/10.1007/s40996-022-00908-8
6. Ott, E.: Chaos in Dynamical Systems. Cambridge University Press, Cambridge (2002)
7. Bucolo, M., Caponetto, R., Fortuna, L., Frasca, M., Rizzo, A.: Does chaos work better than noise? IEEE Circuit Syst. Mag. **2**(3), 4–19 (2002)
8. Kaveh, A., Sheikholeslami, R., Talatahari, S., Keshvari-Ilkhichi, M.: Chaotic swarming of particles: a new method for size optimization of truss structures. Adv. Eng. Softw. **67**, 136–47 (2014)
9. Erramilli, A., Singh, R., Pruthi, P.: Modeling Packet Traffic with Chaotic Maps. KTH, Stockholm (1994)
10. Peitgen, H.-O., Jürgens, H., Saupe, D.: Chaos and Fractals: New Frontiers of Science. Springer, Berlin (2006)

Chapter 5
Chaotic Biogeography Based Optimization

5.1 Introduction

Meta-heuristic algorithms have special features that have attracted increasing attention from academic and industrial communities. Some of these prominent features are:

- No require for derivatives of the objective function
- More flexibility in a wide range of optimization problems
- Ability to solve continuous, discrete, binary or combined problems
- Less sensitive to increasing problem dimension and number of design variables
- Speed of operation compared to gradient based methods
- High speed of operation to solve permutation, conversion and combination problems

Challenges in recent years have led to the introduction of a large number of meta-heuristics with different inspirations and arguments [1]. On the other hand, the problems that are selected for optimization are classified in different categories based on complexity theory. These groupings according to Fig. 5.1 are:

- Polynomial-time problems that increase the dimensions of the problem, the time required to investigate it grows as a polynomial.
- Non-Polynomial-time problems with increasing problem dimensions, time growth is more than polynomial and in some cases it can be exponential.
- NP-Hard problems, with the increase in the dimensions of the problem, the growth of the time required for the investigation is extremely high, and in problems such as the TSP traveling salesman problem, it is factorial.
- NP-Complete problems are problems in which the time growth is between groups 2 and 3.

A. Kaveh and H. Yousefpoor, *Chaotic Meta-heuristic Algorithms for Optimal Design of Structures*, Studies in Computational Intelligence 1129,
https://doi.org/10.1007/978-3-031-48918-1_5

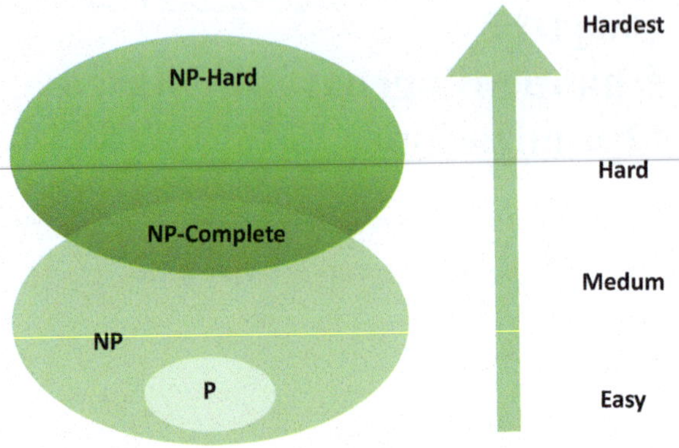

Fig. 5.1 Classification of optimization problems based on complexity theory

We will not have any particular problem in optimization for easy and polynomial problems, but for non-linear, non-convex, NP and NP-Hard problems, the optimization process is slow and stops in some cases.

The main reasons for this problem are the imbalance between the exploration and exploitation stages. In meta-heuristic algorithms, in the exploration phase, the algorithm selects prominent points and indexes of the search space and settles in them, and then in the exploitation phase, it carefully examines the neighborhood of these points. But each of these steps is selected based on random distributions such as Normal, Cauchy, Levy distribution, etc., which plays a key role in the lack of success and lack of balance between the two key steps of exploration and exploitation. Today, the proposed new generation algorithms to improve the efficiency is the Estimation of Distribution Algorithm (EDA_s), in which the suitable distribution function is proposed for the variables. Socha and Dorigo have presented the Ant Colony Optimization for continuous domains (ACO_R), based on the estimation of the distribution function. By performing rotation and transfer in the space of decision variables based on the correlation coefficient, it is possible to form independent linear orthogonal bases like the Gram-Schmidt inner multiplication method and then estimate the appropriate distribution function with the decision variables Fig. 5.2. The condition of linear independence for the vectors related to the decision variables is checked based on the Gram-Schmidt determinant according to Eq. (5.1).

$$for \quad \{v_1, v_2, \ldots, v_n\} \Rightarrow \det[\langle v_i, v_j \rangle] = 0, \quad 1 \le i, j \le n \tag{5.1}$$

Although performing these mathematical operations greatly improves the access to optimal answers, it will occupy a large part of the processing time. Therefore, in recent research, for quick access, instead of estimating the distribution function, chaotic accelerators can be applied as a suitable alternative in the random distributions section of algorithms. With this replacement, the problem of premature convergence in algorithms and also a large number of pests that threatened meta-heuristic algorithms are solved. The method of replacement is done with one of the following scenarios: replacing the chaos functions in the random distribution of the exploration

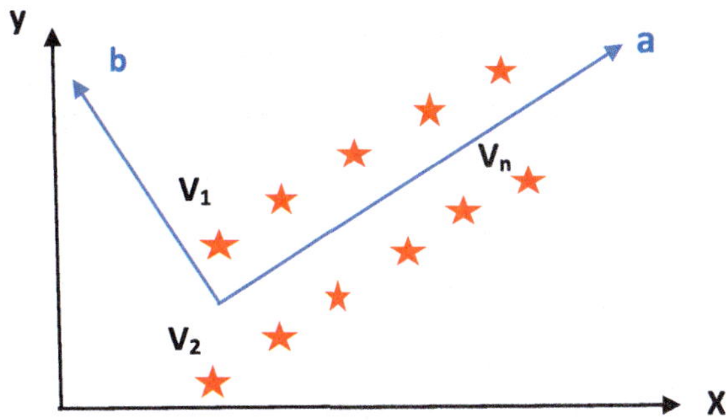

Fig. 5.2 Rotation and transfer based on the correlation coefficient

stage, replacing the chaos functions in the random distribution of the exploitation stage and replacing the chaos functions in both the exploration and exploitation stages simultaneously. The source of inspiration in the Biogeography-Based Optimization (BBO) is the distribution of plant and animal species in different geographical habitats. Animal species are trying to migrate from crowded habitats to secluded habitats and monopolize food sources. Therefore, in this regard, first the Suitability Index Variable (SIV) is determined and the Habitat Suitability Index (HIS) is evaluated, then the migration factor from secluded habitats λ and the migration factor to crowded habitats μ are extracted and finally the migration of the variables with a certain probability is done. Also, in order to create diversity in habitats, mutation is done with a certain probability. In this algorithm, SIV represents the position of decision variables, HSI is the numerical value of the objective function, λ is the immigration rate, and μ is the emigration rate.

5.2 Standard Biogeography-Based Optimization (BBO)

This algorithm was presented in 2008 by Simon [2]. The algorithm has chosen its main inspiration from the distribution of plants and especially animals in different geographical habitats. Basically, animal and plant species are trying to use food resources exclusively, but their request will not be possible in most cases. Therefor they have to have food, water and… Therefore, an ecosystem is formed that becomes the food of other species with the formation of food chains. But this tendency to monopolize has other consequences, and that is that animals insist on migrating to more secluded places. Therefore, a settlement is known as a suitable place when it has a smaller population. Therefore, animal species usually choose a more secluded place from among several options. But where it has richer food resources, it will naturally have more population. Therefore, a factor called the HIS habitat suitable index is introduced, which determines the better location. In optimization problems, this coefficient is the same as the objective function of the problem, and in minimization problems, the lower the coefficient, the better. But the habitats for which this coefficient is higher are in fact very large and because of the competitors, the species want to leave. Therefore, two very common terms of migration are considered: migration to habitat that named Immigration, which Expresses acceptance of immigration, which is expressed as a percentage by the λ, and migration from habitat, that named emigration, which Its percentage state is also expressed by the μ. Habitat location is displayed with SIV that indicating suitability index variable or decision variables in the selected space. Now the variables migrate from the most populated habitats and their migration coefficient is high μ to the habitats with high migration coefficient λ. Also, in order to create variety in the search space, mutations are applied to the components of the variables with a certain probability. The basic steps of this algorithm for standard mode are presented in the following.

5.3 Basic Steps of Biogeography-Based Optimization

Step 1. Formation of primary habitats: In the search space, a set of primary habitats is randomly formed and sorted based on the objective function. At this stage we access the initial population. The number of this initial population is represented by npop.
Step 2. The values of the dimensionless coefficients λ and the migration μ are determined for the habitats.
Step 3. For each selected habitat location such as *i*, repeat Steps 4 to 8 to the initial population.
Step 4. For each variable such as *r* in habitat i, repeat Steps 5 to 8 for the number of array variables.
Step 5. With the probability λ_i in the variable X_{ir}, we make the changes according to Steps 6 to 8.
Step 6. The origin of X_{jr} migration is determined using the values of μ vectors based on random selection in a discrete distribution (roulette wheel).
Step 7. The location of the new habitat is done by migrating from X_{jr} to X_{ir} as follows:

$$if\ \ \mathbf{rand} \leq lambda(i) \Rightarrow X_{ir}^{new} = X_{ir} + \alpha_k \cdot \times (X_{jr} - X_{ir}) \tag{5.2}$$

$$\alpha_k = 0.9 \tag{5.3}$$

In the original version, an alpha value of 0.9 was suggested.
Step 8. With a certain probability, jump changes with a specific random distribution (preferably 'normal distribution') are performed on the selected variable X_{ir}:

$$if\ \ \mathbf{rand} \leq pmutation \Rightarrow X_{ir}^{mut} = X_{ir}^{new} + sigma \times randn \tag{5.4}$$

$$sigma = 0.02 \times (VarMax - VarMin) \tag{5.5}$$

Sigma comprises a percentage of the decision space. In the original version, this value is 2%.
Step 9. Migrate the answers from the previous answers and after evaluating and sorting, select the number of npop from the best of them for the next step.
Step 10. Check the termination conditions and repeat the operation from Step 3 if necessary.

5.4 Chaos Enhanced Biogeography-Based Optimization (CBBO)

In this algorithm, the location of the new habitat is based on two important strategies [3–5]. In the first strategy, migration from X_{jr} to X_{ir} takes place for each location such as i and for each variable such as K in location i with a probability of λ corresponding to i. In the second strategy, random mutation changes are applied to the variables of the decision variable. These two play the same role in the exploration and exploitation process. Therefore, by replacing the chaos map in random choices related to these steps, we will see a significant improvement in the performance of the algorithm. The proposed scenarios for this replacement are as follows:

Scenario 1. Placement of the chaos map in the first stage of the migration strategy of variables: In this case, the first chaos map *CHM1* is replaced in Eq. (5.2), and the result will be Eq. (5.6):

$$if \quad \mathbf{CHM1} \leq lambda(i) \Rightarrow X_{ir}^{new} = X_{ir} + \alpha_k \cdot \times (X_{jr} - X_{ir}) \tag{5.6}$$

Scenario 2. Placing the chaos map in the second strategy phase of random mutation changes: In this case, the second chaos map **CHM2** is applied in Eq. (5.4), and the result will be Eq. (5.7):

$$if \quad \mathbf{CHM2} \leq pmutation \Rightarrow X_{ir}^{mut} = X_{ir}^{new} + sigma \times randn \tag{5.7}$$

Scenario 3. Placing the chaos map in both stages simultaneously: In this case, the two chaotic maps are replaced simultaneously in Eqs. (5.2) and (5.4).

5.5 Truss Weight Optimization with Static Constraints

In this investigation of the effectiveness of chaos functions in Biogeography-Based optimization (CBBO), Logistic, Tenet and Gossin chaos functions have been chosen [6, 7]. By embedded chaos maps and formation of chaotic algorithms, the competition between chaotic functions is formed to access the best optimal weight. In order to introduce the best performance of the chaos algorithm, we need three scenarios.

In these scenarios, chaos functions are embedded at the locations of the distributions related to exploration, exploitation, or both simultaneously. In each example, 9 chaotic models are formed and compared with the standard state. To form the statistical space and increase diversity in the research, each model is independently evaluated 30 times. Also, to analyze the statistical results, the best weight, the best average and the coefficient of variation are provided. In a number of studies, the application of chaos functions has increased the coefficient of variation. This performance shows sudden jumps from local optima to global optima and an increase in the difference between the best and worst answer. Due to the wide scope of investigations, there is a strong competition between chaos functions and as a result, the possibility of approaching the global optimum increases. This optimum cannot be far from the crosshairs of chaotic mutations. In each investigation, the chaos function compatible with the algorithm and the scenario corresponding to the best improvement are selected. We can also find out about the weakness of the algorithm in the exploration or exploitation stages.

5.5.1 Formulation of the Structural Optimization Problems

In optimizing the size of truss structures, the main goal is to find the optimal value for the cross-sectional area of the members A_i, in order to optimize the weight of the structure W, and fulfill the limitations related to the response of the structure. Therefore, the optimal design problems are expressed according to the following equations:

$$\begin{array}{ll} Find & X = [x_1, x_2, \ldots, x_n] \\ to\ optimal & Mer(X) = f_{penalty}(X) \times W(X) \\ subject\ to & \begin{cases} \sigma_{\min} \le \sigma_i \le \sigma_{\max} & i = 1, 2, \ldots, nm \\ \delta_{\min} \le \delta_j \le \delta_{\max} & j = 1, 2, \ldots. nn \end{cases} \end{array} \tag{5.8}$$

where X is the vector embrace the decision variable; the variable x_i is selected between upper and lower bound; n is the number of element, $Mer(X)$ is the merit function; $W(X)$ is the weight of structure; $f_{penalty}(X)$ is the penalty function which results from the violations of the constraints; nm is the number of members forming the topology of structure; nn is the number of nodes except support nodes; σ_i and δ_i are the stress of members and nodal displacement, respectively; *min* and *max* mean the lower and upper bound of constraints, respectively. The weight of structure can be expressed as:

$$W(x) = \sum \gamma_i \cdot A_i \cdot L_i \tag{5.9}$$

where γ_i is the material density of member i, L_i is the length of member i, and A_i is the cross-sectional area of member i. The penalty function can be defined as:

$$f_{penalty}(X) = \left[1 + \varepsilon_1 \cdot \sum_{i=1}^{np} (\phi^k_{\sigma(i)} + \phi^k_{\delta(i)})\right]^{\varepsilon_2} \quad (5.10)$$

where np is the number of multiple loadings; ε_1 is taken as unity; ε_2 is set from 1.5 and increases linearly to 6; φ_σ and φ_δ are the summation of stress and nodal displacement penalize, which are expressed as following equations:

$$\begin{cases} \phi_\sigma = \sum_{i=1}^{nm} \max\left(\left|\frac{\sigma_i}{\overline{\sigma}_i}\right| - 1\right) \\ \phi_\delta = \sum_{i=1}^{nn} \max\left(\left|\frac{\delta_i}{\overline{\delta}_i}\right| - 1\right) \end{cases} \quad (5.11)$$

where σ_i, δ_i are the stress and nodal displacement which are divided by the allowable amount.

5.5.2 Introduction of Selected Chaos Map

By iterative formation of chaos functions, chaos dynamic series are determined. By embedding in meta-heuristic algorithms, these series can converge towards global answers and have a good potential to avoid falling into the trap of local optima. With chaotic meta-heuristic algorithms, the non-linear and non-convex behavior of the objective function in the optimization of truss structures can be easily controlled and adjusted. Among the advantages of the series formed by the mediation of chaos functions are non-repetition and non-convergence towards a certain limit. The standard mode of Biogeography-Based optimization (CBBO), will have premature convergence when dealing with non-linear and non-convex problems. The reason for that is the random distribution related to the phases of the algorithm, including the migration of biological species, immigration for habitat and emigration from settlements. By substituting chaos series in the place of random distributions, local optima jump to global optima. The *Logistic chaos function* [6] converges from a local minimum to the global value and will be effective for modifying the exploration conditions of the algorithm. But the *Gaussian chaos function* [7] is formed in the range of local optima and will improve the exploitation conditions of the algorithm. Also, the *tent chaos function* will be suitable for improving both conditions at the same time. Therefore, by choosing these functions, the weakness of algorithms of any kind is improved. The relationships related to these chaos functions have been introduced in the previous chapter.

5.5.3 *Numerical Examples of Optimal Truss Design*

To determine the efficiency of chaotic models, and to choose the most appropriate chaos function, the competition between 9 answers is formed, and the best chaotic state is selected. To determine the improvement values, the best chaotic model is compared with the standard case. In example 1, the simultaneous optimization of the cross section and the geometric shape of the truss structure is considered. Due to the diversity of the response space, the optimization results are very close to the global optimum. The examples provided are as follows:

- The 52-bar planar truss;
- The 72-bar spatial truss;
- The 120-bar dome shaped truss.

5.5.3.1 A 52-Bar Planar Truss

The topology and nodal numbering of the 52-bar planer truss are shown in Fig. 5.3. The material density is 7860 kg/m^3 and the modulus of elasticity is 207 GPa. The members are subjected to stress limitation 180 MPa in both tension and compression. For loading, the intensity of the loads in the X and Y directions is 100 and 200 kN, respectively, and both loads are applied simultaneously. Also, truss members are classified into 12 groups according to geometric symmetry.

The cross-section areas of the members were chosen from the 64 discrete values of the AISC code. In order to ensure the performance of chaos map and algorithms, as well as to increase the accuracy and sensitivity of calculations, each of the modes has been performed independently 30 times and the results related to the best response and the average value of responses are presented in Table 5.1. The graph of these results is also shown in Fig. 5.4. The coefficient of variation of the responses, which indicates the stability and robustness of the responses, has been calculated and another criterion for the efficiency of chaos map and algorithms has been obtained. After reviewing the optimization results for different combinations of algorithms with chaos map and comparing it with the initial and standard mode, a significant improvement in reducing the weight of the 52-bar planar truss results. The enhanced results for Chaotic Biogeography-Based optimization (CBBO), by Logistic chaos map and Scenario 1, a weight of 1903.183 kg has the optimal response.

5.5.3.2 A 72-Bar Spatial Truss

The topology and nodal numbering of the 72-bar spatial truss are shown in Fig. 5.5. The material density is 0.1 lb/in^3 and the modulus of elasticity is 10,000 ksi. Permissible members' stress of 25 ksi is assumed. The maximum displacement for each of the upper nodes is limited at 0.25 in in all directions. The two loading conditions are applied according to Table 5.2. Also, truss members are classified into 16 groups

Fig. 5.3 Schematic of a 52-bar planar truss

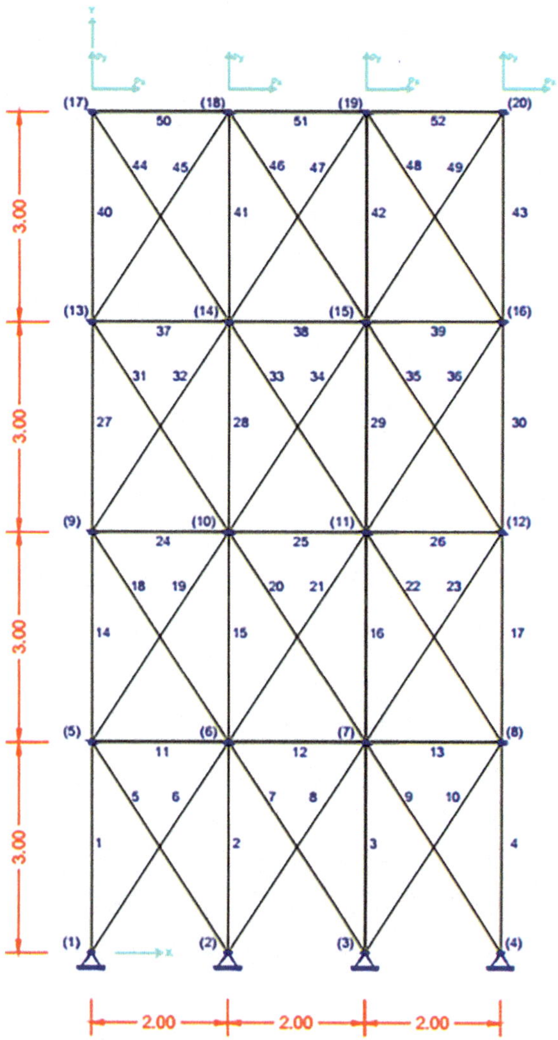

according to considering the geometric symmetry. The cross section of the members, which are the same as the decision variables, is considered in the range of 0.1–4 in^2. In order to ensure the performance of chaos map and algorithms, as well as to increase the accuracy and sensitivity of calculations, each of the modes has been performed independently 30 times and the results related to the best response and the average value of responses are presented in Table 5.3. A graph of these results is also shown in Fig. 5.6.

The coefficient of variation of the responses, which indicates the stability and robustness of the responses, has been calculated and another criterion for the efficiency of chaos map and algorithms has been obtained. Examining the optimization

Table 5.1 Optimal design comparison for the 52-bar planar truss

Number group	Element group	Standard	Logistic map			Tent map			Gauss map		
		BBO	CBBO 21	CBBO 22	CBBO 23	CBBO 31	CBBO 32	CBBO 33	CBBO 41	CBBO 42	CBBO 43
1	A_1–A_4	4658.055	4658.055	4658.055	4658.055	4658.055	4658.055	4658.055	4658.055	4658.055	4658.055
2	A_5–A_{10}	1161.288	1161.288	1161.288	1161.288	1161.288	1161.288	1161.288	1161.288	1161.288	1161.288
3	A_{11}–A_{13}	494.193	494.193	506.451	494.193	388.386	494.193	494.193	494.193	494.193	494.193
4	A_{14}–A_{17}	3303.219	3303.219	3303.219	3303.219	3303.219	3303.219	3303.219	3303.219	3303.219	3303.219
5	A_{18}–A_{23}	1008.385	939.998	1008.385	939.998	1008.385	939.998	939.998	939.998	939.998	939.998
6	A_{24}–A_{26}	494.193	494.193	363.225	506.451	363.225	506.451	506.451	494.193	494.193	494.193
7	A_{27}–A_{30}	2290.318	2238.705	2290.318	2290.318	2341.931	2238.705	2238.705	2238.705	2238.705	2290.318
8	A_{31}–A_{36}	1008.385	1008.385	1008.385	1008.385	1008.385	1008.385	1008.385	1008.385	1008.385	1008.385
9	A_{37}–A_{39}	506.451	506.451	494.193	494.193	494.193	506.451	494.193	494.193	494.193	506.451
10	A_{40}–A_{43}	1374.191	1283.868	1283.868	1374.191	1283.868	1374.191	1283.868	1374.191	1283.868	1283.868
11	A_{44}–A_{49}	1161.288	1161.288	1161.288	1161.288	1161.288	1161.288	1161.288	1161.288	1283.868	1161.288
12	A_{50}–A_{52}	494.193	494.193	494.193	494.193	494.193	494.193	506.451	494.193	494.193	494.193
Best weight		1928.199	**1903.183**	1915.237	1916.571	1915.086	1912.281	1903.761	1911.124	1923.448	1908.052
Mean weight		1934.301	**1905.848**	1922.059	1920.642	1919.648	1916.484	1907.117	1915.212	1932.948	1915.827
Coefficient variation		0.213541	**0.11059**	0.2310	0.19371	0.19245	0.19287	0.18258	0.19394	0.42939	0.26973

The bold items in the table indicate the optimal values for the best weight, the best mean and the best coefficient of variation

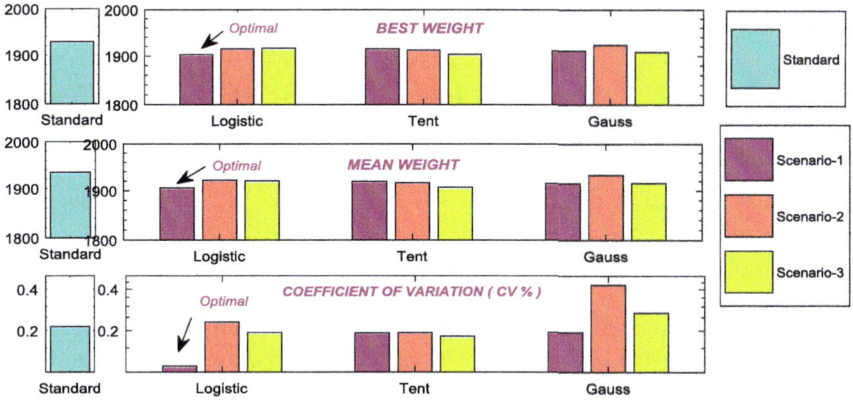

Fig. 5.4 Optimization results in standard and chaos map for the 52-bar planar truss

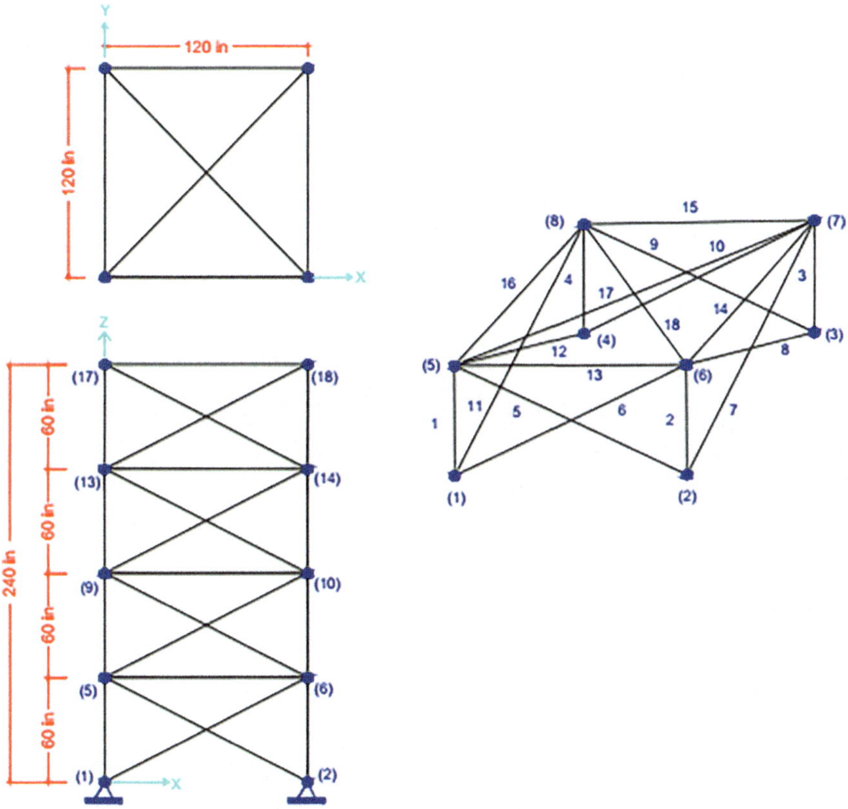

Fig. 5.5 Schematic of a 72-bar spatial truss and typical story

Table 5.2 Loading conditions for the 72- bar spatial truss

Node	Load case 1			Load case 2		
	F_x (kips)	F_y (kips)	F_z (kips)	F_x (kips)	F_y (kips)	F_z (kips)
17	5	5	– 5	0	0	– 5
18	0	0	0	0	0	– 5
19	0	0	0	0	0	– 5
20	0	0	0	0	0	– 5

results for different combinations of algorithms with chaos map and comparing it with the initial and standard modes, shows a significant improvement in reducing the weight of the 72-bar spatial truss. The enhanced results for Chaotic Biogeography-Based optimization (CBBO), by Tent chaos map and Scenario 2, a weight of 379.1046 lb has the optimal response.

5.5.3.3 A 120-Bar Dome Shaped Truss

The topology and nodal numbering of the 120-bar dome shaped truss shown in Fig. 5.7. The material density is 0.288 lb/in^3 and the modulus of elasticity is 30,450 lb/in^2. Permissible members' tensile and compressive stresses are proposed in accordance with AISC ASD code.

The maximum deformation for each node in all directions is set at 0.1969 in. All non-abutment nodes are affected by vertical loading, with intensities of – 13.49 kips in node one and – 6.744 kips in spheres 2– 14 and 2.248 kips in the other nodes, respectively. Also, truss members are classified into seven groups according to their shape and geometric symmetry. The minimum cross-sectional area of the members, which is the lower limit of the decision variables, is 0.775 in^2, and the maximum cross-sectional area is 20 in^2. The allowable stress in tension and pressure is determined based on the following equations.

$$\sigma_i = \begin{cases} 0.6F_y \; for \;\; \sigma_i \geq 0 \\ \sigma_i^- \quad for \;\; \sigma_i < 0 \end{cases} \tag{5.12}$$

For compressive stresses, we have the following equations:

$$\sigma_i^- = \begin{cases} (1 - \frac{\lambda^2}{2C^2})F_y/FS \; for \;\; \lambda < C \\ \frac{12\pi^2 E}{23\lambda^2} \qquad\qquad for \;\; \lambda \geq C \end{cases} \tag{5.13}$$

In this equation we have:

$$FS = (\frac{5}{3} + \frac{3\lambda}{8C} - \frac{\lambda^3}{8C^3}); \, C = \sqrt{(\frac{2\pi^2 E}{F_y})};$$

Table 5.3 Optimal design comparison for the 72-bar spatial truss

Number group	Element group	Standard	Logistic map			Tent map			Gauss map		
		BBO	CBBO 21	CBBO 22	CBBO 23	CBBO 31	CBBO 32	CBBO 33	CBBO 41	CBBO 42	CBBO 43
1	A_1–A_4	1.8755	1.8992	1.8802	1.9132	1.8937	1.8645	1.8835	1.8647	1.881	1.8452
2	A_5–A_{12}	0.50019	0.51062	0.51064	0.50212	0.50588	0.5143	0.51642	0.51522	0.51018	0.51079
3	A_{13}–A_{16}	0.1	0.1	0.1	0.1	0.1	0.1	0.10001	0.10041	0.1	0.1
4	A_{17}–A_{18}	0.1	0.10005	0.1	0.10001	0.1	0.1	0.1	0.1	0.1	0.1
5	A_{19}–A_{22}	1.2283	1.2869	1.2738	1.3116	1.2592	1.2636	1.2746	1.3013	1.2667	1.2861
6	A_{23}–A_{30}	0.51921	0.51281	0.5225	0.51336	0.50992	0.51517	0.51512	0.51116	0.51312	0.51819
7	A_{31}–A_{34}	0.1	0.10003	0.1	0.1	0.1	0.1	0.10001	0.1	0.1	0.1
8	A_{35}–A_{36}	0.10001	0.1	0.1	0.1	0.1	0.1	0.1	0.1	0.1	0.1
9	A_{37}–A_{40}	0.54598	0.52566	0.50153	0.5123	0.55376	0.50825	0.50462	0.53185	0.5063	0.52267
10	A_{41}–A_{48}	0.51858	0.51928	0.51224	0.51451	0.50933	0.51842	0.52049	0.51829	0.5203	0.51846
11	A_{49}–A_{52}	0.1	0.1	0.1	0.1	0.1	0.1	0.1	0.1	0.1	0.10001
12	A_{53}–A_{54}	0.1	0.1	0.1	0.10006	0.1	0.1	0.1	0.10031	0.1	0.10001
13	A_{55}–A_{58}	0.15613	0.15719	0.15919	0.15802	0.15586	0.15675	0.1582	0.15691	0.15834	0.15692
14	A_{59}–A_{66}	0.55284	0.53316	0.52832	0.53792	0.54951	0.54401	0.53486	0.53719	0.54204	0.53389
15	A_{67}–A_{70}	0.41652	0.41098	0.40762	0.41476	0.40999	0.41102	0.40668	0.40135	0.40501	0.42103
16	A_{71}–A_{72}	0.57242	0.5657	0.56023	0.56957	0.58208	0.56662	0.57327	0.56899	0.58249	0.57746
Best weight		379.7566	379.1749	379.2879	379.2577	379.393	**379.1046**	379.2523	379.1542	379.1461	379.245
Mean weight		380.0406	379.4195	380.151	379.5769	379.5586	379.5592	379.589	379.4161	**379.3653**	379.459
Coefficient variation		0.095228	0.048454	0.19388	0.057394	**0.03664**	0.091087	0.071742	0.060816	0.057399	0.04024

The bold items in the table indicate the optimal values for the best weight, the best mean and the best coefficient of variation

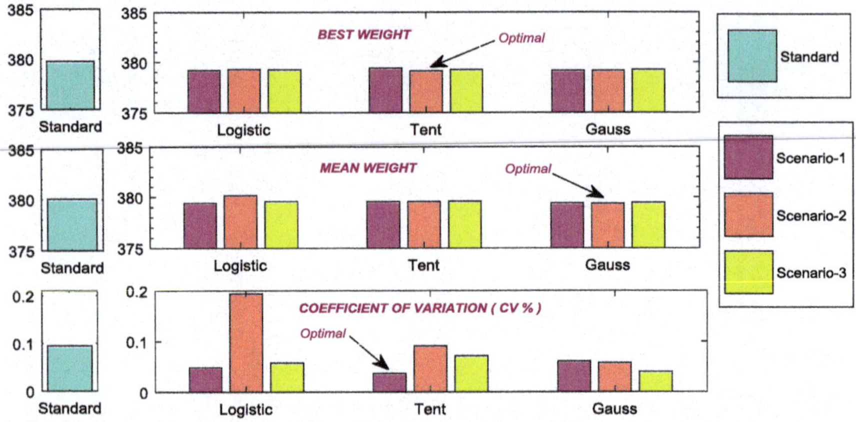

Fig. 5.6 Optimization results in standard mode and chaos map for the 72-bar spatial truss

$$\lambda = \frac{kl}{r}; r = aA^b; a = 0.4993; b = 0.6777 \tag{5.14}$$

The components used include E modulus of elasticity, Fy the yield stress of steel, C the ratio of the limit slenderness which separates the elastic and inelastic buckling region in comparison with the existing slandering λ. Also, k is the effective length factor of the l and r is the radius of rotation of the limb. In order to ensure the performance of chaos map and algorithms, as well as to increase the accuracy and sensitivity of calculations, each of the modes has been performed independently 30 times and the results related to the best response and the average value of responses are presented in Table 5.4. A graph of these results is shown in Fig. 5.8.

Also, the coefficient of variation of the responses, which indicates the stability and robustness of the responses, has been calculated and another criterion for the efficiency of chaos map and algorithms has been obtained. The minimum and maximum cross-sectional area of all members is 0.775 in^2 and 20 in^2 respectively. By examining the optimization results for different combinations of algorithms chaos map and comparing it with standard mode, a significant improvement in reducing the weight of the 120-bar dome shaped truss is achieved. The enhanced results for Chaotic Biogeography-Based optimization (CBBO), by Gauss chaos map and Scenario 3, a weight of 33,240.56 lb has the optimal response.

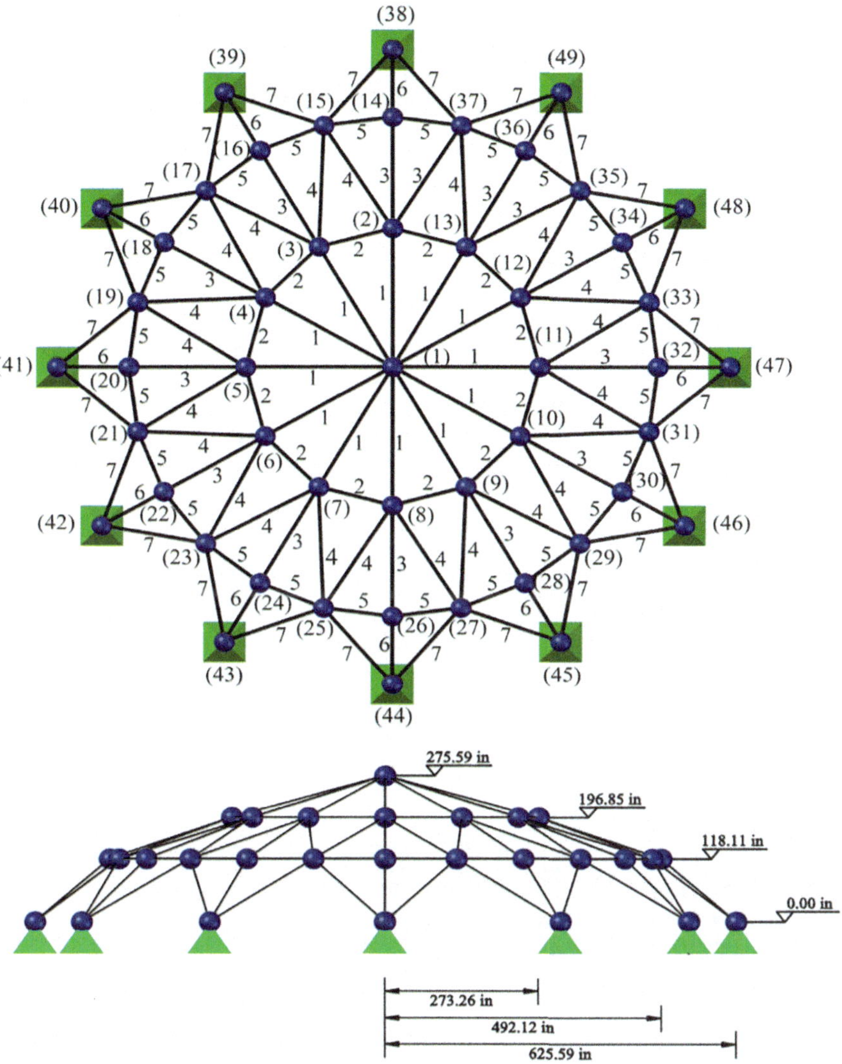

Fig. 5.7 Schematic of a 120-bar dome shaped truss

5.6 Truss Weight Optimization with Multiple Frequency Constraints

In the analysis of dynamic loads such as earthquake and wind load, the natural frequency of the structure is compared with the frequency of the incoming load. The main purpose of this comparison is to prevent the destructive phenomenon of resonance. Therefore, in order to prevent this phenomenon, the natural frequency of

Table 5.4 Optimal design comparison for the 120-bar dome shaped truss

Number group	Standard	Logistic map			Tent map			Gauss map		
	BBO	CBBO 21	CBBO 22	CBBO 23	CBBO 31	CBBO 32	CBBO 33	CBBO 41	CBBO 42	CBBO 43
1	3.02488	3.02491	3.02606	3.02546	3.02326	3.02568	3.02338	3.02458	3.02403	3.0248
2	14.7418	14.6484	14.7318	14.6298	14.5975	14.4199	14.3207	14.5192	14.5978	14.7029
3	5.08802	5.11281	5.03887	5.1684	4.98821	5.21854	4.99201	5.16703	5.01339	5.09729
4	3.13496	3.13109	3.12917	3.13049	3.12794	3.12873	3.13001	3.12914	3.12836	3.1266
5	8.45746	8.48733	8.52677	8.48379	8.6573	8.50638	8.77634	8.54883	8.59981	8.48967
6	3.34338	3.34421	3.31076	3.28598	3.29958	3.37219	3.35202	3.28589	3.3528	3.32243
7	2.49574	2.49555	2.49746	2.49778	2.49866	2.4939	2.49759	2.49587	2.49675	2.49588
Best weight	33,250.77	33,245.09	33,241.69	33,244.71	33,241.34	33,246.55	33,249.08	33,240.99	33,241.9	**33,240.56**
Mean weight	33,255.41	33,249.7	33,251.17	33,252.45	33,246.48	33,266.1	33,271.86	33,249.82	**33,246.2**	33,246.65
Coefficient variation	0.014163	**0.000666**	0.027271	0.019698	0.022279	0.059382	0.072742	0.021735	0.085582	0.020825

The bold items in the table indicate the optimal values for the best weight, the best mean and the best coefficient of variation

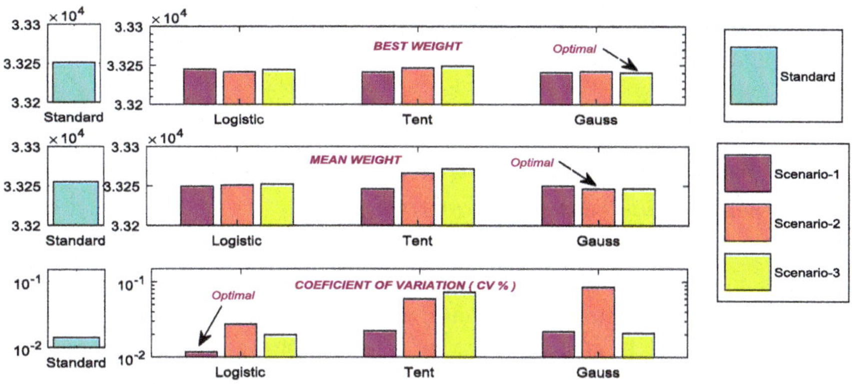

Fig. 5.8 Optimization results in standard mode and chaos map for the 120-bar dome shaped truss

structures must be limited in a certain range. By calculating the structure's natural frequencies and oscillation modes, all information about the structure's dynamic behavior can be accessed. In most of the vibration problems that have low frequency, the response of the structure depends on the basic frequencies and the mode shape, and by applying the frequency limit, the dynamic behavior of the structure can be easily controlled. The effects of some incompatible and destructive fashions can be reduced by using these relationships. In addition to being non-linear and non-convex, the frequency constraint relations have complex implicit variables. Most of the meta-heuristic algorithms in the standard mode are caught in the local optimal trap for frequency-constrained optimization, and premature convergence causes the optimization to stop. Therefore, the embedding of chaos functions provides suitable conditions for jumping from local optima to global optima. In the Biogeography-Based optimization (CBBO), embedding chaos functions instead of random values of exploration and exploitation steps significantly improves the results. In this investigation, *Gaussian*, *Liebovitch* and *Piecewise* chaos functions with triple scenarios have replaced the random selection of Immigration, Emigration and mutation of animal species from densely populated to sparsely populated settlements [8]. The reason for this migration is to monopolize food resources. In order to increase diversity in the statistical space, each model evaluated 20 times independently and the best weight, the best mean and coefficient of variation are the results. By selecting the best response from among the 9 improved chaotic models, the appropriate chaotic function is selected and compared with the standard mode. Also, useful information is obtained about the weakness of the algorithm and how to improve the results. Investigations show that Liebovitch chaos map converges local optima to global optima with a very high probability, and this function will be suitable for improving exploitation conditions [9]. The Gaussian chaos map is likely to be in the range of local optima, and it is suitable for improving the exploitation conditions. Finally, the piecewise chaos map simultaneously improves both conditions [10]. Therefore, by choosing these functions, the weakness of algorithms of any kind is improved. The relationships of these chaos functions have been introduced in the previous chapter.

5.6.1 Formulation of the Structural Optimization with Multi-frequency Constraints

In structural optimization problems the main objective is to minimize the weight of the structures under some constraints. In this class of optimization problems, the design variables can be considered as: the cross-sectional areas of elements and/or the coordinates of nodes. The optimization problem for a truss structure can be stated as Eq. (5.15).

$$\begin{aligned} &Find && \{X\} = [x_1, x_2, \ldots, x_{ng}] \\ &to\ minimize && W(\{X\}) = \sum_{i=1}^{nm} \rho_i \cdot A_i \cdot L_i \\ &subjected\ to && \begin{cases} g_j(\{X\}) \leq 0, & j = 1, 2, \ldots, nc \\ x_{i\,\min} \leq x_i \leq x_{i\,\max} \end{cases} \end{aligned} \tag{5.15}$$

The most common and convenient method in metaheuristic algorithms to handle constraints is to use penalty functions. In this method, a constrained optimization problem is transformed into an unconstrained one by multiplying a certain value to objective function based on the amount of constraint violation appears in a problem. Therefore, the merit (or pseudo-objective) function which should be minimized is defined as:

$$Mer(X) = W(X) \times f_{penalty}(X) \tag{5.16}$$

where Mer(X) is the merit function; W(X) is the objective function (weight of truss structure); $f_{penalty}(X)$ is the penalty function which results from the violations of the constraints corresponding to the response of the structure:

$$f_{penalty}(X) = 1 + \gamma_p \cdot \sum_{j=1}^{m} \max(o, g_j(X)) = 1 + \gamma_p \times G(X) \tag{5.17}$$

where γ_p is penalty parameter, and G(X) is the constraint function. In dynamic constraint problem of a structure, the behavioral constraints are the restricted natural frequencies that are defined as follows:

$$
if\begin{cases} \omega_n(X) \le \omega_{upper,n} \Rightarrow g_n(X) = \dfrac{\omega_n(X)}{\omega_{upper,n}} - 1 \le 0 \quad for\ some\ natural\ frequencies\ n \\ \omega_m(X) \ge \omega_{lower,m} \Rightarrow g_m(X) = 1 - \dfrac{\omega_m(X)}{\omega_{lower,m}} \le 0 \quad for\ some\ natural\ frequencies\ m \end{cases} \tag{5.18}
$$

where ω_n and ω_m are the nth and mth natural frequency, respectively; $\omega_{\text{lower,n}}$ and $\omega_{\text{upper,m}}$ is lower bound and upper bound of the nth and mth natural frequency, respectively. As can be seen from Eq. (5.16), the penalty function method consists of a penalty parameter where the feasible solutions are dependent on right tuning of these parameters. By setting small values of the penalty parameter, the pseudo-objective function becomes less sensitive to optimization constraints and the optimization process may converge to a violated solution. Conversely, setting the penalty parameter to a large value led to amplify the effect of optimization constraints and the optimization process may get stuck in a local optimum.

5.6.2 Numerical Examples of Optimal Truss Design

To determine the efficiency of chaotic models and to choose the most appropriate chaos function, a competition between nine chaotic modes is formed, and the best chaotic mode is selected. To compare the improvement values, the best chaotic mode is compared with the standard model. In order to form the statistical population, each model has been implemented 20 times and the best weight, the best average and the best coefficient of variation have been presented. In Example 2, the optimization of the cross section and geometric shape of truss structures has been selected. Due to the variety of answers, the optimization results are very close to the global optimum. The examples provided are as follows:

- The planar 10-bar truss;
- The simply supported 37-bar planar truss;
- 200-bar planar truss structure.

5.6.2.1 A Planar 10-Bar Truss

The planar 10-bar truss as depicted in Fig. 5.9 is a well-known benchmark problem in the field of weight optimization of the structures with frequency constraints. This truss only considers "optimizing the size of the sections" and the geometric shape of the structure is constant during the optimization process. The non-structural concentrated mass of 454 kg affected the free nodes. The mechanical properties of the structure are: The density of materials is 2770 kg/m^3, the modulus of elasticity is 69800 MPa, the frequency limits of the structure in the first, second and third modes are greater than 7, 15 and 20 Hz, respectively. The cross-sectional area of each

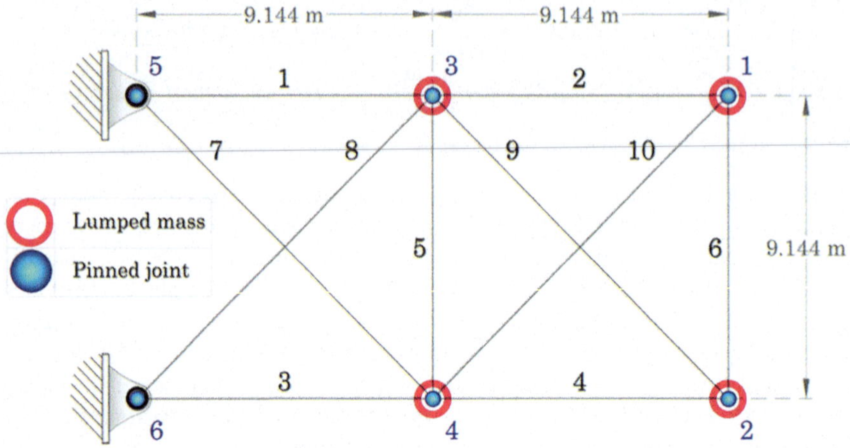

Fig. 5.9 Schematic of a planar 10-bar truss

member is selected as an independent decision variable. Therefore, ten design variables with a lower limit of 0.645 cm^2 and an upper limit of 50 cm^2 are considered. In order to ensure the performance of chaotic functions and algorithms, as well as to increase the accuracy and sensitivity of calculations, each of the modes has been performed independently 30 times and the results related to the best response and the average value of responses are presented in Statistical Table 5.5. Also, the coefficient of variation of responses, which is a measure of the robustness of responses, has been calculated and used to compare the efficiency of chaos maps and standard algorithms. For quick access to optimization information, a bar chart of each component is shown in Fig. 5.10. Examining the optimization results for different combinations of algorithms with chaos maps and comparing it with the standard mode, shows a significant improvement in reducing the weight of the planar 10-bar truss. The enhanced results for Chaotic Biogeography-Based optimization (CBBO), by Piecewise chaos map and Scenario 1, a weight of 524.4289 kg has the optimal response.

5.6.2.2 A Simply Supported 37-Bar Planar Truss

A 37-member planar truss with a simple abutment as shown in Fig. 5.11 is a well-known benchmark problem for optimizing the weight and shape of a truss with a frequency limit. This truss considers both optimizations of the section size and geometric coordinates of the nodes and the geometric shape of the structure is determined during the optimization process. The lower edge member is considered to be a rod member with a rectangular cross-section and an area of 40 cm^2. The other members are modeled as rod members. These decision variables are grouped according to symmetry in the geometric shape. The y-axis coordinates form the

Table 5.5 Optimal design comparison for the planar 10-bar truss

Number group	Standard	Gauss map			Liebovitch map			Piecewise map		
	BBO	CBBO 21	CBBO 22	CBBO 23	CBBO 31	CBBO 32	CBBO 33	CBBO 41	CBBO 42	CBBO 43
A_1	34.5564	34.8793	35.5817	34.9762	34.6127	37.6617	36.7075	36.9987	35.1201	34.9749
A_2	14.2734	14.9897	14.205	15.4707	14.8705	18.2947	9.86943	9.85285	15.6044	14.3579
A_3	35.1557	35.5114	34.8935	35.213	35.2541	36.9491	38.9168	38.2403	35.0251	34.8426
A_4	13.9081	13.7242	14.9285	14.6599	13.4575	9.80946	18.5843	18.706	14.2226	13.9957
A_5	0.645	0.645	0.645	0.645	0.645725	4.724	4.73147	4.63981	0.645	0.645
A_6	4.63538	4.61221	4.57229	4.52705	4.64977	4.06059	4.07577	4.09576	4.56582	4.63656
A_7	24.6965	25.4532	23.0268	24.3054	25.8644	22.5267	20.1708	19.799	25.0472	24.6398
A_8	23.8608	22.6975	24.7044	22.4814	22.6206	20.3961	21.4089	22.0158	22.0126	23.755
A_9	12.298	11.6621	12.7437	12.1875	11.8274	11.2569	13.6454	13.5926	11.9642	12.2718
A_{10}	12.6946	12.9075	11.8492	12.8948	13.1017	13.6455	11.4488	11.4245	13.0779	12.6345
Best weight (kg)	524.7873	524.8238	524.581	524.6373	525.1022	525.3685	524.7549	**524.4289**	524.6891	524.602
Mean weight (kg)	527.5939	528.6819	**525.8308**	527.3456	529.0053	526.7618	526.9934	526.4199	526.0595	527.093
Coefficient variation	0.3737	0.45636	0.38532	0.50091	0.4229	0.45557	**0.31314**	0.51217	0.3407	0.42592
ω_1 (HZ)	6.99995	6.99963	6.99992	6.99986	6.99996	6.95839	6.95156	6.95006	6.99858	6.99878
ω_2 (HZ)	16.1452	16.1709	16.192	16.1841	16.1334	17.6039	17.579	17.5765	16.1656	16.155
ω_3 (HZ)	20	20.0004	19.9999	20	20.0009	20	20.0001	20	20	20

The bold items in the table indicate the optimal values for the best weight, the best mean and the best coefficient of variation

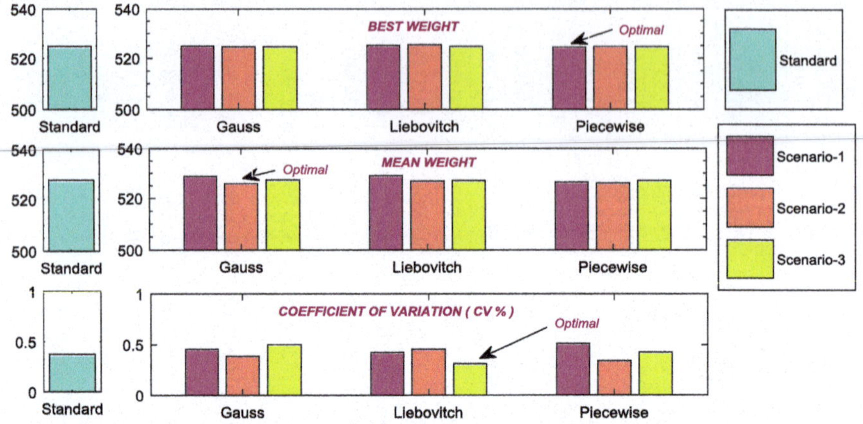

Fig. 5.10 Optimization results in standard mode and chaos map for the planar 10-bar truss

shape variables in the upper year, taking into account the symmetry. In the lower axis, a non-structural concentrated mass of 10 kg affects all free nodes. The mechanical properties of the structure are: The density of the materials is 7800 kg/m^3, the modulus of elasticity is 21,000 MPa, the frequency limits of the structure in the first, second and third modes are greater than 20, 40 and 60 Hz, respectively. For the cross section of the members, the lower limit is 1 cm^2 and the upper limit is 10 cm^2. The total size and shape decision variables are 19 variables, of which 14 variables are related to the size and cross-sectional area of the members and 5 variables are related to the shape of the structure and the coordinates of the nodes. There are also three frequency limits. In order to ensure the performance of turbulent functions and algorithms, as well as to increase the accuracy and sensitivity of calculations, each of the independent modes has been performed 20 times and the results related to the best response and the average value of responses. Is presented in statistical Table 5.6. Also, the coefficient of variation of responses, which is a measure of the robustness and robustness of responses, has been calculated and used to compare the efficiency of turbulence functions and algorithms. For quick access to optimization information, a bar chart of each component is shown in Fig. 5.12.

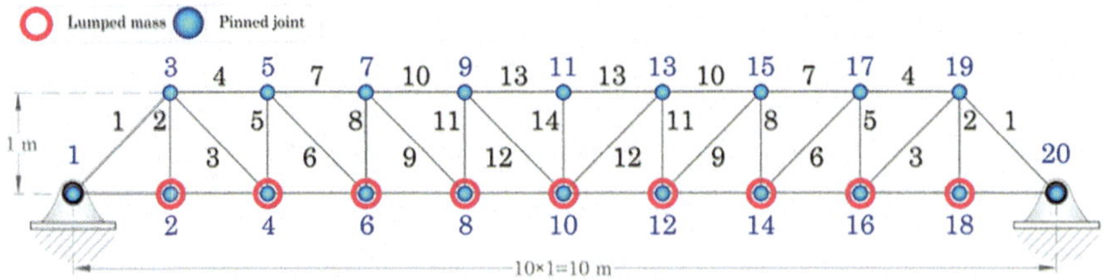

Fig. 5.11 Schematic of A simply supported 37-bar planar truss

Table 5.6 Optimal design comparison for the simply supported 37-bar planar truss

Decision variable	Standard	Gauss map			Liebovitch map			Piecewise map		
	BBO	CBBO 21	CBBO 22	CBBO 23	CBBO 31	CBBO 32	CBBO 33	CBBO 41	CBBO 42	CBBO 43
Y_3, Y_{19} (m)	0.92124	0.91183	0.90878	0.88386	0.88784	0.9248	0.95292	0.89298	0.90013	0.90086
Y_5, Y_{17} (m)	1.3043	1.2926	1.2825	1.2578	1.2704	1.2939	1.3197	1.2694	1.264	1.2855
Y_7, Y_{15} (m)	1.5237	1.4954	1.469	1.4705	1.4413	1.4899	1.5215	1.4562	1.4566	1.475
Y_9, Y_{13} (m)	1.6932	1.6585	1.5969	1.6073	1.5842	1.6421	1.6954	1.6181	1.588	1.5954
Y_{11} (m)	1.7691	1.7406	1.6618	1.68	1.6548	1.7175	1.7775	1.6931	1.6677	1.6694
A_1, A_{27}	2.9587	2.8999	2.9039	2.9567	3.039	2.8465	2.9251	2.9013	2.8843	3.0217
A_2, A_{26}	1.0121	1	1	1	1.0011	1	1.0026	1.0053	1	1
A_3, A_{24}	1	1	1	1	1	1	1	1	1	1
A_4, A_{25}	2.7443	2.7328	2.7292	2.7655	2.8857	2.6762	2.7025	2.6171	2.8668	2.684
A_5, A_{23}	1.27	1.2056	1.1212	1.1768	1.2258	1.212	1.1812	1.2425	1.1642	1.2441
A_6, A_{21}	1.1419	1.2182	1.2589	1.1278	1.2761	1.1879	1.1822	1.1974	1.1571	1.1798
A_7, A_{22}	2.4747	2.5859	2.7975	2.6121	2.8203	2.4442	2.6931	2.7402	2.6823	2.716
A_8, A_{20}	1.4714	1.4464	1.3292	1.478	1.2518	1.3402	1.3923	1.3513	1.4457	1.3303
A_9, A_{18}	1.4447	1.4177	1.5556	1.5555	1.4231	1.5166	1.4488	1.4408	1.4748	1.5398
A_{10}, A_{19}	2.3073	2.5614	2.7148	2.6302	2.7263	2.6409	2.244	2.7018	2.8073	2.701
A_{11}, A_{17}	1.3411	1.2304	1.1674	1.2537	1.1804	1.2231	1.3744	1.2839	1.2707	1.2091
A_{12}, A_{15}	1.4017	1.469	1.3482	1.34	1.3744	1.4378	1.3421	1.4091	1.2824	1.294
A_{13}, A_{16}	2.3301	2.3721	2.6028	2.7185	2.6072	2.6883	2.3382	2.7051	2.7335	2.6116
A_{14}	1	1	1	1	1	1	1	1	1	1
Best weight (kg)	360.023	359.9641	359.9327	360.0342	360.0508	359.9468	359.9858	360.0344	360.0302	**359.9232**
Mean weight (kg)	360.4218	360.3283	360.4144	360.2804	360.491	360.4123	360.2645	360.384	**360.1752**	360.3598

(continued)

Table 5.6 (continued)

Decision variable	Standard	Gauss map			Liebovitch map			Piecewise map		
	BBO	CBBO 21	CBBO 22	CBBO 23	CBBO 31	CBBO 32	CBBO 33	CBBO 41	CBBO 42	CBBO 43
Coefficient variation	0.079996	0.065435	0.084862	0.051335	0.069324	0.12003	0.053823	0.076587	**0.030952**	0.09874
ω_1 (HZ)	19.9989	19.9998	20.0006	19.9988	19.9945	19.9995	19.9999	20.0022	19.9975	20.0006
ω_2 (HZ)	40.0055	40.0068	40.0029	40.0012	40.0038	40.0018	40.0004	40.0001	40.0041	40.002
ω_3 (HZ)	59.997	59.9991	59.9958	59.9984	60	60.0001	59.9989	60.0048	59.9905	60.0013

The bold items in the table indicate the optimal values for the best weight, the best mean and the best coefficient of variation

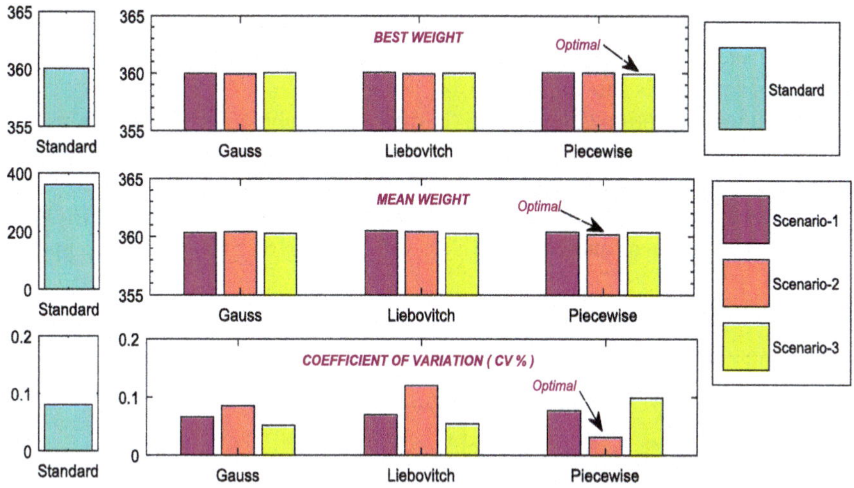

Fig. 5.12 Optimization results in standard mode and chaos map for the simply supported 37-bar planar truss

Examining the optimization results for different combinations of algorithms with turbulence functions and comparing it with the standard mode, shows a significant improvement in reducing the weight of A simply supported 37-bar planar truss. The enhanced results for Chaotic Biogeography-Based optimization (CBBO), by Piecewise chaos map and Scenario 3, a weight of 359.9232 kg has the optimal response.

5.6.2.3 A 200-Bar Planar Truss Structure

The 200-bar planar truss structure as shown in Fig. 5.13 is a well-known benchmark problem for weight optimization with frequency limits. This truss only considers "optimizing the size of the sections" and the geometric shape of the structure is constant during the optimization process. The decision variables related to the sections size of the members are classified into 29 groups according Table 5.7. Non-structural concentrated mass in nodes 1–5 and in the amount of 100 kg affect the structure. The mechanical characteristics of the structure are: material density 7860 kg/m^3, modulus of elasticity 210,000 MPa, frequency limits of the structure in the first, second and third modes are greater than 5, 10 and 15 Hz, respectively. For the cross section of the members, the lower limit range is 0.1 cm^2 and the upper limit is 25 cm^2. In order to ensure the performance of turbulent functions and algorithms, as well as to increase the accuracy and sensitivity of calculations, each of the modes has been performed independently 30 times and the results related to the best response and the average value of responses are presented in Statistical Table 5.8. Also, the coefficient of change of responses, which is a measure of the robustness and robustness of responses, has been calculated and used to compare the efficiency of turbulence functions and algorithms. For quick access to optimization information, a bar chart of each component is shown in Fig. 5.14.

Examining the optimization results for different combinations of algorithms with turbulence functions and comparing it with the standard mode, shows a significant and significant improvement in reducing the weight of the 200-bar planar truss structure. The enhanced results for Chaotic Biogeography-Based optimization (CBBO), by Liebovitch chaos map and Scenario 2, a weight of 2161.566 kg has the optimal response.

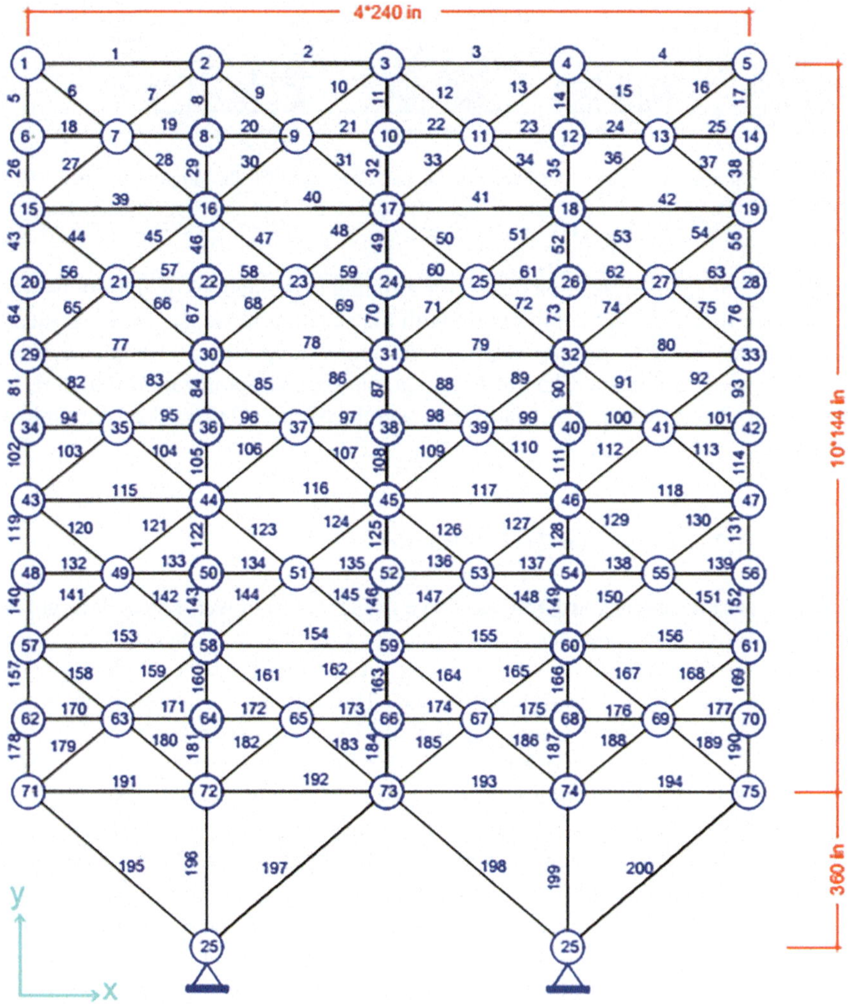

Fig. 5.13 Schematic of A 200-bar planar truss structure

Table 5.7 Classification of members in the 200-bar planar truss structure

Group	Element group	Group	Element group
A_1	1, 2, 3, 4	A_{16}	82, 83, 85, 86, 88, 89, 91, 92, 103, 104, 106, 107, 109, 110, 112, 113
A_2	5, 8, 11, 14, 17	A_{17}	115, 116, 117, 118
A_3	19, 20, 21, 22, 23, 24	A_{18}	119, 122, 125, 128, 131
A_4	18, 25, 56, 63, 94, 101, 132, 139, 170, 177	A_{19}	133, 134, 135, 136, 137, 138
A_5	26, 29, 32, 35, 38	A_{20}	140, 143, 146, 149, 152
A_6	6, 7, 9, 10, 12, 13, 15, 16, 27, 28, 30, 31, 33, 34, 36, 37	A_{21}	120, 121, 123, 124, 126, 127, 129, 130, 141, 142, 144, 145, 147, 148, 150, 151
A_7	39, 40, 41, 42	A_{22}	153, 154, 155, 156
A_8	43, 46, 49, 52, 55	A_{23}	157, 160, 163, 166, 169
A_9	57, 58, 59, 60, 61, 62	A_{24}	171, 172, 173, 174, 175, 176
A_{10}	64, 67, 70, 73, 76	A_{25}	178, 181, 184, 187, 190
A_{11}	44, 45, 47, 48, 50, 51, 53, 54, 65, 66, 68, 69, 71, 72, 74, 75	A_{26}	158, 159, 161, 162, 164, 165, 167, 168, 179, 180, 182, 183, 185, 186, 188, 189
A_{12}	77, 78, 79, 80	A_{27}	191, 192, 193, 194
A_{13}	81, 84, 87, 90, 93	A_{28}	195, 197, 198, 200
A_{14}	95, 96, 97, 98, 99, 100	A_{29}	196, 199
A_{15}	102, 105, 108, 111, 114		

5.7 Concluding Remarks

The Biogeography-Based optimization (BBO), is inspired by the distribution of plants and animals in geographical habitats. Animal and plant species are trying to use food resources exclusively, but they do not achieve this goal due to the limitation of food resources. Therefore, they are trying to migrate to more thinly populated settlements. Based on the suitability of each habitat, migration to or from it may take place. In this algorithm, μ and λ coefficients have been introduced for these two types of migration. This part actually constitutes the exploitation phase of the algorithm. Also, the probability of mutating to get the new settlement position constitutes the exploration phase. The random distribution used in these phases is of a uniform type, so the imbalance between exploration and exploitation in standard model is natural. For this reason, in non-linear and non-convex problems, the algorithm falls into the trap of local optima. The random distributions used in these two important phases can be replaced by chaotic dynamic functions and achieve a significant improvement. By embedding Chaos functions in Biogeography-Based optimization (CBBO), the algorithm is saved from the local optimal trap and converges from the local optimal to the global optimal by chaotic mutative. As a result, the problem of premature convergence of the algorithm is solved and the balance between the stages of exploration

Table 5.8 Optimal design comparison for the 200-bar planar truss structure

Number group	Standard	Gauss map			Liebovitch map			Piecewise map		
	BBO	CBBO 21	CBBO 22	CBBO 23	CBBO 31	CBBO 32	CBBO 33	CBBO 41	CBBO 42	CBBO 43
A_1	0.305051	0.337935	0.310077	0.29401	0.336019	0.32702	0.329683	0.305579	0.339197	0.333796
A_2	0.57205	0.453834	0.539997	0.534342	0.465465	0.468673	0.473839	0.498175	0.603354	0.489167
A_3	0.1	0.1	0.1	0.1	0.100457	0.1	0.1	0.1	0.101506	0.100096
A_4	0.100018	0.1	0.1	0.1	0.1	0.1	0.1	0.1	0.110655	0.1
A_5	0.489079	0.51327	0.557429	0.49114	0.505159	0.524138	0.542997	0.556806	0.527832	0.519342
A_6	0.856072	0.851604	0.858854	0.809025	0.827887	0.871746	0.877371	0.852999	0.83589	0.840704
A_7	0.112521	0.1	0.1	0.1	0.10784	0.104385	0.1	0.1	0.109322	0.109566
A_8	1.5329	1.42789	1.36436	1.4238	1.36033	1.48665	1.32845	1.43401	1.35929	1.51955
A_9	0.1	0.1	0.1	0.1	0.1	0.1	0.1	0.1	0.1	0.108562
A_{10}	1.78963	1.65166	1.48945	1.54246	1.56791	1.5346	1.64057	1.51209	1.45185	1.65421
A_{11}	1.08428	1.10458	1.13825	1.18533	1.15682	1.15596	1.10129	1.16839	1.15997	1.18596
A_{12}	0.100202	0.116814	0.166822	0.100081	0.1	0.151898	0.144528	0.100333	0.151454	0.110585
A_{13}	2.7528	2.96355	3.24081	3.15643	2.92281	3.10391	3.09351	3.0256	2.79664	2.96099
A_{14}	0.103808	0.102233	0.1	0.1	0.100073	0.113654	0.1	0.106846	0.1	0.100052
A_{15}	3.47482	3.41838	3.47643	3.12356	3.20347	3.09549	3.62024	3.17184	3.23233	3.24774
A_{16}	1.62531	1.50286	1.55389	1.5154	1.65362	1.58429	1.57494	1.61447	1.67503	1.6147
A_{17}	0.100015	0.390016	0.316922	0.106285	0.313072	0.2098	0.333148	0.352225	0.303855	0.439443
A_{18}	5.19525	4.78338	5.17335	5.00884	5.26224	5.46341	5.02611	5.12982	5.28917	5.04063
A_{19}	0.1	0.104153	0.100618	0.1	0.1	0.1	0.1	0.101033	0.101771	0.100298
A_{20}	5.46498	5.26415	5.8343	5.92097	5.47554	5.32215	5.05431	5.28166	5.26163	5.78842
A_{21}	2.06255	2.16061	2.07679	2.07584	2.16809	1.93415	1.93467	2.20869	2.09015	2.0866

(continued)

Table 5.8 (continued)

Number group	Standard	Gauss map			Liebovitch map			Piecewise map		
	BBO	CBBO 21	CBBO 22	CBBO 23	CBBO 31	CBBO 32	CBBO 33	CBBO 41	CBBO 42	CBBO 43
A_{22}	0.693316	0.889075	0.570566	0.874356	0.895691	0.648907	0.901021	0.710971	0.499314	0.81607
A_{23}	7.30237	7.94523	7.7416	8.05833	7.75629	7.8271	7.91158	7.12617	7.80177	7.55209
A_{24}	0.350178	0.1	0.595366	0.139422	0.198174	0.243167	0.167734	0.44238	0.157816	0.263212
A_{25}	8.68997	8.02606	8.19748	7.68748	7.97954	8.18425	8.95896	7.58997	8.63953	7.84443
A_{26}	3.13466	2.95016	2.8177	2.89462	3.10129	2.80952	3.04749	2.90733	2.78296	3.09452
A_{27}	10.6634	10.3895	9.67134	10.1566	9.44989	10.8694	10.3208	10.0495	10.0233	9.17087
A_{28}	19.2649	20.1795	20.652	20.7226	20.2072	20.791	20.5901	20.4812	20.8281	20.2513
A_{29}	14.9734	12.8967	12.1079	12.1302	12.5048	11.3466	11.0312	12.9812	12.0183	13.1391
Best weight	2183.932	2163.597	2164.12	2162.292	2161.652	**2161.566**	2165.3	2163.372	2165.019	2165.683
Mean weight	2192.087	**2165.625**	2173.617	2172.499	2169.272	2168.778	2169.94	2167.179	2170.96	2175.701
Coefficient variation	0.46106	**0.093395**	0.29618	0.36184	0.22424	0.20266	0.26866	0.12069	0.4198	0.31631
ω_1 (HZ)	5.00253	5.00014	5.00052	5.00006	4.99951	5.00022	5.00024	4.99999	5.0004	5.00019
ω_2 (HZ)	12.432	12.5433	12.4652	12.2092	12.5049	12.4891	12.5369	12.3564	12.7179	12.603
ω_3 (HZ)	15.2498	15.1312	15.211	15.097	15.0455	15.1455	15.1891	15.1762	15.2633	15.2066

The bold items in the table indicate the optimal values for the best weight, the best mean and the best coefficient of variation

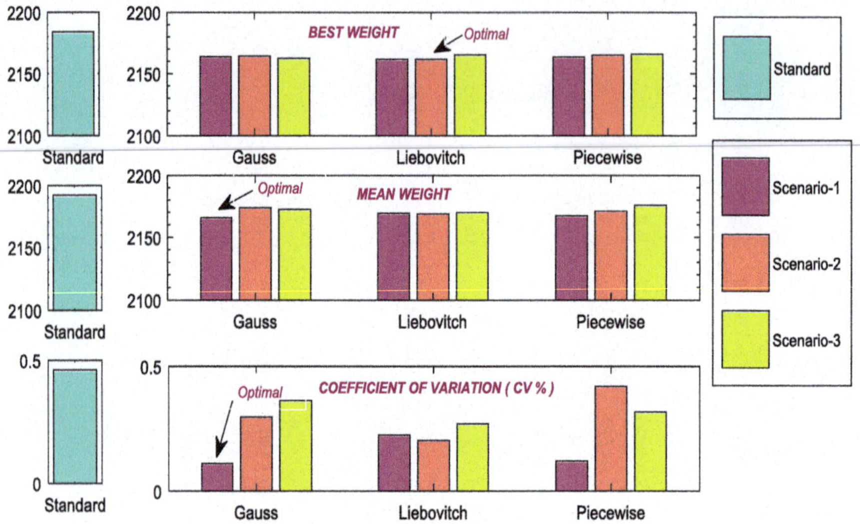

Fig. 5.14 Optimization results in standard mode and selection of chaos map for the 200-bar planar truss structure

and exploitation is achieved. In this investigation; Logistic, Tenet, Gauss, Liebovitch and Piecewise chaos functions are considered for embedding in Biogeography-Based optimization (BBO). Embedding chaos functions can be done with three scenarios. Therefore, the chaotic modes compete with each other and the best optimal response is compared with the standard mode.

References

1. Kaveh, A.: Advances in Meta-Heuristic Algorithms for Optimal Design of Structures, 2nd edn., 3rd edn. Springer, Cham, Switzerland, 2021. https://doi.org/10.1007/978-3-319-05549-7
2. Simon, D.: Biogeography-based optimization. IEEE Trans. Evol. Comput. **12**(6), 702–713 (2008). https://doi.org/10.1109/TEVC.2008.919004
3. Kaveh, A., Yousefpoor, H.: Chaos-embedded meta-heuristic algorithms for optimal design of truss structures. Scientia Iranica A **29**(6), 2868–2885 (2022). https://doi.org/10.24200/sci.2022.59812.6441
4. Kaveh, A., Yosefpoor, H.: Chaotically enhanced meta-heuristic algorithms for optimal design of truss structures with frequency constraints. Periodica Polytechnica Civil Eng. **66**(3), 900–921 (2022). https://doi.org/10.3311/PPci.20220
5. Kaveh, A., Zarfam, P., Aziminejad, A., Yosefpoor, H.: Comparison of four chaotic meta-heuristic algorithms for optimal design of large-scale Truss structures. Iranian J. Sci. Technol. Trans. Civil Eng. https://doi.org/10.1007/s40996-022-00908-8
6. Ott, E.: Chaos in Dynamical Systems. Cambridge University Press, Cambridge (2002)
7. Bucolo, M., Caponetto, R., Fortuna, L., Frasca, M., Rizzo, A.: Does chaos work better than noise? IEEE Circuit Syst. Mag. **2**(3), 4–19 (2002)

8. Kaveh, A., Sheikholeslami, R., Talatahari, S., Keshvari-Ilkhichi, M.: Chaotic swarming of particles: a new method for size optimization of truss structures. Adv. Eng. Softw. **67**, 136–47 (2014)
9. Erramilli, A., Singh, R., Pruthi, P.: Modeling Packet Traffic with Chaotic Maps. KTH, Stockholm (1994)
10. Peitgen, H.-O., Jürgens, H., Saupe, D.: Chaos and Fractals: New Frontiers of Science. Springer, Berlin (2006)

Chapter 6
Chaotic Differential Evolution

6.1 Introduction

Our world is constantly changing and no condition remains constant over time, so any living being that intends to survive in such an environment must be able to adapt to the surrounding conditions [1]. Nature itself chooses creatures that match its criteria. This adaptation process is known as evolution. Evolution is the result of phenomena such as natural selection, reproduction, crossover, mutation and symbiosis. The role of each operator briefly includes the following:

- Natural selection: the criteria of adaptation of each living organism is determined by the objective function, and the organisms that are more adapted to the criteria of nature will have a better chance of survival.
- Reproduction: By exchanging the information of gene chains between pairs, more compatible chromosomes are obtained in the new population. In Darwin's theory, compatible chromosomes are obtained in the generation itself, but in Lamarck's theory, it is believed that the existing generation acquires it and transfers it to the next generation.
- Mutation: This operator causes the genetic representation of population members to change according to the rules of probability. Therefore, a series of unplanned changes occur and errors in the genetic transcript are obtained.

Changing the generations of population members and creating new generations is done based on crossover and mutation operators. Evolutionary algorithms are an imitation of what happened in the real world. For millions of years, single-celled non-living molecules have evolved into the most complex creatures in the universe. In nature, to record the optimization information related to living organisms, ladder molecules of spiral DNA strands have been used and all genetic information related to the sequence of generations is available in it. Genes are nature's handwriting to record the evolutionary stages of living organisms. The main branches of the DNA molecule are sugar and phosphate, and its bases are composed of four types of organic

A. Kaveh and H. Yousefpoor, *Chaotic Meta-heuristic Algorithms for Optimal Design of Structures*, Studies in Computational Intelligence 1129,
https://doi.org/10.1007/978-3-031-48918-1_6

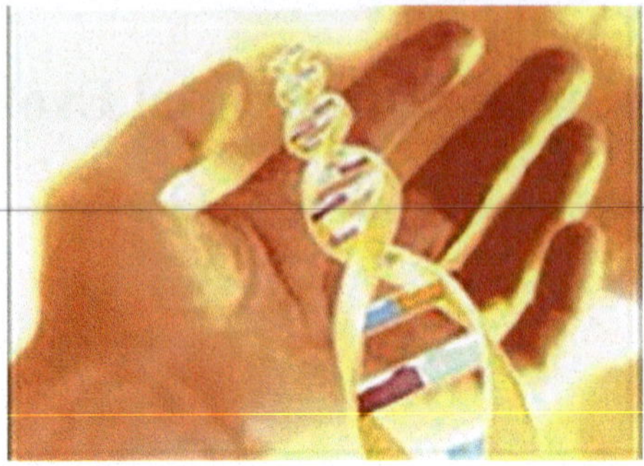

Fig. 6.1 The spiral structure of the DNA molecule

bases called cytosine, guanine, thymine and adenine. An example of it is shown in Fig. 6.1.

The molecular structure of DNA in humans contains more than three billion basic compounds, the type and sequence of which are more than 99% common to all humans. The molecular structure of DNA in humans contains more than three billion basic compounds, the type and sequence of which are more than 99% similar for all humans. In fact, each person's unique information, such as fingerprints, eye pupil color, hair color, etc., is stored in less than one percent of the DNA structure. Current organisms are the result of thousands of repetitions of a huge optimization algorithm, the purpose of which is to increase the survival power of living organisms. The Differential Evolution (DE) is among the most successful evolutionary algorithms and it adopts a unique method to generate answers and new generation population. In this algorithm, unlike other evolutionary algorithms, the mutation is done first and then the crossover is formed. Also, the difference between the points of the answer space will be optimally used and there will be no extraneous or removable information. Embedded chaos functions in this algorithm make a remarkable improvement in the optimization results. Consequently, three chaos functions, consist Logistic, Tent and Gaussian functions have been challenged with different scenarios. In the rest of this chapter, after introducing the differential evolution algorithm, the main steps of the standard algorithm will be presented, and then the chaotic mode of the differential evolution algorithm will be formed with three scenarios. Finally, with well-known optimization examples of static constraints and multi-frequency constraints, the efficiency of several chaos functions will be challenged.

6.2 Standard Differential Evolution (DE)

The differential evolution (DE) algorithm is a population-based metaheuristic algorithm developed by Price and Storn [2]. In DE, the process of finding the new solutions is unique. The difference of the DE algorithm and other evolutionary algorithms consists of forming the offspring population. Although, the way of applying mutation and selecting the step length for mutation is in accordance to a considered distribution. In the DE algorithm, a momentary solution is formed first through the mutation operation and then a new solution is produced through the cross-over operation. In the DE algorithm, instead of using a known probabilistic distribution for the mutation operation, the distance between the solutions is utilized to produce the mutation steps. The differences between the solutions of the population have important information about the objective function and the problem at hand, and therefore its utilization may improve the search processes. In a metaheuristic algorithm, as the search process progresses, the solutions converge to a single solution. Thus, the difference between the solutions of the population gets smaller and smaller. It should be noted that the population size plays a dominant role in the value of the distance in the initial population. In fact, as the population size is increased, the distances between the search agents decrease and vice versa. The distance between solutions of the population shows useful information about the diversity of the solutions of the population. In fact, by knowing the distance between the solutions of the population, one can tune the step-size of the movement and in this way regulate the intensity of the mutation operation. If the distance is bigger, one reaches the promising regions faster by utilizing larger step-sizes. On the other hand, for smaller distances, smaller step-sizes can search the promising regions with higher accuracy. Therefore, in the DE algorithm, the step-size and direction of the mutation operator are calculated as follows: In the first step, two solutions of the population are chosen randomly. Then, the difference of the two chosen solutions is determined and taken as the difference vector. The coefficient of the difference vector is considered as the mutation vector. The use of difference vectors has some benefits. The useful information of the solutions of the population is fully explored. Such information can be very useful for directing the search process. Considering the theorem on central limit, by increasing the size of the population, length of the jump steps tends to a random value having a normal distribution. In order to carry out the proposed relations, some arrays can be chosen from the initial and the mutant populations, as shown in Fig. 6.2.

$$\begin{cases} X = \{x_1, x_2, \ldots, x_{n\text{var}}\} \\ Y = \{y_1, y_2, \ldots, y_{n\text{var}}\} \Rightarrow Z = \{z_1, z_2, \ldots, z_{n\text{var}}\} \\ \alpha = \beta \times (b - c) \end{cases} \tag{6.1}$$

$$Z_j = \begin{cases} y_i \ if \ j \in J \\ x_i \ if \ j \notin J \end{cases} \tag{6.2}$$

Let us consider an empty set called J. For generating the set J, first J_o is randomly chosen from the numbers 1 to *nvar*. Then, the chosen number (J_o) is added to the set J. Now, the following operations are repeated for all values up to 1 *nvar*: A uniformly distributed random number like r_j is generated between the range zero and one:

$$Z_j = \begin{cases} y_i \ r \le P_{cr} \quad or \quad j \in j_0 \\ x_i \ otherwise \end{cases} \tag{6.3}$$

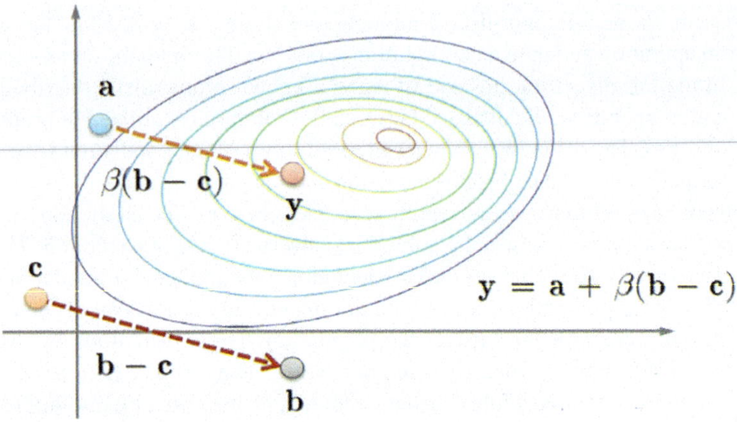

Fig. 6.2 Mutation process in differential evolution (DE)

For *r* being less than or equal to *Pcr* (the probability of intersection), one adds the number *j* to the set *J*. The basic steps in standard mode of this algorithm are provided in the subsequent subsection.

6.3 Basic Steps of Differential Evolution

Step 1. The algorithm parameters are defined consisting of the number of design variables, *nvar,* upper and lower bounds of the design variables, the maximum number of objective function evaluations, population size (npop), upper and lower limits of the scale factor Beta (*Betamax* and *betamin*, respectively), which were set to be 0.2 and 0.8 in the classic version, respectively. Ultimately, the crossover probability *Pcr* is obtained.
Step 2. Generation of the initial population: Similar to almost all of the population-based metaheuristic algorithms, the initial population is produced randomly.
Step 3. The steps below are repeated until the termination criterion is fulfilled:
The following steps are carried out for all solutions of the population:— A temporary solution is produced employing the jump operator. Where β is calculated as:

$$Beta = bet\_\min + \mathbf{rand}(\text{var}size) \cdot \times (beta\_\max - beta\_\min) \tag{6.4}$$

The intersection operator is used and a new solution is generated and calculated using the objective function. Then, if the new solution is improved compared with the current one, it is replaced by the current solution, otherwise the current solution is preserved.

$$\begin{aligned} &if \quad j = j_0 \quad or \quad \mathbf{rand} \le Pcr \\ &Z(j) = Y(j) \\ &Else \\ &Z(j) = X(j) \end{aligned} \tag{6.5}$$

The best solution found so far is then taken as the output.

6.4 Chaos Differential Evolution (CDE)

Similar to other evolutionary methods, this algorithm consists of two main stages consisting of mutation and cross-over [3–5]. The main difference between the DE algorithm belongs to the order of the two mentioned stages, where first the mutation operation and afterwards the cross-over operation is performed. It also uses the difference between solutions as much as possible until it converges while escaping the local optimums. Obviously, the cross-over and mutation operators have an important role in exploitation and exploration phases, respectively. Thus, by incorporating chaotic functions in random components related to these phases, a remarkable improvement can be obtained, as shown in Fig. 6.2. Three different scenarios are considered as follows:

Scenario 1. The chaotic map **is used** in the process of the mutation operator: Here, the first chaotic function **CHM1** is substituted in Eq. (6.4), and the following equation is resulted:

$$Beta = bet\_\min + \mathbf{CHM1} \cdot \times (beta\_\max - beta\_\min) \tag{6.6}$$

Scenario 2. The chaotic map **is placed** in the stage of the cross-over operator: Here, the second chaotic map **CHM2** is substituted in Eq. (6.5), and the following equation is resulted:

$$\begin{aligned} &if \quad j = j_0 \quad or \quad \mathbf{CHM2} \le Pcr \\ &Z(j) = Y(j) \\ &Else \\ &Z(j) = X(j) \end{aligned} \tag{6.7}$$

Scenario 3. The chaotic map is placed simultaneously in both the cross-over and mutation operators: In this case, the two chaotic maps are simultaneously substituted in Eqs. (6.4) and (6.5).

6.5 Truss Weight Optimization with Static Constraints

To form chaos algorithms, three chaos functions are selected and embedded in differential evolution (CDE). This function includes Logistic, Tent and Gaussian chaos function. Then each of these functions will be formed with three different scenarios. In these scenarios, chaos functions respectively replace the distributions related to exploration, exploitation or both stages simultaneously. 9 chaotic modes and standard mode will compete with each other to determine the best efficiency. In order to increase the diversity in the statistical space, each model is run independently for 30 times and the statistical population corresponding to the best and average values and the coefficient of variation is extracted. By researching the efficiency of chaotic models and triple scenarios, a considerable improvement in the obtained results is observed and it is possible to determine the chaos function compatible with the algorithm that creates the trade-off between exploitation and exploration phases with the best efficiency. We can also learn about the weakness and strength of the algorithm in the exploration or exploitation phases. Research results show that by using a Logistic map, it is very probable to move from a local optimum to a global one [6]. Thus, the Logistic map is used to improve the exploration capability of the algorithms. The Gaussian map is likely to be in the local minimum range and is usually employed for improving the exploitation capability [7]. Ultimately, the Tenet map can improve simultaneously both the exploration and exploitation phases. Thus, by choosing these maps, the drawbacks belonging to any kind of algorithms are removed.

6.5.1 *Formulation of the Structural Optimization Problem*

Truss optimization problems include objective function, design constraints, and bounds of the design variables [8]. In structural optimization problems, the structural weight is usually chosen as the objective function and all the design constraints should be met resulting in the lowest possible weight. The mathematical formulation of these problems can be stated as follows:

$$\begin{array}{lll} Find & A = \{A_1, A_2, A_3, ..., A_n\} & \\ to\ minimize & W(A) = \sum_{i=1}^{n} \gamma_i \times A_i \times L_i & \\ & g_j(A) \leq 0; & j = 1, 2, 3, \ldots, m \\ subjected\ to & h_k(A) \leq 0; & k = 1, 2, 3, \ldots, p \\ & \{A_L\} \leq \{A\} \leq \{A_U\} & \end{array} \tag{6.8}$$

In the above equation, A denotes the vector of cross-sectional area of the members, W stands for the total structural weight, n is the number of structural members, g_i and h_i are design constraints that can define limitations on member slenderness ratios, member stresses and nodal displacements. A_l and A_u are, respectively, the lower bound and upper bound of the design variables. Note that the selected algorithms are utilized to solve unconstrained optimization problems. In this research, the penalty function method is adopted to transform the original constrained optimization problem to an unconstrained one [9]. In the penalty function method, if none of the design constraints are violated, the value of the penalty function is set to one; Otherwise the value of the penalty function is determined with respect to the severity of the design constraint violations (see Eqs. (6.9)–(6.13)):

$$\sigma_i \le \sigma^{\max} \Rightarrow V_i = \max(0, \frac{\sigma_i}{\sigma^{\max}} - 1); \quad i = 1, 2, 3, \ldots, n \tag{6.9}$$

$$\delta_j \le \delta^{\max} \Rightarrow V_j = \max(\frac{\delta_j}{\delta^{\max}} - 1); \quad j = 1, 2, 3, \ldots, n \tag{6.10}$$

$$\lambda_k \le \lambda^{\max} \Rightarrow V_k = \max(0, \frac{\lambda_k}{\lambda^{\max}} - 1); \quad k = 1, 2, 3, \ldots, p \tag{6.11}$$

$$F_{penalty}(A) = 1 + \gamma_p \times \left\{ \sum_{i=1}^{m} V_i + \sum_{j=1}^{n} V_j + \sum_{k=1}^{p} V_k \right\} \tag{6.12}$$

$$to\ Minimize \quad Mer(A) = W(A) \times F_{penalty}(A) \tag{6.13}$$

Equations (6.9), (6.10) and (6.11) respectively correspond to stress, displacement and slenderness constraints. Here, the penalty function is considered as in Eq. (6.12) and the penalized objective function is constructed using the penalty function (see Eq. (6.13)).

6.5.2 Numerical Examples

The main objective is to find the optimal cross-sectional areas of the members of truss structures having the least weight to satisfy some limitations on allowable stress, allowable nodal displacements and slenderness rations of the members based on the used code of practice. Well known optimization benchmarks are selected and the performance of the algorithms are improved using chaotic maps and the proposed scenarios. The studied benchmarks are as follows:

- The 25-bar spatial truss;
- The 47-bar power transmission tower;
- The 52-bar planar truss;
- The 72-bar spatial truss;
- The 120-bar dome-shaped truss;
- The 200-bar planar truss.

6.5.2.1 The 25-Bar Spatial Truss

The schematic of the 25-bar spatial truss is illustrated in Fig. 6.3. The modulus of elasticity of the members and the material density are taken as 10,000 ksi and 0.1 lb/in^3, respectively. The structure includes 25 members, which are divided into eight groups according to Table 6.1. The allowable tensile and compressive stresses have also been determined.

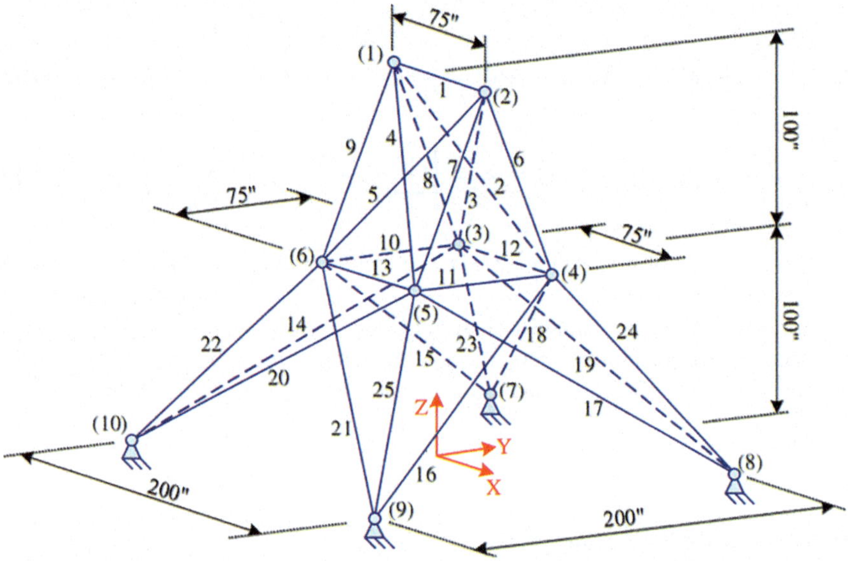

Fig. 6.3 Schematic of the 25-bar spatial truss

Table 6.1 Grouping of members and stress limitations of the 25-member spatial truss problem

Number group	Element group	Compressive stress limitation (ksi)	Tensile stress limitation (ksi)
1	A_1	35.092	40
2	A_2–A_5	11.590	40
3	A_6–A_9	17.305	40
4	A_{10}–A_{11}	35.092	40
5	A_{12}–A_{13}	35.092	40
6	A_{14}–A_{17}	6.759	40
7	A_{18}–A_{21}	6.959	40
8	A_{22}–A_{25}	11.082	40

Table 6.2 Loading conditions for the 25-bar spatial truss

Node point	Load case 1			Load case 2		
	F_x (kips)	F_y (kips)	F_z (kips)	F_x (kips)	F_y (kips)	F_z (kips)
1	0	20	− 5	1	10	− 5
2	0	− 20	− 5	0	10	− 5
3	0	0	0	0.5	0	0
4	0	0	0	0.5	0	0

The truss should be designed for two different loading conditions as summarized in Table 6.2. Also, the maximum nodal displacement in all directions is limited to ± 0.35 in. The range of choice of members' cross-section or range of decision variables is in the range of 0.01–3.4 in^2.

For ensuring the performance of the chaotic maps and considered algorithms, and also to take into account the stochastic nature of the optimization process, each mode has independently been optimized 30 times, and the results of the best run and the average of the obtained results are reported in Table 6.3. The obtained results are illustrated in Fig. 6.4. The coefficient of variation of the solutions, that shows the stability and robustness of the algorithms, are determined and another criterion for the performance of chaotic functions and algorithms are obtained.

Investigating the results obtained for combinations of algorithms with chaotic maps and comparing with standard modes indicate a considerable improvement in reduction of the optimal weight of the 25-bar spatial truss. The enhanced results for Chaotic Differential Evolution (CDE), by Logistic chaotic map and Scenario 1, a weight of 544.0047 lb is the optimal solution.

6.5.2.2 The 47-Bar Power Transmission Tower

The schematic of the 47-bar power transmission tower is depicted in Fig. 6.5. The material density is taken 0.3 lb/in^3 and the elasticity modulus of the material is considered as 30,000 ksi. Both buckling and stress limitations have to be met for all members. The allowable compressive and tensile stresses are 15 and 20 ksi, respectively. The Euler buckling stress of the members is obtained by the following equation:

$$\sigma_i^e = \frac{-KEA_i}{L_i^2}; \quad i = 1, 2, \ldots, 47 \tag{6.14}$$

Here, K is a constant value chosen with respect to the geometry of the section, E is the elasticity modulus, A_i denotes the member cross-sectional area and L_i denotes the member length. Here, k is set to be 3.96.

The structure is subjected to three loading cases. The load cases are as follows: As the first load case, loads of 6 kips and 14 kips are applied in the positive direction of

Table 6.3 Comparison of optimization results for the 25-bar spatial truss problem

Number group	Element group	Standard	Logistic map			Tent map			Gauss map		
		DE	CDE 21	CDE 22	CDE 23	CDE 31	CDE 32	CDE 33	CDE 41	CDE 42	CDE 43
1	A_1	0.01	0.01	0.01	0.01	0.01	0.01	0.01	0.01	0.01	0.01
2	A_2–A_5	1.9091	1.9552	1.975	1.8934	2.0153	2.0011	1.9589	1.96	1.9407	1.9489
3	A_6–A_9	3.0428	2.9925	2.9513	3.1195	2.9267	2.9356	2.9815	2.9256	3.0204	2.9971
4	A_{10}–A_{11}	0.01	0.01	0.01	0.01	0.01	0.01	0.01	0.01	0.01	0.01
5	A_{12}–A_{13}	0.01	0.01	0.01	0.01	0.01	0.01	0.01	0.01	0.01	0.01
6	A_{14}–A_{17}	0.68637	0.70134	0.68342	0.67751	0.6969	0.66042	0.67563	0.70805	0.67406	0.68339
7	A_{18}–A_{21}	1.7272	1.6939	1.7001	1.6928	1.6752	1.6882	1.7004	1.7231	1.7014	1.7044
8	A_{22}–A_{25}	2.6339	2.6257	2.6601	2.6229	2.655	2.6984	2.662	2.6285	2.6544	2.6499
Best weight		545.5119	**544.0047**	544.2655	544.2452	544.2191	544.473	544.2781	544.1498	544.5356	544.6218
Mean weight		546.274	544.818	**544.7989**	544.8724	545.1222	545.1124	544.9173	544.9133	544.857	545.0729
CV		0.10739	0.12407	**0.06474**	0.077288	0.094271	0.075905	0.088811	0.096847	0.05118	0.075308

The bold items in the table indicate the optimal values for the best weight, the best mean and the best coefficient of variation

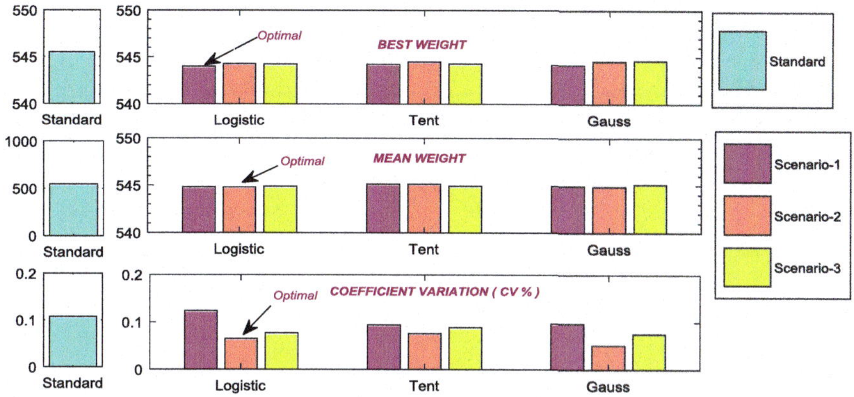

Fig. 6.4 Optimization results in standard mode and selection of chaotic maps for the 25-bar spatial truss

the X-axis and in the negative direction of the Y-axis, respectively, at both nodes 17 and 22. As the second load case, loads of 6 kips and 14 kips are, respectively, applied in the positive direction of the X-axis and in the negative direction of the Y-axis, at node 17. As the third load case, loads of 6 kips and 14 kips are, respectively, applied in the positive direction of the X-axis and in the negative direction of the Y-axis, at node 22. The first load case is the oblique loads of both the power transmission lines. In addition, the second and third load cases denote the state where one of the two power transmission lines breaks. The 47 members of the truss are grouped into 27 different groups with respect to geometric symmetry of the structure. The design variables (the cross-sectional areas of the member groups) should be selected from the 64 discrete values provided by AISC. For ensuring the performance of the chaotic maps and considered algorithms, and also to take into account the stochastic nature of the optimization process, each mode has been carried out independently 30 times, and the results of the best run and the average of the runs are reported in Table 6.4. The obtained results are illustrated in Fig. 6.6. The coefficient of variation of the solutions, which shows the robustness and stability of the algorithms, is computed and another criterion for the performance of chaotic maps and considered algorithms is investigated. By investigating the results of optimization obtained for different combinations of algorithms with chaotic map and comparing them with the standard version, a considerable improvement in reducing the structural of the 47-bar truss is obtained. The enhanced results for Chaotic Differential Evolution (CDE), by Logistic chaotic map and Scenario 2, a weight of 2359.297 lb is the optimal solution.

6.5.2.3 The 52-Bar Planar Truss

The schematic of the 52-bar planer truss is depicted in Fig. 6.7. The density of the used material is 7860 kg/m^3 and the elasticity modulus is 207 GPa. All members of

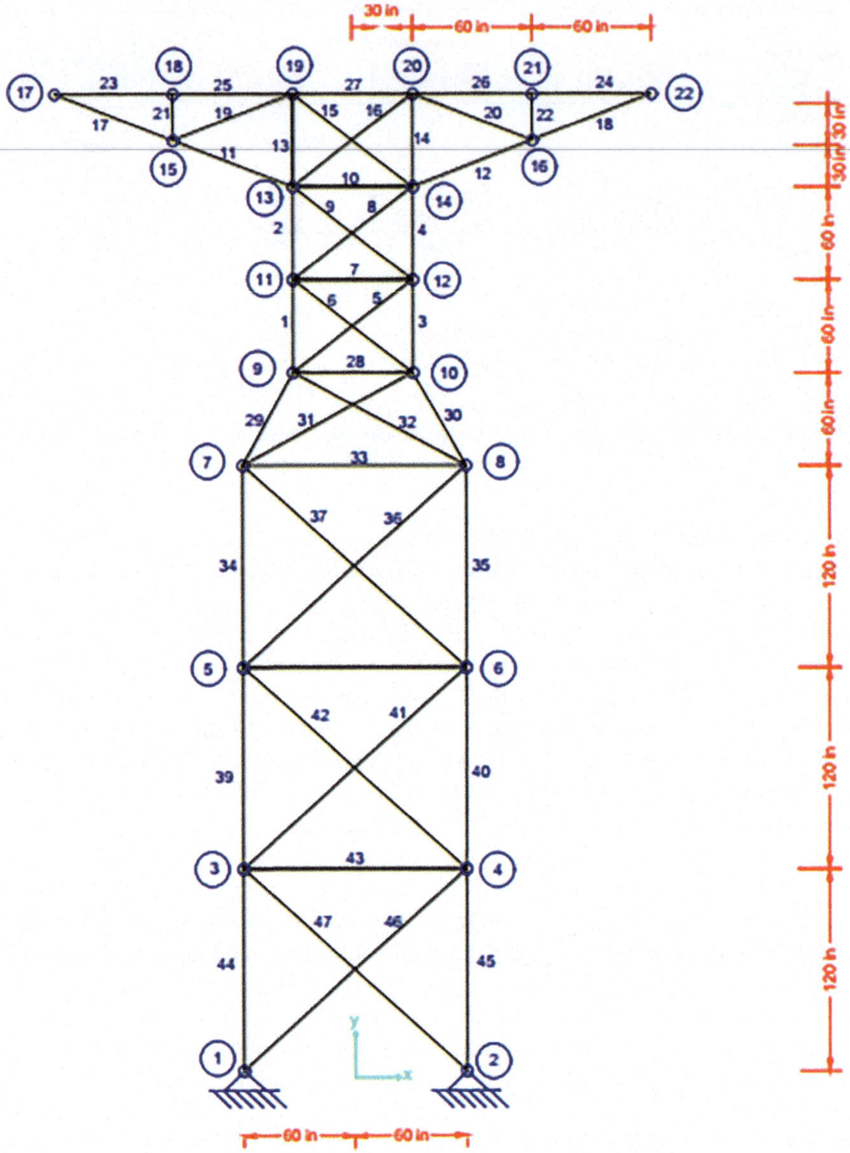

Fig. 6.5 Schematic of the 47-bar power transmission tower

the truss are subjected to a stress limitation of 180 MPa in both compression and tension. For loading, the intensity of the loads in the X and Y directions are 100 and 200 kN, respectively, and both loads are applied simultaneously. The 52 members of the truss are categorized into 12 groups considering the geometric symmetry. The cross-sectional areas of all members should be selected from the list of 64 discrete

Table 6.4 Comparison of the optimization results for the 47-bar power transmission tower problem

Number group	Element group	Standard	Logistic map			Tent map			Gauss map		
		DE	CDE 21	CDE 22	CDE 23	CDE 31	CDE 32	CDE 33	CDE 41	CDE 42	CDE 43
1	A_1–A_3	3.84	3.84	3.84	3.84	3.84	3.84	3.84	3.84	3.84	3.84
2	A_2–A_4	3.47	3.47	3.38	3.38	3.38	3.38	3.38	3.47	3.47	3.47
3	A_5–A_6	0.766	0.785	0.766	0.766	0.785	0.994	0.766	0.766	0.766	0.766
4	A_7	0.111	0.196	0.111	0.111	0.141	0.111	0.141	0.111	0.111	0.141
5	A_8–A_9	0.785	0.994	0.785	0.825	0.994	0.994	0.994	0.994	0.785	0.994
6	A_{10}	1.99	1.99	1.85	1.91	1.99	1.99	1.99	1.99	1.99	1.99
7	A_{11}–A_{12}	2.13	2.13	2.13	2.13	2.13	2.13	2.13	2.13	2.13	2.13
8	A_{13}–A_{14}	1.228	1.228	1.228	1.228	1.228	1.228	1.228	1.228	1.228	1.228
9	A_{15}–A_{16}	1.62	1.563	1.563	1.62	1.563	1.563	1.563	1.563	1.62	1.563
10	A_{17}–A_{18}	2.13	2.38	2.13	2.13	2.13	2.13	2.13	2.13	2.13	2.13
11	A_{19}–A_{20}	0.141	0.111	0.111	0.121	0.196	0.141	0.111	0.111	0.141	0.141
12	A_{21}–A_{22}	0.196	0.111	0.111	0.111	0.141	0.141	0.111	0.111	0.196	0.141
13	A_{23}–A_{24}	1.8	1.8	1.8	1.8	1.8	1.8	1.8	1.8	1.8	1.8
14	A_{25}–A_{26}	1.8	1.8	1.8	1.8	1.8	1.8	1.8	1.8	1.8	1.8
15	A_{27}	1.62	1.457	1.457	1.457	1.457	1.457	1.457	1.457	1.62	1.457
16	A_{28}	0.563	0.442	0.543	0.543	0.442	0.766	0.766	0.766	0.563	0.602
17	A_{29}–A_{30}	3.65	3.63	3.63	3.63	3.63	3.63	3.63	3.63	3.59	3.61
18	A_{31}–A_{32}	1.563	1.563	1.457	1.51	1.457	1.457	1.457	1.457	1.563	1.563
19	A_{33}	0.391	0.563	0.26	0.26	0.307	0.307	0.307	0.391	0.391	0.563
20	A_{34}–A_{35}	3.09	3.09	3.09	3.09	3.09	3.09	3.09	3.09	3.09	3.05
21	A_{36}–A_{37}	1.457	1.457	1.228	1.318	1.457	1.457	1.266	1.457	1.457	1.457

(continued)

Table 6.4 (continued)

Number group	Element group	Standard	Logistic map			Tent map			Gauss map		
		DE	CDE 21	CDE 22	CDE 23	CDE 31	CDE 32	CDE 33	CDE 41	CDE 42	CDE 43
22	A_{38}	0.25	0.111	0.25	0.25	0.307	0.442	0.391	0.307	0.25	0.196
23	A_{39}–A_{40}	3.93	3.73	3.74	3.74	3.84	3.75	3.84	3.763	3.71	3.71
24	A_{41}–A_{42}	1.563	1.563	1.563	1.563	1.563	1.563	1.563	1.563	1.563	1.563
25	A_{43}	0.141	0.111	0.111	0.111	0.111	0.111	0.111	0.141	0.141	0.111
26	A_{44}–A_{45}	4.818	4.54	4.59	4.59	4.586	4.538	4.681	4.528	4.618	4.53
27	A_{46}–A_{47}	1.457	1.457	1.457	1.457	1.457	1.457	1.457	1.457	1.457	1.563
Best weight		2440.885	2418.221	**2359.297**	2378.322	2410.082	2418.725	2401.401	2408.070	2408.230	2418.837
Mean weight		2452.271	2429.356	**2361.247**	2391.521	2422.831	2431.297	2425.571	2412.418	2423.812	2422.536
Coefficient variation		0.24109	0.22631	**0.10804**	0.19193	0.2332	0.2383	0.37785	0.1892	0.2578	0.19376

The bold items in the table indicate the optimal values for the best weight, the best mean and the best coefficient of variation

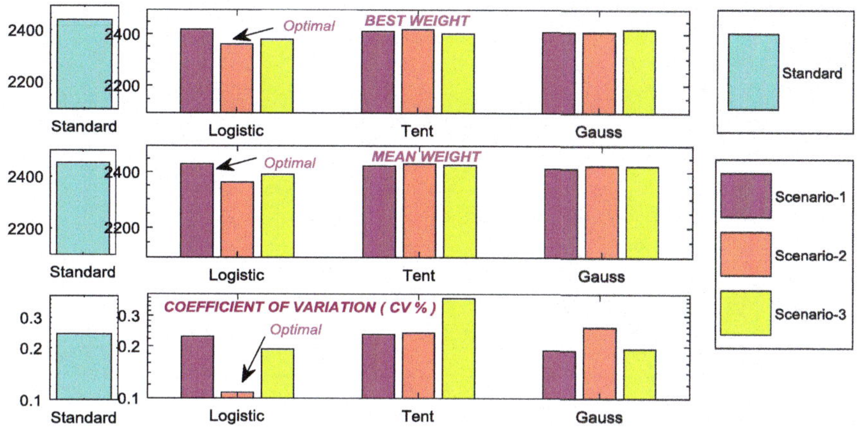

Fig. 6.6 Optimization results in standard mode and selection of chaos map for the 47-bar power transmission tower

values provided by AISC. For ensuring the performance of the chaotic maps and considered algorithms, and also to take into account the stochastic nature of the optimization process, each mode has been carried out independently 30 times, and the results of the best run and the average of the runs are reported in Table 6.5. The obtained results are illustrated in Fig. 6.8.

The coefficient for the variation of the obtained solution, that demonstrate the robustness of the algorithms, has been computed and another tool for evaluating the performance of the chaotic maps and algorithms are investigated. After reviewing the results of optimization for different combinations of algorithms with chaotic maps and comparing them with the initial and standard version, a considerable improvement in reducing the structural weight of the 52-bar planar truss results. The enhanced results for Chaotic Differential Evolution (CDE), by Logistic chaotic map and Scenario 2, a structural weight of 1899.319 kg is the optimal solution.

6.5.2.4 The 72-Bar Spatial Truss

The schematic of the 72-bar spatial truss is depicted in Fig. 6.9. The density of material is 0.1 lb/in^3 and the elasticity modulus is assumed as 10,000 ksi. Permissible stress of all members is taken as 25 ksi. The maximum displacement of the upper nodes is restricted to 0.25 in in all directions. The structure is subjected to two different load cases, as shown in Table 6.6. The 72 members of the truss are categorized into 16 groups with respect to the symmetry. The cross-sectional areas of the members are taken as the design variables of the problem, and they can vary in the range of 0.1–4 in^2.

For ensuring the performance of the chaotic maps and considered algorithms, and also to take into account the stochastic nature of the optimization process, each mode

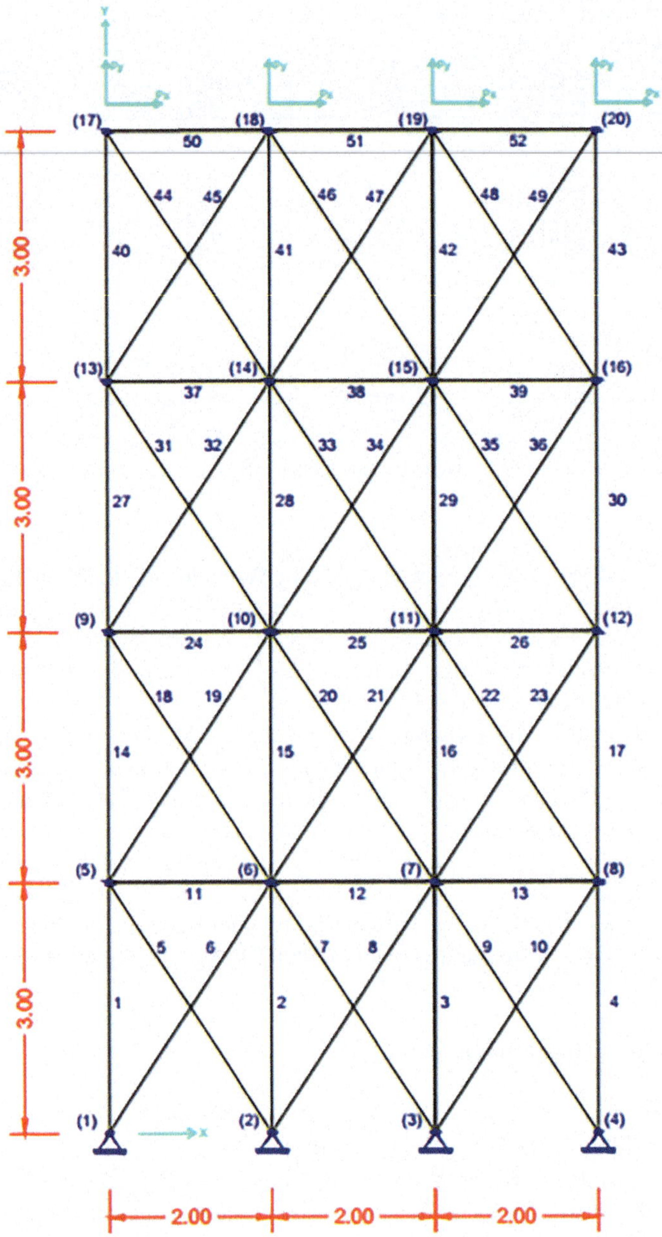

Fig. 6.7 Schematic of the 52-bar planar truss

Table 6.5 Comparison of optimization results for the 52-bar planar truss problem

Number group	Element group	Standard	Logistic map			Tent map			Gauss map		
		DE	CDE 21	CDE 22	CDE 23	CDE 31	CDE 32	CDE 33	CDE 41	CDE 42	CDE 43
1	A_1–A_4	4658.055	4658.055	4658.055	4658.055	4658.055	4658.055	4658.055	4658.055	4658.055	4658.055
2	A_5–A_{10}	1161.288	1161.288	1161.288	1161.288	1161.288	1161.288	1161.288	1161.288	1161.288	1161.288
3	A_{11}–A_{13}	494.193	506.451	494.193	506.451	506.451	494.193	506.451	363.225	494.193	506.451
4	A_{14}–A_{17}	3303.219	3303.219	3303.219	3303.219	3303.219	3303.219	3303.219	3303.219	3303.219	3303.219
5	A_{18}–A_{23}	939.998	939.998	939.998	1008.385	1008.385	1008.385	939.998	939.998	939.998	1008.385
6	A_{24}–A_{26}	506.451	506.451	506.451	363.225	641.289	494.193	494.193	494.193	494.193	506.451
7	A_{27}–A_{30}	2290.318	2238.705	2238.705	2290.318	2238.705	2238.705	2238.705	2238.705	2238.705	2290.705
8	A_{31}–A_{36}	1008.385	1008.385	1008.385	1008.385	1008.385	1008.385	1008.385	1008.385	1008.385	1008.385
9	A_{37}–A_{39}	506.451	388.386	388.386	494.193	388.386	494.193	494.193	388.386	506.451	388.386
10	A_{40}–A_{43}	1283.868	1283.868	1283.868	1283.868	1283.868	1283.868	1283.868	1283.868	1374.191	1283.868
11	A_{44}–A_{49}	1283.868	1161.288	1161.288	1161.288	1161.288	1161.288	1283.868	1161.288	1161.288	1161.288
12	A_{50}–A_{52}	494.193	506.451	506.451	494.193	494.193	494.193	494.193	792.256	494.193	494.193
Best weight		1929.473	1899.897	**1899.319**	1915.237	1917.306	1914.233	1924.026	1907.777	1911.702	1915.852
Mean weight		1934.848	1902.485	**1901.848**	1925.991	1921.822	1918.911	1932.106	1915.991	1920.107	1921.159
CV		0.15059	0.11727	**0.1035**	0.35484	0.2378	0.12287	0.53945	0.31484	0.3217	0.15397

The bold items in the table indicate the optimal values for the best weight, the best mean and the best coefficient of variation

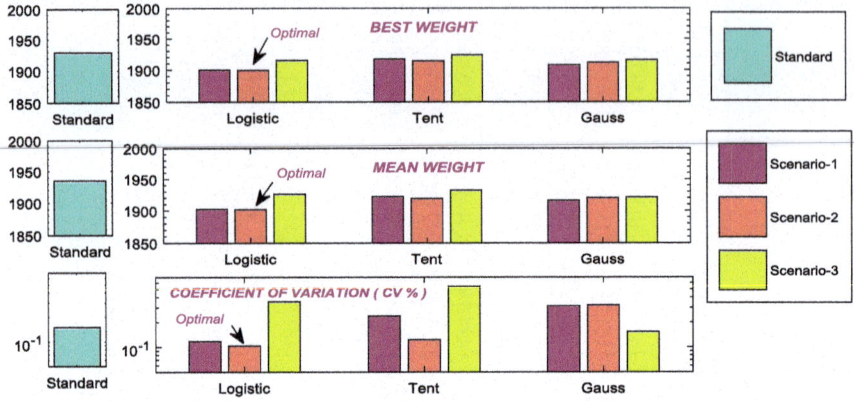

Fig. 6.8 Optimization results in standard and chaotic map for the 52-bar planar truss problem

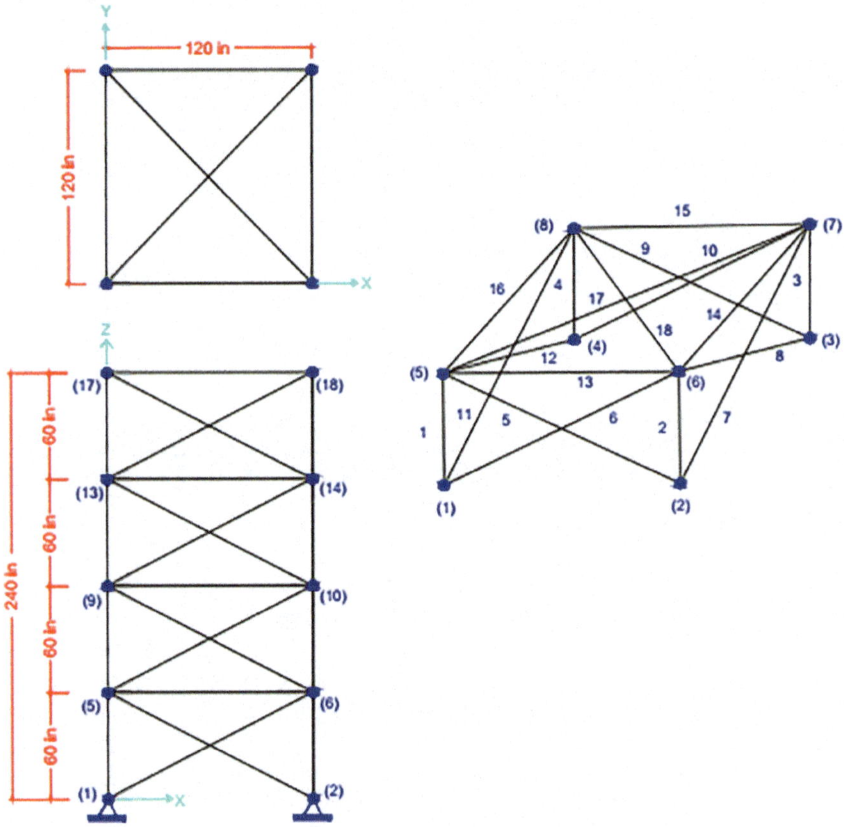

Fig. 6.9 Schematic of the 72-bar spatial truss and typical story

Table 6.6 Loading conditions for the 72-bar spatial truss

Node	Load case 1			Load case 2		
	F_x (kips)	F_y (kips)	F_z (kips)	F_x (kips)	F_y (kips)	F_z (kips)
17	5	5	– 5	0	0	– 5
18	0	0	0	0	0	– 5
19	0	0	0	0	0	– 5
20	0	0	0	0	0	– 5

has been carried out independently 30 times, and the results of the best run and the average of the runs are reported in Table 6.7. The obtained results are illustrated in Fig. 6.10. The coefficient of variation of the obtained solutions, which shows the robustness and stability of the optimization results, are computed and another criterion for evaluating the performance of chaotic maps and considered algorithms has been investigated. Observing the results of optimization for different combinations of algorithms with chaotic maps and comparing them with the initial and standard version, a considerable improvement in reducing the structural weight of the 72-bar truss. The enhanced results for Chaotic Differential Evolution (CDE), by Tent chaotic map and Scenario 1, a structural weight of 379.0187 lb is the optimal response.

6.5.2.5 The 120-Bar Dome Shaped Truss

The 120-bar dome-shaped truss depicted in Fig. 6.11 is considered as the next design example. The material density is 0.288 lb/in^3 and the elasticity modulus is taken as 30,450 lb/in^2. The permissible compressive and tensile stresses are considered based on the provisions of the AISC ASD code. For all nodes of the truss, the maximum deformation in all directions is taken as 0.1969 in. All free nodes of the truss are affected by vertical loading, as – 13.49 kips at node one, and – 6.744 kips at nodes 2 to 14, and 2.248 kips at the remaining nodes. The 120 members of the dome are categorized into seven groups with respect to their shapes and the existing symmetry.

The minimum member cross-sectional areas, as the lower limit of the design variables, is 0.775 in^2, while the maximum cross-sectional area is 20 in^2. The allowable compressive and tensile stresses are considered according to the following equations:

$$\sigma_i = \begin{cases} 0.6F_y \ for \ \ \sigma_i \geq 0 \\ \sigma_i^- \quad \ for \ \ \sigma_i < 0 \end{cases} \tag{6.15}$$

The allowable compressive stresses are determined by the following equations:

$$\sigma_i^- = \begin{cases} (1 - \frac{\lambda^2}{2C^2})F_y/FS \ for \ \ \lambda < C \\ \frac{12\pi^2 E}{23\lambda^2} \qquad\qquad\quad for \ \ \lambda \geq C \end{cases} \tag{6.16}$$

Table 6.7 Comparison of optimization results for the 72-bar spatial truss problem

Number group	Element group	Standard	Logistic map			Tent map			Gauss map		
		DE	CDE 21	CDE 22	CDE 23	CDE 31	CDE 32	CDE 33	CDE 41	CDE 42	CDE 43
1	A_1–A_4	1.9345	1.8866	1.8949	1.8883	1.8997	1.8945	1.8804	1.8649	1.8843	1.8913
2	A_5–A_{12}	0.51721	0.51071	0.51046	0.5094	0.51342	0.51292	0.51355	0.51742	0.51499	0.51212
3	A_{13}–A_{16}	0.1	0.10001	0.10002	0.10026	0.1	0.1	0.1	0.10006	0.1	0.10004
4	A_{17}–A_{18}	0.1	0.10004	0.10004	0.10009	0.1	0.1	0.1	0.1	0.10011	0.1
5	A_{19}–A_{22}	1.3123	1.2926	1.2838	1.2897	1.2753	1.2616	1.2663	1.2908	1.2603	1.2822
6	A_{23}–A_{30}	0.49009	0.51042	0.51614	0.50892	0.51251	0.50796	0.5132	0.51034	0.51448	0.50879
7	A_{31}–A_{34}	0.10023	0.1	0.1	0.1	0.1	0.10003	0.1	0.10016	0.1	0.1
8	A_{35}–A_{36}	0.1	0.1	0.1	0.10047	0.1	0.1	0.1	0.1	0.10001	0.10002
9	A_{37}–A_{40}	0.5185	0.51864	0.50295	0.51953	0.50878	0.51879	0.51124	0.512	0.53427	0.52011
10	A_{41}–A_{48}	0.51046	0.50669	0.51861	0.51159	0.5186	0.51527	0.51528	0.51534	0.51193	0.51747
11	A_{49}–A_{52}	0.10091	0.10004	0.1	0.1	0.10008	0.10041	0.1	0.1	0.1	0.1
12	A_{53}–A_{54}	0.1	0.10001	0.1	0.1	0.1	0.10027	0.10011	0.1	0.10017	0.1
13	A_{55}–A_{58}	0.15657	0.15651	0.15701	0.15669	0.15711	0.15669	0.15672	0.15687	0.15651	0.15667
14	A_{59}–A_{66}	0.5644	0.54784	0.54111	0.54679	0.53611	0.54009	0.54227	0.54404	0.54086	0.54402
15	A_{67}–A_{70}	0.37973	0.41384	0.40093	0.40456	0.40886	0.42958	0.41598	0.40582	0.41437	0.41109
16	A_{71}–A_{72}	0.56581	0.57146	0.5708	0.58226	0.56886	0.55951	0.57224	0.56597	0.56443	0.56335
Best weight		379.6738	379.1342	379.2588	379.1903	**379.0187**	379.0244	379.2372	379.0853	379.1792	379.351
Mean weight		379.7721	379.3436	379.4705	379.4347	379.2736	379.313	379.4492	**379.2044**	379.3444	379.506
Coefficient variation		**0.021252**	0.04711	0.046402	0.046123	0.06374	0.04605	0.040675	0.041523	0.051209	0.02554

The bold items in the table indicate the optimal values for the best weight, the best mean and the best coefficient of variation

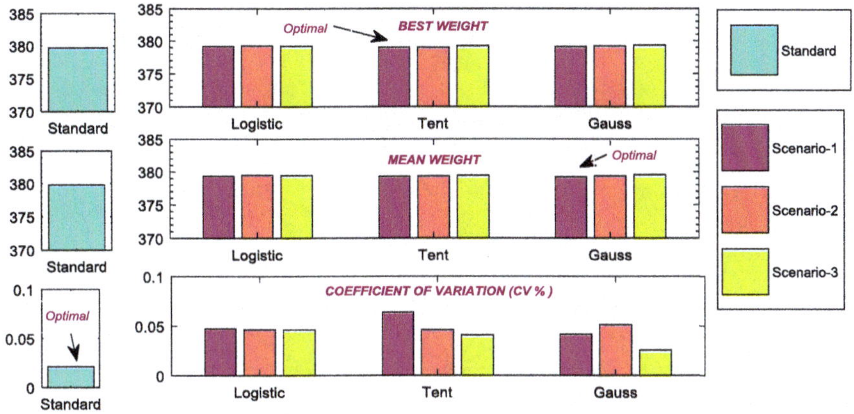

Fig. 6.10 Optimization results in standard mode and chaotic map modes for the 72-bar spatial truss problem

where:

$$FS = (\frac{5}{3} + \frac{3\lambda}{8C} - \frac{\lambda^3}{8C^3}); C = \sqrt{(\frac{2\pi^2 E}{F_y})};$$
$$\lambda = \frac{kl}{r}; r = aA^b; a = 0.4993; b = 0.6777 \tag{6.17}$$

In the above equations, E is the elasticity modulus, F_y is the yield stress of steel, C is the critical slenderness ratio which separates the inelastic and elastic buckling regions, k is the effective length factor and r is the radius of gyration. For ensuring the performance of the chaotic maps and considered algorithms, and also to take into account the stochastic nature of the optimization process, each mode has been carried out independently 30 times, and the results of the best run and the average of the runs are reported in Table 6.8. The obtained results are illustrated in Fig. 6.12.

The coefficient for variation of the obtained solutions, which demonstrate the robustness of the optimization results, are computed and a different criterion is obtained to assess the performance of the chaotic maps and the proposed algorithms. The lower and upper bounds of the cross-sectional areas of the members are respectively taken as 0.775 and 20 in^2. Observing the obtained results for different combinations of algorithms with chaotic maps and comparing them with the initial and standard version, a considerable reduction of the structural weight of the 120-bar dome is observed. The enhanced results obtained for Chaotic Differential Evolution (CDE), by Gauss chaotic map and Scenario 1, a structural weight of 33,241.97 lb is the optimal solution.

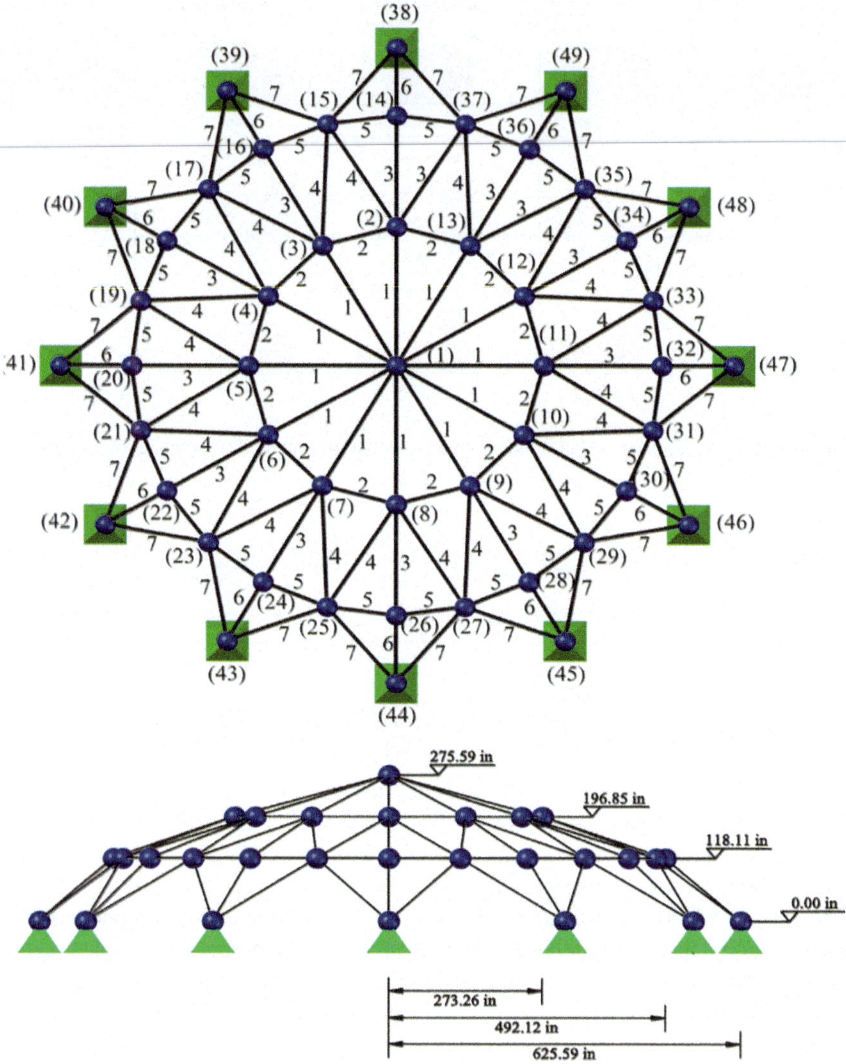

Fig. 6.11 Schematic of the 120-bar dome

6.5.2.6 The 200-Bar Planar Truss Structure

For the 200-bar planar truss structure, the topology and nodal ordering are illustrated in Fig. 6.13. The material density is 0.283 lb/in^3 and its module of elasticity is taken 30,000 ksi. The allowable stress of members is taken as 10 klb. No deformation restriction is considered for optimization. Three different load cases are applied to the structure as follows: As the first load case, a lateral load of 1 kips is applied in positive X-axis at nodes 1, 6, 15, 20, 29, 34, 43, 48, 57, 62 and 71. The second load

Table 6.8 Comparison of the optimization results for the dome-shaped 120-bar truss problem

Number group	Standard	Logistic map			Tent map			Gauss map		
	DE	CDE 21	CDE 22	CDE 23	CDE 31	CDE 32	CDE 33	CDE 41	CDE 42	CDE 43
1	3.02546	3.02444	3.02437	3.02442	3.02501	3.02428	3.02451	3.02453	3.02441	3.02431
2	14.75	14.6478	14.8242	14.7554	14.7987	14.744	14.7644	14.8004	14.7668	14.8321
3	5.12625	5.07877	5.06158	5.07062	5.12698	5.08473	5.07825	5.0796	5.09108	5.0545
4	3.12647	3.1223	3.13276	3.13154	3.12503	3.13087	3.13371	3.12682	3.12704	3.12766
5	8.43044	8.54635	8.47336	8.49377	8.42358	8.49819	8.48115	8.4641	8.4842	8.46608
6	3.34835	3.32704	3.28793	3.30141	3.31089	3.29047	3.30139	3.30583	3.28594	3.31401
7	2.49503	2.49679	2.49671	2.49645	2.4957	2.49655	2.49642	2.49638	2.49776	2.49717
Best weight	33,249.27	33,245.51	33,247.03	33,244.61	33,242.61	33,246.37	33,247.7	**33,241.97**	33,242.83	33,246.59
Mean weight	33,253.34	33,249.32	33,249.16	33,248.4	**33,247.52**	33,248.2	33,248.55	33,247.96	33,247.64	33,248.5
Coefficient variation	0.007028	0.007815	0.005418	0.007266	0.009784	0.005294	**0.0026**	0.014293	0.012065	0.005855

The bold items in the table indicate the optimal values for the best weight, the best mean and the best coefficient of variation

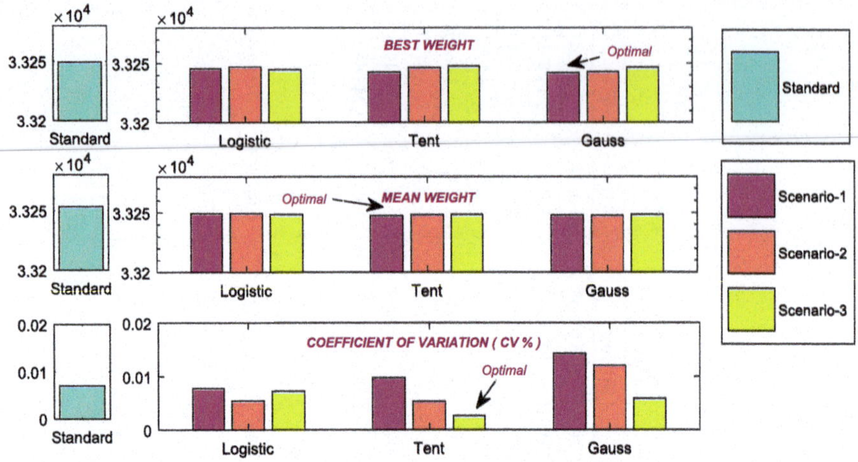

Fig. 6.12 Results of the optimization in standard mode and chaotic map for the 120-bar dome truss problem

case include a gravity load of 10 kips is applied in the negative Y-axis at nodes 1, 2, 3, 4, 5, 6, 8, 10, 12, 14, 15, 16, 17, 18, 19 and... 71, 72, 73, 74 and 75. The third load case consists of both the above-mentioned load cases, simultaneously.

The truss members are categorized into 29 groups according Table 6.9. The minimum cross-sectional area of the truss members, being considered as the lower bound of the design variables, is set to be 0.1 in^2 and the maximum cross-sectional area is as 16 in^2. For ensuring the performance of the chaotic maps and considered algorithms, and also to take into account the stochastic nature of the optimization process, each mode has been carried out independently 30 times, and the results of the best run and the average of the runs are reported in Table 6.10. The obtained results are illustrated in Fig. 6.14. The coefficient of variation of the obtained solutions, which demonstrates the robustness of the algorithms, is computed and another criterion for evaluating the performance of chaotic maps and algorithms is achieved. The enhanced results obtained for Chaotic Differential Evolution (CDE), by Gauss chaotic map and Scenario 3, a structural weight of 25,944.98 lb is found as optimal.

6.6 Concluding Remarks

The Differential Evolution (DE), uses a unique new method of generating responses. In most evolutionary algorithms, children population are first generated using the cross-over operator, and then mutation is applied to them, but DE produces a temporary solution through the mutation operation and is stabilized by the intersection of that response. In the process of optimization, the distance between the population solutions gets smaller and smaller and converges towards a single answer. Therefore,

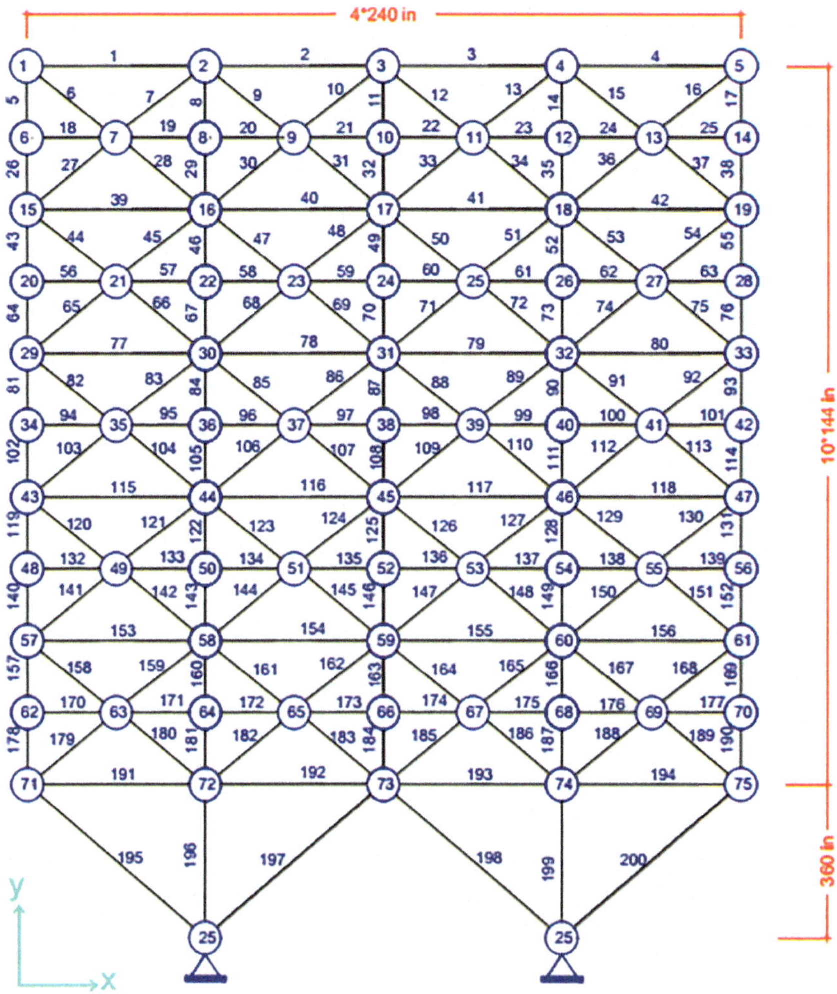

Fig. 6.13 Schematic of the 200-bar planar truss structure

by using the distance of the population solutions, one can adjust the step length and the intensity of the mutation. If the distance is longer, one can reach the promising regions faster by selecting large steps, and on the contrary, we need small steps for small distances. Therefore, by defining the difference vector, the algorithm selects a coefficient of it for mutation. For cross-over operator, a number of solution of the initial and mutant populations are selected. For the mutation, determining the β scale parameter is very important, and the generation of new answers is achieved by the intersection method using a binomial. With all the advantages that this algorithm has, it is still threatened by some pests, and that is the random distribution used in the two phases of exploitation and exploration. By embedding chaotic dynamic

Table 6.9 Member groups of the 200-bar planar truss structure

Group	Element group	Group	Element group
A_1	1, 2, 3, 4	A_{16}	82, 83, 85, 86, 88, 89, 91, 92, 103, 104, 106, 107, 109, 110, 112, 113
A_2	5, 8, 11, 14, 17	A_{17}	115, 116, 117, 118
A_3	19, 20, 21, 22, 23, 24	A_{18}	119, 122, 125, 128, 131
A_4	18, 25, 56, 63, 94, 101, 132, 139, 170, 177	A_{19}	133, 134, 135, 136, 137, 138
A_5	26, 29, 32, 35, 38	A_{20}	140, 143, 146, 149, 152
A_6	6, 7, 9, 10, 12, 13, 15, 16, 27, 28, 30, 31, 33, 34, 36, 37	A_{21}	120, 121, 123, 124, 126, 127, 129, 130, 141, 142, 144, 145, 147, 148, 150, 151
A_7	39, 40, 41, 42	A_{22}	153, 154, 155, 156
A_8	43, 46, 49, 52, 55	A_{23}	157, 160, 163, 166, 169
A_9	57, 58, 59, 60, 61, 62	A_{24}	171, 172, 173, 174, 175, 176
A_{10}	64, 67, 70, 73, 76	A_{25}	178, 181, 184, 187, 190
A_{11}	44, 45, 47, 48, 50, 51, 53, 54, 65, 66, 68, 69, 71, 72, 74, 75	A_{26}	158, 159, 161, 162, 164, 165, 167, 168, 179, 180, 182, 183, 185, 186, 188, 189
A_{12}	77, 78, 79, 80	A_{27}	191, 192, 193, 194
A_{13}	81, 84, 87, 90, 93	A_{28}	195, 197, 198, 200
A_{14}	95, 96, 97, 98, 99, 100	A_{29}	196, 199
A_{15}	102, 105, 108, 111, 114		

series with triple scenarios, a significant improvement is achieved and the algorithm is saved from premature convergence. In chaotic states, we can get rid of the local optimal trap and converge to the global optimal by creating a mutation. By embedding Logistic, Tenet and Gaussian chaotic functions with three scenarios, a great challenge between chaos functions has been achieved and significant results have been improved. In each model, nine chaotic modes compete with each other and then the best solution is compared with the standard mode to determine the amount of improvement.

Table 6.10 Comparison of optimization results for the 200-bar planar truss problem

Number group	Standard	Logistic map			Tent map			Gauss map		
	DE	CDE 21	CDE 22	CDE 23	CDE 31	CDE 32	CDE 33	CDE 41	CDE 42	CDE 43
A_1	0.109804	0.111968	0.133334	0.1518	0.139758	0.128249	0.124777	0.145959	0.123874	0.137444
A_2	1.0013	1.0019	1.00513	0.9361	0.970259	0.955915	0.951117	0.997033	0.950766	1.10377
A_3	0.110432	0.129351	0.104934	0.1	0.124648	0.1	0.112	0.101723	0.118156	0.1
A_4	0.1	0.1	0.1	0.1004	0.102849	0.1	0.1	0.1	0.107598	0.103773
A_5	1.98799	2.00228	2.18996	1.9674	1.97256	2.06544	2.05272	2.08527	2.09164	2.06069
A_6	0.272905	0.287364	0.290136	0.3234	0.305247	0.281629	0.308542	0.293969	0.293036	0.285309
A_7	0.144468	0.1	0.1	0.1	0.1	0.111118	0.176904	0.1	0.141453	0.107926
A_8	3.01138	3.19851	3.29609	3.1721	3.1163	3.15725	3.07922	3.13249	3.16747	3.11814
A_9	0.160321	0.1	0.1	0.1152	0.101798	0.102197	0.1	0.1	0.1	0.108558
A_{10}	4.02464	4.17349	4.18262	4.1405	4.11543	4.1284	4.1019	4.14145	4.07637	4.11268
A_{11}	0.468984	0.430061	0.435545	0.4046	0.408911	0.408231	0.432817	0.415525	0.451286	0.436898
A_{12}	0.136734	0.1	0.1	0.1	0.104182	0.10279	0.154718	0.1	0.137208	0.11479
A_{13}	5.37855	5.4345	5.38746	5.3787	5.41479	5.34336	5.42221	5.53738	5.5261	5.45544
A_{14}	0.167681	0.178761	0.12	0.1545	0.100585	0.1	0.108904	0.103312	0.161605	0.147573
A_{15}	6.64142	6.37694	6.37768	6.3976	6.60039	6.3413	6.69565	6.69513	6.64046	6.69777
A_{16}	0.541264	0.543705	0.527312	0.5555	0.522813	0.532863	0.557485	0.527096	0.555244	0.580085
A_{17}	0.342541	0.271354	0.412044	0.4355	0.324989	0.388815	0.211266	0.363476	0.201965	0.250868
A_{18}	7.98224	7.79065	7.90307	8.1938	7.948	7.782	8.03003	8.19935	7.87956	8.20042
A_{19}	0.100675	0.1	0.197329	0.1142	0.129617	0.111	0.103753	0.1	0.109357	0.103911
A_{20}	8.83395	8.82831	8.98214	8.9821	8.91169	8.7748	8.9213	8.92501	8.93009	8.9805
A_{21}	0.815249	0.889663	0.884391	0.8934	0.832012	0.846524	0.755394	0.863188	0.758023	0.889963

(continued)

Table 6.10 (continued)

Number group	Standard	Logistic map			Tent map			Gauss map		
	DE	CDE 21	CDE 22	CDE 23	CDE 31	CDE 32	CDE 33	CDE 41	CDE 42	CDE 43
A_{22}	0.176725	0.491397	0.1	0.2633	0.710391	0.62645	0.844355	0.5882	0.626019	0.851129
A_{23}	10.7815	10.72	11.4916	11.3254	11.188	10.8632	11	11.4158	11.5141	11.3492
A_{24}	0.178354	0.15608	0.1	0.1	0.101125	0.13308	0.1	0.1	0.1	0.108883
A_{25}	11.6241	11.8274	12.6137	12.3304	12.3104	11.8517	11.9514	12.1701	12.2788	12.3597
A_{26}	0.95791	1.16881	0.850457	1.011	1.259	1.3699	1.30515	1.17538	1.23006	1.36937
A_{27}	7.89651	6.82736	7.41096	6.5203	6.44418	5.83602	5.41386	5.5353	5.80837	5.5083
A_{28}	11.9779	11.5108	10.9749	11.3814	10.5292	11.5232	11.3177	11.3113	11.5237	10.144
A_{29}	13.3582	13.8913	13.7803	13.8988	14.5201	14.1102	14.2151	14.1187	14.0046	14.638
Best weight	26,238.97	26,103.37	26,047.30	26,123.09	25,959.04	26,020.50	25,988.41	26,087.70	26,205.10	**25,944.98**
Mean weight	26,369.75	26,259.53	26,186.69	26,178.49	26,147.29	26,090.43	26,303.55	26,303.14	26,259.55	**26,067.28**
Coefficient variation	0.37892	0.44071	0.32072	**0.14413**	0.46014	0.28446	0.29325	0.85912	0.38258	0.29596

The bold items in the table indicate the optimal values for the best weight, the best mean and the best coefficient of variation

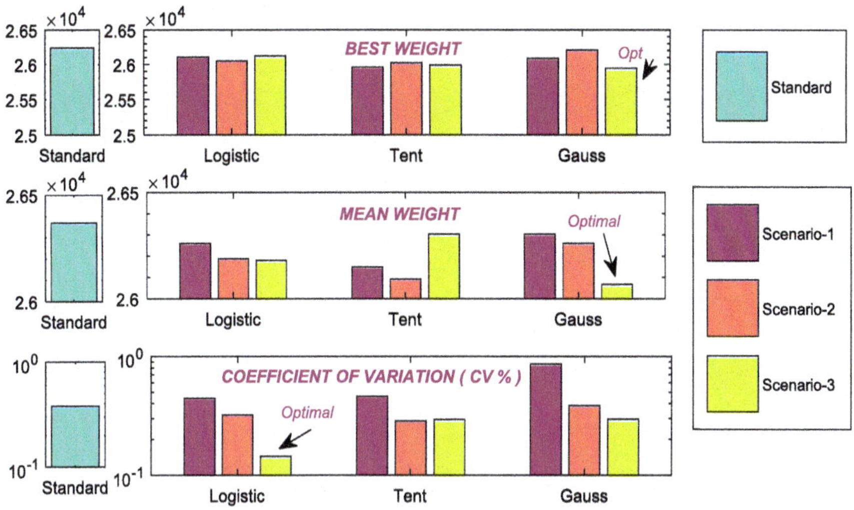

Fig. 6.14 The results of optimization in standard and chaotic map for the 200-bar planar truss problem

References

1. Holland, J.H.: Adaptation in Natural and Artificial Systems: An Introductory Analysis with Applications to Biology, Control, and Artificial Intelligence. MIT Press, Ann Arbor (1975)
2. Price, K.V., Storn, R.M.: Differential Evolution: A Practical Approach to Global Optimization. Springer, Berlin, Heidelberg (2005). https://doi.org/10.1007/3-540-31306-0
3. Kaveh, A., Yousefpoor, H.: Chaos-embedded meta-heuristic algorithms for optimal design of truss structures. Scientia Iranica A **29**(6), 2868–2885 (2022). https://doi.org/10.24200/sci.2022.59812.6441
4. Kaveh, A., Yosefpoor, H.: Chaotically enhanced meta-heuristic algorithms for optimal design of truss structures with frequency constraints. Periodica Polytechnica Civil Eng. **66**(3), 900–921 (2022). https://doi.org/10.3311/PPci.20220
5. Kaveh, A., Zarfam, P., Aziminejad, A., Yosefpoor, H.: Comparison of four chaotic meta-heuristic algorithms for optimal design of large-scale truss structures. Iran. J. Sci. Technol. Trans. Civil Eng. https://doi.org/10.1007/s40996-022-00908-8
6. Ott, E.: Chaos in Dynamical Systems. Cambridge University Press, Cambridge (2002)
7. Bucolo, M., Caponetto, R., Fortuna, L., Frasca, M., Rizzo, A.: Does chaos work better than noise? IEEE Circuit Syst. Mag. **2**(3), 4–19 (2002)
8. Kaveh, A.: Advances in Meta-heuristic Algorithms for Optimal Design of Structures, 2nd edn., 3rd edn. Springer, Cham, Switzerland (2021). https://doi.org/10.1007/978-3-319-05549-7
9. Kaveh, A., Sheikholeslami, R., Talatahari, S., Keshvari-Ilkhichi, M.: Chaotic swarming of particles: a new method for size optimization of truss structures. Adv. Eng. Softw. **67**, 136–47 (2014)

Chapter 7
Chaotic Water Evaporation Optimization

7.1 Introduction

The combination of meta-heuristic algorithms with chaos functions brings significant improvement in non-linear and non-convex problems. Since these functions replace the random distributions of the exploration and exploitation stages, the need for meta-heuristic algorithms to create a balance between these two stages is greatly improved [1]. In recent research, algorithms from different groups such as evolutionary and swarm intelligence algorithms in combination with chaos functions have been significantly improved. In this chapter, a new algorithm from the sub-field of physics called Water Evaporation Optimization (WEO) is examined. The inspiration of the algorithm is the evaporation of small water molecules on a solid surface with different wet states. Also, in the structure of this algorithm, the dynamic behavior of water during its transformation from liquid to gas and surface evaporation is considered. Physically, by applying special conditions, the surface in contact with water can change from a hydrophobicity state to a hydrophily state. Different strategies have been investigated for this algorithm, which have been introduced under the title of single layer strategy and drop strategy. In this study, Logistic, Gaussian and Piecewise chaos functions have been used for optimizations with static selection constraints and well-known structural problems for evaluation. Logistic, Gaussian, and Piecewise chaos functions have been chosen to form chaotic algorithms in static loading. Also, well-known structural problems have been investigated to determine the effectiveness of chaotic algorithms. For optimization with multi-frequency constraints, logistic and Gaussian chaos functions have resulted in significant improvement and we did not need to check the third chaos function. In most cases, chaotic modes have a significant improvement compared to the standard mode. The results of embedded chaos functions show the fact that chaos accelerators, with different scenarios, have been able to overcome the weaknesses of algorithms and achieve a significant improvement in creating a balance between search and extraction. By creating chaotic mutations,

A. Kaveh and H. Yousefpoor, *Chaotic Meta-heuristic Algorithms for Optimal Design of Structures*, Studies in Computational Intelligence 1129,
https://doi.org/10.1007/978-3-031-48918-1_7

these functions have jumped from local optima to global optima, thus saving algorithms from the trap of local optima. In chaotic states, premature convergence of algorithms has been resolved and, in some cases, the Number of Function Evaluation (NFE) to achieve convergence has been reduced. In the rest of this chapter, after introducing the Water Evaporation Optimization (WEO), the basic steps of the algorithm will be presented in the standard mode, and then the Chaotic Water Evaporation Optimization (CWEO), will be formed with three scenarios. In the end, the effectiveness of chaos functions will be challenged with well-known optimization examples of static constraints and multi-frequency constraints.

7.2 Standard Water Evaporation Optimization (WEO)

The meta-heuristic algorithm of Water Evaporation Optimization (WEO) was presented by Kaveh et al. in 2016 [2]. The physical phenomenon of evaporation of tiny water molecules occurs in different forms when the solid surface is in different wet states. If this phenomenon is combined with the dynamic simulation of water molecules, we will see interesting results. Based on the laboratory tests, the investigated surface changes from hydrophobicity to hydrophily, but the rate of evaporation of water molecules in this sudden change will not have a uniform reduction condition. Also, the rate of evaporation of water molecules increases first, and after reaching a maximum volume, the rate of evaporation decreases. This intelligently behavior of water molecules is one of the inspirations in the search space. The second inspiration is about the spherical and layered form of the water molecule. The second inspiration is about the spherical form of the water drop and the conditions for its transformation into a spread layer form. With experimental investigations, the results show the fact that when the wetness of the surface of the desired layer is not enough, the water molecule will appear in the form of drops. But in cases where the wetness of the investigated surface increases to an acceptable level, the water molecule spreads as a single layer. These two modes play a constructive role in the formation of exploration and exploitation steps.

7.3 Basic Steps in Standard Water Evaporation Optimization (WEO)

Step 1. Selecting the parameters of the algorithm, which are: determining the number of water molecules, which is the initial population nWM, maximum number of repetitions of the "maxNITs" algorithm, minimum and maximum probability of evaporation of the single-layer state MEPmin and MEPmax, which we usually choose 0.03 and 0.6, respectively, minimum and maximum probability of evaporation of droplet state DEPmin and DEPmax which we usually choose 0.6 and 1 respectively. This algorithm with n number of proposed answers which constitute the same water molecules is randomly selected from the search space and the matrix of molecules it forms water. Then, the responses are applied to objective functions and penalized objective functions, and the corresponding vectors of those functions are formed.

Step 2. The formation of the evaporation phase based on the single-layer strategy of water, which is carried out in the form of Eq. (7.1) and the probability in the range of 0.03–0.6:

$$if \quad NFE_s \leq \frac{\max NFE_s}{2} \rightarrow \begin{cases} if \ \mathbf{rand}_{ij} < \exp(E_{sub}(i)) \Rightarrow MEP_{ij} = 1 \\ if \ \mathbf{rand}_{ij} \geq \exp(E_{sub}(i)) \Rightarrow MEP_{ij} = 0 \end{cases} \tag{7.1}$$

The formation of the evaporation phase based on the water drop strategy, which is carried out in the form of Eq. (7.2) probability relations in the range of 0.6–1:

$$if \quad NFE_s > \frac{\max NFE_s}{2} \rightarrow \begin{cases} if \ \mathbf{rand}_{ij} < J(\theta_i) \Rightarrow DEP_{ij} = 1 \\ if \ \mathbf{rand}_{ij} \geq J(\theta_i) \Rightarrow DEP_{ij} = 0 \end{cases} \tag{7.2}$$

Step 3. Determining the range of step-size is determined according to Eq. (7.3):

$$Stepsize = rand \otimes (WM[permute1(i)(j)] - WM[permute2(i)(j)]) \tag{7.3}$$

Step 4 The formation of new water molecules (offspring) that arise according to Eq. (7.4):

$$\begin{cases} if \ NFE_s \leq \frac{\max NFE_s}{2} \Rightarrow newWM = WM + Stepsize \times MEP \\ if \ NFE_s > \frac{\max NFE_s}{2} \Rightarrow newWM = WM + Stepsize \times DEP \end{cases} \tag{7.4}$$

The new evaluated molecules are replaced if they are better than the previous answers.

Step 5. Termination conditions are checked and if necessary, the operation from Step 2 is repeated.

7.4 Chaos Enhanced Water Evaporation Optimization (CWEO)

In this algorithm, the formation of the evaporation phase is based on two important strategies, which include the single-layer and water droplet strategies [3–5]. Therefore, these two steps will play an important role in exploration and exploitation. By replacing the chaos maps in the random selections related to these steps, we will witness a significant improvement in the performance of the algorithms. The suggested scenarios for this replacement are as follows:

Scenario 1. Replacing the chaos map in the evaporation phase based on a monolayer strategy: In this case, the first chaos map CHM1 is replaced in Eq. (7.1) and the result of Eq. (7.5) is as follows:

$$if \quad NFE_s \le \frac{\max NFE_s}{2} \rightarrow \begin{cases} if\ \mathbf{CHM1}_{ij} < \exp(E_{sub}(i)) \Rightarrow MEP_{ij} = 1 \\ if\ \mathbf{CHM1}_{ij} \ge \exp(E_{sub}(i)) \Rightarrow MEP_{ij} = 0 \end{cases} \tag{7.5}$$

Scenario 2. Replacing the chaos function in the evaporation phase based on the drip strategy: In this case, the second chaos map CHM2 is replaced in Eq. (7.2) and the result of Eq. (7.6) is as follows:

$$if \quad NFE_s > \frac{\max NFE_s}{2} \rightarrow \begin{cases} if\ \mathbf{CHM2}_{ij} < J(\theta_i) \Rightarrow DEP_{ij} = 1 \\ if\ \mathbf{CHM2}_{ij} \ge J(\theta_i) \Rightarrow DEP_{ij} = 0 \end{cases} \tag{7.6}$$

Scenario 3. Placing the chaos function in both steps simultaneously: In this case, two chaos maps are simultaneously replaced in Eqs. (7.1) and (7.2).

7.5 Truss Weight Optimization with Static Constraints

To improve the optimization results in the Water Evaporation Optimization (CWEO), the Logistic, Gossin and Piecewise chaos functions have been chosen. By embedding chaos functions and forming chaos algorithms, the competition between chaos functions to reach the best optimal weight starts. To reach the best performance of the chaos algorithm, three scenarios are formed. In these scenarios, the chaos functions replace the distributions related to exploration, exploitation or both phases simultaneously. In each example, 9 chaotic models are formed and compared with the standard state. To form the statistical space and increase diversity in the research, each model is independently evaluated 30 times. Also, to analyze the statistical results, the best weight, the best average and the coefficient of variation are provided. In a number of studies, the application of chaos functions has increased the coefficient of variation. This performance indicates sudden jumps and increasing the difference between the best and worst answers. Due to the wide scope of investigations, there is a strong competition between chaos functions and as a result, the possibility of approaching the global optimum increases. This optimum cannot be far from the crosshairs of chaotic mutations. In each investigation, the chaos function compatible with the algorithm and the scenario corresponding to the best improvement are selected. We can also find out the strengths and weaknesses of the algorithm in the stages of exploration or exploitation.

7.5.1 Statement of the Structural Optimization Problems

The purpose of optimizing structures is to determine the minimize weight of the structure, while satisfy some constraints on the element stress and nodal displacement. The general form related to these problems can be expressed as follows:

$$\begin{array}{ll} Find & \{X\} = [x_1, x_2, ..., x_{ng}] \\ to\ minimize & W(\{X\}) = \sum_{i=1}^{nm} \rho_i \cdot A_i \cdot L_i \\ subjected\ to & \begin{cases} g_j(\{X\}) \leq 0, & j = 1, 2, ..., nc \\ x_{i\,\min} \leq x_i \leq x_{i\,\max} \end{cases} \end{array} \tag{7.7}$$

where {X} is the vector belong the decision variable; ng is the Dimensions of variables; W({X}) presents weight of the structure; nm is the number of elements; A_i, ρ_i and L_i denote the cross-sectional area, material density and the length of the ith member, respectively. X_{imin} and x_{imax} are the lower and upper bound of the decision variable x_i, respectively. G({X}) denotes design constraints; and nc is the number of constraints. The penalty function is used to apply the constraints. Therefore, the objective function is redefined to the following equation:

$$\begin{array}{l} f(\{X\}) = f_{penalty}(\{X\}) \times W(\{X\}) \\ f_{penalty}(\{X\}) = (1 + \varepsilon_1 \cdot \upsilon)^{\varepsilon_2} \\ \upsilon = \sum_{j=1}^{nc} \max[0, g_j(\{X\})] \end{array} \tag{7.8}$$

where V denotes the sum of the violations from constraints and parameter ε_1 and ε_2 are selected based on a balance between exploration and exploitation. In this study, the value of ε_1 is set as unity, and the parameter ε_2 start from 1.5 and increases linearly to 6 in all test example.

7.5.2 Introduction of Selected Chaos Map

In most meta-heuristic algorithms, premature convergence causes computational stagnation. In such cases, the algorithm is trapped in local optimization and there is no way to escape it. Therefore, after several repetitions, we will see the optimization process stop. To escape from this unfavorable situation, chaos functions create suitable conditions that by creating disorder in the search space, it is possible to jump to most positions in the search space. Therefore, in the first stage, it is spread in scattered areas of the search space and then the neighborhood of those points is carefully reviewed. Chaos maps have no traces of random behaviors; however, they cause very chaotic behaviors in the search space. The most important features of these functions are sensitivity to initial conditions and non-periodic and ergodic behavior can be introduced.

For these functions, there is no inverse, and the use of turbulent series to evolve variables in COA probe algorithms has significant advantages over other methods. How to apply chaos maps to meta-heuristic algorithms is shown in Flowchart Fig. 7.1. By selecting chaos maps in deterministic searches compared to random search, more speed and convergence towards the overall response is obtained [5]. In this research, the Logistic and Gaussian chaos maps have been selected. In the Logistic chaos map, it most likely converges from local minima to general minima. Therefore, this function is suitable for improving the exploration conditions of algorithms. The Gaussian chaos map is very likely to be within the local minimums and is suitable for improving exploitation conditions. Finally, Piecewise map simultaneously improves both conditions.

Therefore, by selecting these chaos maps, the weakness of algorithms of any kind will be repaired and global optimum will not have the opportunity to escape the chaos functions. The numerical distribution of these chaos map for a period of 100 iterations is shown in Fig. 7.2. In the following, we will examine the equations of chaos maps.

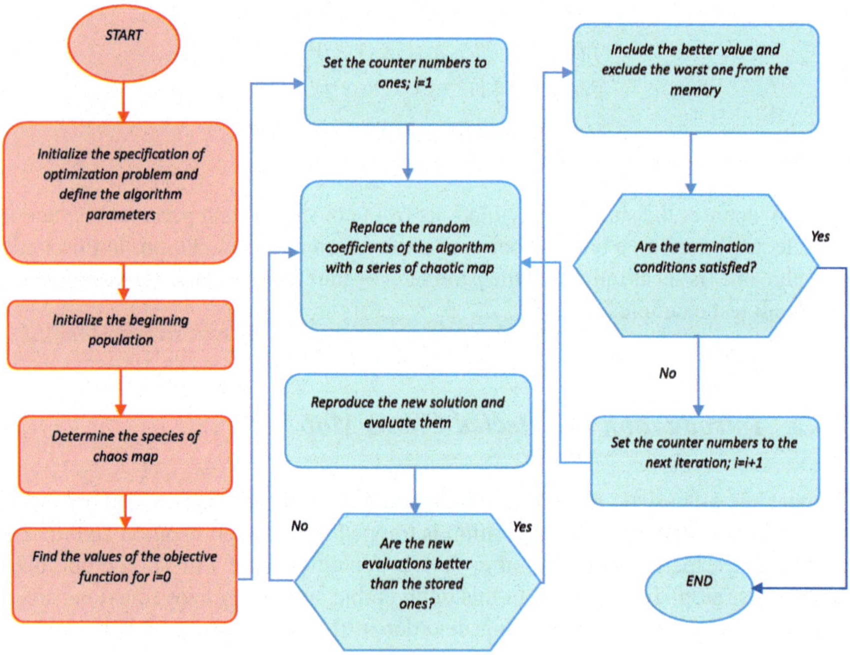

Fig. 7.1 The flowchart for the chaos Algorithm

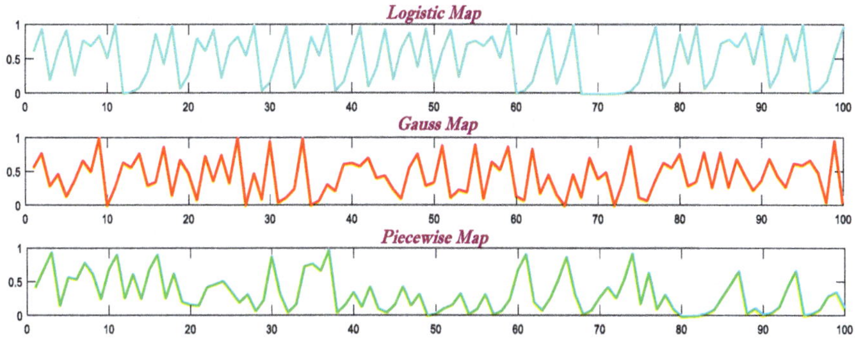

Fig. 7.2 The Chaotic value distribution during 100 iterations

7.5.2.1 Logistics Map

This map appears in nonlinear dynamic behaviors related to biological populations [6]. The sentences of chaotic sequences in the logistic map are obtained according to the following equations:

$$CHM_{k+1} = \mathrm{a} \times CHM_k(1 - CHM_k) \tag{7.9}$$

$$CHM_{k+1} \in (0, 1); CHM_0 \in (0, 1); CHM_0 \notin (0, 0.25, 0.5, 0.75, 1) \tag{7.10}$$

In the performed studies, a = 4 has been considered.

7.5.2.2 Gauss Map

The following equation shows the Eq. (7.11) [7] between the sentences of chaotic sequences in the Gauss function:

$$CHM_{k+1} = \begin{cases} 0 & CHM_k = 0 \\ \frac{1}{CHM_k} - \left[\frac{1}{CHM_k}\right] & CHM_k \neq 0 \end{cases} \tag{7.11}$$

7.5.2.3 Piecewise Map

The piecewise map is the same as the Liebovitch map of three criteria, and the value of P is considered as the control parameter [9]. The range of changes of P is in the range of 0–0.5, which in this study we have considered 1/3, and the following category of equations indicates the relationship between the sentences of chaotic sequences in this function that expressed by Eq. (7.12):

$$CHM_{k+1} = \begin{cases} \frac{CHM_k}{p} & 0 \le CHM_k < p \\ \frac{CHM_k - p}{0.5 - p} & p \le CHM_k < \frac{1}{2} \\ \frac{1 - p - CHM_k}{0.5 - p} & \frac{1}{2} \le CHM_k < 1 - p \\ \frac{1 - CHM_k}{p} & 1 - p \le CHM_k < 1 \end{cases} \quad (7.12)$$

7.5.3 *Numerical Examples of Optimal Truss Design*

To validate chaotic meta-heuristic algorithms, common examples of truss optimization are selected as benchmark problems. The examined samples were selected from both discrete and continuous groups. By examining the competition between chaos functions and triple scenarios, the improved results are compared with the standard case. In fact, 10 independent answers compete with each other. Due to the variety of reviews, the optimal responses will be largely close to the global optimality. The examples provided are as follows:

- The 25-bar spatial truss
- The 47-bar power transmission tower
- The 200-bar planar truss structure
- The 582-bar tower truss structure

7.5.3.1 A 25-Bar Spatial Truss

The topology and nodal numbering of the 25-bar spatial truss is as shown in Fig. 7.3. The material density is 0.1 lb/in^3 and the modulus of elasticity of the members is 10,000 ksi. The structure includes 25 members, which are divided into eight groups according to Table 7.1. Also, allowable compressive and tensile stresses have been determined for each type.

This spatial truss is subjected to two Loading conditions as shown in Table 7.2. Also, the maximum displacement limitations for the nodes in all directions of the axes is $\pm$ 0.35 in. The range of choice of members' cross-section or range of decision variables is in the range of 0.01–3.4 in^2.

In order to ensure the performance of chaotic maps and algorithms, as well as to increase the accuracy and sensitivity of calculations, each of the modes has been performed independently 30 times and the results related to the best response and the average value of responses are presented in Table 7.3. The graph of these results is also shown in Fig. 7.4. The coefficient of variation of the responses, which indicates the stability and robustness of the responses, has been calculated and another criterion for the efficiency of chaos functions and algorithms has been obtained. Examining the optimization results for different combinations of algorithms with chaos map and comparing it with standard modes, shows a significant improvement in reducing the weight of the 25-bar spatial truss. The enhanced results for Chaotic Water Evaporation

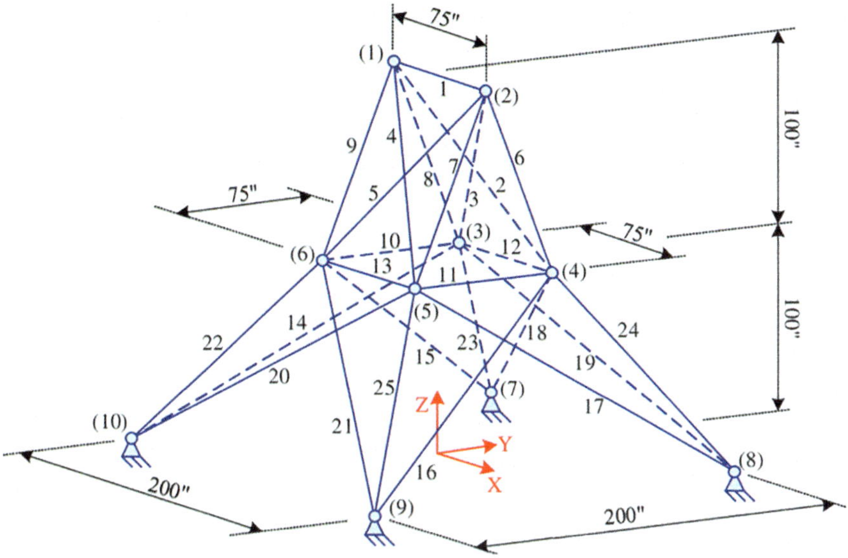

Fig. 7.3 Schematic of a twenty-five-bar spatial truss

Table 7.1 Grouping of members and stresses limitation for the 25-member spatial truss

Number group	Element group	Compressive stress limitation (ksi)	Tensile stress limitation (ksi)
1	A_1	35.092	40
2	A_2–A_5	11.590	40
3	A_6–A_9	17.305	40
4	A_{10}–A_{11}	35.092	40
5	A_{12}–A_{13}	35.092	40
6	A_{14}–A_{17}	6.759	40
7	A_{18}–A_{21}	6.959	40
8	A_{22}–A_{25}	11.082	40

Table 7.2 Loading condition for the 25-member spatial truss

Node point	Load case 1			Load case 2		
	F_x (kips)	F_y (kips)	F_z (kips)	F_x (kips)	F_y (kips)	F_z (kips)
1	0	20	− 5	1	10	− 5
2	0	− 20	− 5	0	10	− 5
3	0	0	0	0.5	0	0
4	0	0	0	0.5	0	0

Optimization (CWEO), by Gauss chaos map and Scenario 1, a weight of 544.041 lb has the optimal response.

7.5.3.2 A 47-Bar Power Transmission Tower

The topology and nodal numbering of the power transmission tower that consists of 47 members and 22 nodes, as shown in Fig. 7.5. The density of material is 0.3 lb/in^3 and the modulus of elasticity of the members is 30,000 ksi. Both stress and buckling limits must be satisfied for all member. The allowable stress of tensile members is 20 ksi and for compressive members 15 ksi is recommended. Euler buckling compressive stress for members is also calculated from the following equation

$$\sigma_i^e = \frac{-KEA_i}{L_i^2}; \quad i = 1, 2, \ldots, 47 \tag{7.13}$$

where K is a constant coefficient that must be selected according to the geometric type of the section, E is the modulus of elasticity of the material, A_i is cross-sectional area of member and L_i is the length of member. Here we consider k to be 3.96. Structural nodes are affected by a combination of triple loading modes. The first group includes loads of 6 kip in the positive direction of the X-axis and 14 kip in the negative direction of the Y-axis, which are acting in nodes 17 and 22. The second group consists of loads of 6 kg in the positive direction of the X-axis and 14 kg in the negative direction of the Y-axis, which acting only at node 17. The third group includes loads of 6 kP in the positive direction of the X-axis and 14 kP in the negative direction of the Y-axis, which acting only in node 22. The first group represents the oblique loads of both power transmission lines and the second and third groups represent the state where one of the two lines breaks. Also, truss members are classified into 27 groups according to geometric symmetry. The cross-section areas of the members, were chosen from the 64 discrete values of the AISC code.

In order to ensure the performance of chaos map and algorithms, as well as to increase the accuracy and sensitivity of calculations, each of the modes has been performed independently 30 times and the results related to the best response and the average value of responses are presented in Table 7.4. The graph of these results is also shown in Fig. 7.6. The coefficient of variation of the responses, which indicates the stability and robustness of the responses, has been calculated and another criterion for the efficiency of chaos maps and algorithms has been obtained. By examining the optimization results for different combinations of algorithms with chaos map and comparing it with the standard mode, a significant improvement in reducing the weight of the 47-bar power transmission tower is achieved. The enhanced results for Chaotic Water Evaporation Optimization (CWEO), by Logistic chaos map and Scenario 2, a weight of 2355.733 lb has the optimal response.

Table 7.3 Optimal design comparison for the 25-bar spatial truss

Number group	Element group	Standard	Logistic map			Gauss map			Piecewise map		
		WEO	CWEO 21	CWEO 22	CWEO 23	CWEO 31	CWEO 32	CWEO 33	CWEO 41	CWEO 42	CWEO 43
1	A_1	0.01	0.01006	0.01	0.01	0.01	0.01003	0.01003	0.01	0.01	0.01
2	A_2–A_5	1.9519	1.9785	1.9742	1.9577	1.9798	1.976	1.9803	1.9673	1.9784	1.9623
3	A_6–A_9	3.0367	2.996	2.989	3.0203	2.9837	2.9956	2.9898	3.0047	2.9923	3.0098
4	A_{10}–A_{11}	0.01	0.01001	0.01	0.01069	0.01	0.01002	0.01	0.01	0.01004	0.01000
5	A_{12}–A_{13}	0.01491	0.01058	0.01	0.01	0.01	0.01	0.01	0.01	0.01002	0.01
6	A_{14}–A_{17}	0.6819	0.6762	0.68787	0.67746	0.68314	0.67825	0.6829	0.68114	0.68348	0.68609
7	A_{18}–A_{21}	1.6886	1.6788	1.6805	1.6813	1.6779	1.6782	1.6788	1.6795	1.6787	1.6809
8	A_{22}–A_{25}	2.651	2.6662	2.6482	2.6532	2.6557	2.661	2.6581	2.6528	2.6557	2.6444
Best weight		545.349	544.623	544.101	544.156	**544.041**	544.294	544.500	544.101	544.414	544.061
Mean weight		545.517	544.846	544.699	544.817	544.647	**544.448**	544.710	544.770	544.795	544.661
CV		0.04798	0.03644	0.07764	0.07557	0.06737	**0.02298**	0.03950	0.07247	0.05003	0.08954

The bold items in the table indicate the optimal values for the best weight, the best mean and the best coefficient of variation

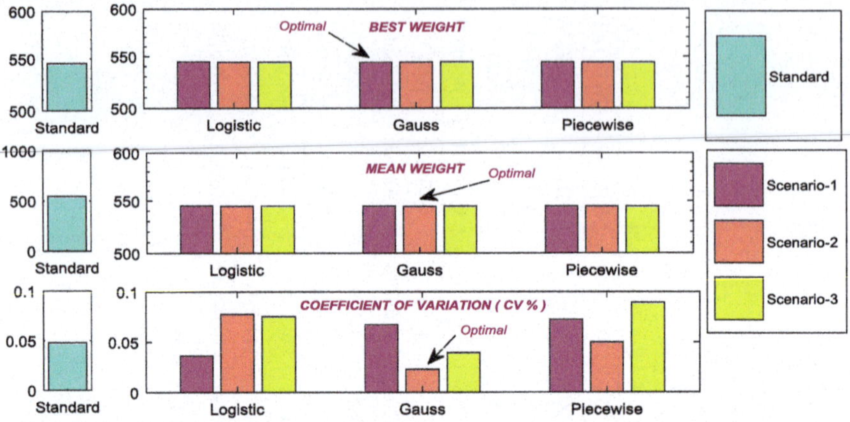

Fig. 7.4 Optimization results in standard mode and selection of chaos map for the 25-member spatial truss

7.5.3.3 A 200-Bar Planar Truss Structure

The topology and nodal numbering of the 200-bar planar truss structure are shown in Fig. 7.7. The material density is 0.283 lb/in^3 and the module of elasticity is 30,000 ksi. Allowable stress of members 10 klb has been assumed. No deformation restrictions are intended for optimization. Truss loading is done in three independent groups. The first group is the lateral load of the structure, which includes a load of 1 kip along the positive axis of Xs and is applied in nodes 1, 6, 15, 20, 29, 34, 43, 48, 57, 62 and 71. The second group covers the gravity load of the structure, which includes a load of 10 kip in the negative direction of the axis of the Ys and in nodes 1, 2, 3, 4, 5, 6, 8, 10, 12, 14, 15, 16, 17, 18, 19 and... 71, 72, 73, 74 and 75 apply. The third group applies both loading groups together. Truss members are also classified into 29 groups according Table 7.5. The minimum cross section of the members, which is the lower limit of the decision variables, is considered 0.1 in^2 and the maximum cross section is 16 in^2. In order to ensure the performance of chaotic maps and algorithms, as well as increasing the accuracy and sensitivity of calculations, each state is independently executed 30 times and the results of the best response and the average amount of responses are presented in statistical Table 7.6 and shown in Fig. 7.4. Also, the coefficient of changes in responses, which indicates the stability and robustness of the responses, has been calculated and another criterion for the efficiency of chaotic functions and algorithms has been achieved. The enhanced results for Chaotic Water Evaporation Optimization (CWEO), by Logistic chaos map and Scenario 3, a weight of 25,855.19 lb has the optimal response (Fig. 7.8).

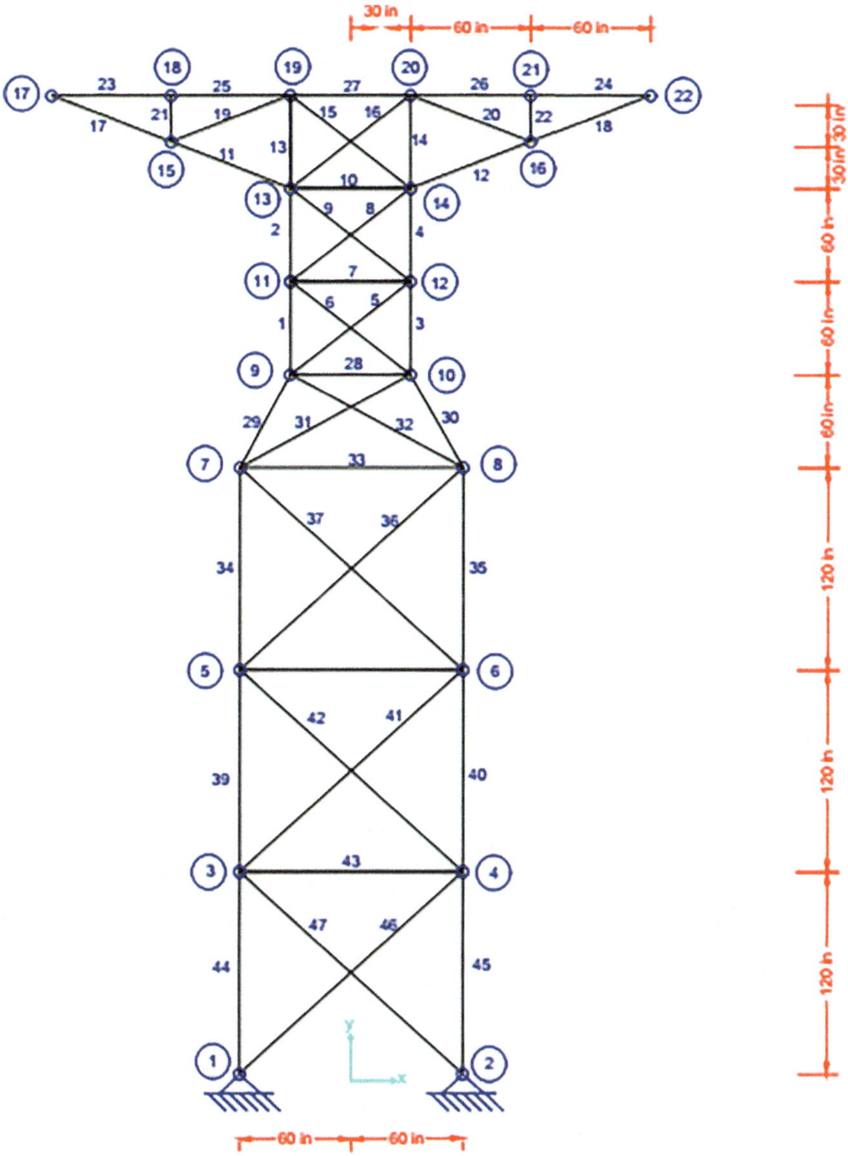

Fig. 7.5 Schematic of a 47-bar power transmission tower

7.5.3.4 A 582-Bar Tower Truss Structure

The topology and nodal numbering of the 582-bar tower truss structure are shown in Fig. 7.9. The optimization of this tower with continuous variables has been done for the first time by Kaveh et al.

Table 7.4 Optimal design comparison for the 47-bar power transmission tower

Number group	Element group	Standard	Logistic map			Gauss map			Piecewise map		
		WEO	CWEO 21	CWEO 22	CWEO 23	CWEO 31	CWEO 32	CWEO 33	CWEO 41	CWEO 42	CWEO 43
1	A_1–A_3	3.84	3.84	3.84	3.84	3.84	3.84	3.84	3.84	3.84	3.84
2	A_2–A_4	3.38	3.47	3.38	3.47	3.38	3.38	3.47	3.47	3.38	3.38
3	A_5–A_6	0.785	0.785	0.766	0.785	0.766	0.994	0.785	0.785	0.766	0.785
4	A_7	0.111	0.111	0.111	0.25	0.141	0.196	0.111	0.141	0.25	0.25
5	A_8–A_9	0.994	0.994	0.785	0.994	0.994	0.994	0.994	0.994	0.994	0.994
6	A_{10}	2.13	1.99	1.85	2.38	1.99	2.13	1.99	1.99	1.99	2.38
7	A_{11}–A_{12}	2.13	2.38	2.13	2.38	2.13	2.38	2.13	2.13	2.13	2.38
8	A_{13}–A_{14}	1.457	1.228	1.228	1.228	1.228	1.266	1.457	1.457	1.266	1.457
9	A_{15}–A_{16}	1.563	1.563	1.563	1.563	1.563	1.563	1.563	1.62	1.563	1.563
10	A_{17}–A_{18}	2.13	2.13	2.13	2.13	2.38	2.38	2.13	2.13	2.13	2.13
11	A_{19}–A_{20}	0.196	0.141	0.111	0.111	0.391	0.111	0.111	0.196	0.141	0.111
12	A_{21}–A_{22}	0.25	0.141	0.111	0.196	0.196	0.111	0.196	0.141	0.111	0.111
13	A_{23}–A_{24}	1.99	1.8	1.8	1.8	1.8	1.8	1.8	1.8	1.8	1.8
14	A_{25}–A_{26}	1.8	1.8	1.8	1.8	1.8	1.99	1.8	1.8	1.8	1.8
15	A_{27}	1.563	1.457	1.457	1.563	1.563	1.563	1.563	1.457	1.563	1.457
16	A_{28}	0.563	0.766	0.543	0.785	0.994	0.563	0.563	0.766	0.766	0.766
17	A_{29}–A_{30}	3.63	3.81	3.63	3.84	3.84	3.84	3.63	3.605	3.63	3.63
18	A_{31}–A_{32}	1.563	1.457	1.457	1.457	1.563	1.563	1.563	1.563	1.563	1.457
19	A_{33}	0.391	0.307	0.25	0.563	0.442	0.307	0.442	0.442	0.442	0.307
20	A_{34}–A_{35}	3.64	3.22	3.09	3.10	3.231	3.18	3.340	3.11	3.59	3.61
21	A_{36}–A_{37}	1.457	1.457	1.228	1.457	1.457	1.266	1.457	1.457	1.457	1.457

(continued)

Table 7.4 (continued)

Number group	Element group	Standard	Logistic map			Gauss map			Piecewise map		
		WEO	CWEO 21	CWEO 22	CWEO 23	CWEO 31	CWEO 32	CWEO 33	CWEO 41	CWEO 42	CWEO 43
22	A_{38}	0.25	0.563	0.361	0.111	0.111	0.307	0.307	0.25	0.196	0.307
23	A_{39}–A_{40}	3.82	3.77	3.68	3.702	3.792	3.69	3.74	3.97	3.59	3.61
24	A_{41}–A_{42}	1.62	1.563	1.563	1.563	1.563	1.563	1.563	1.563	1.563	1.563
25	A_{43}	0.111	0.111	0.111	0.111	0.141	0.141	0.141	0.141	0.25	0.111
26	A_{44}–A_{45}	4.58	5.11	4.55	4.681	4.528	4.541	4.724	4.58	4.631	4.547
27	A_{46}–A_{47}	1.563	1.457	1.457	1.457	1.563	1.457	1.563	1.563	1.563	1.457
Best weight		2496.667	2477.052	**2355.733**	2446.254	2467.389	2442.094	2467.030	2461.222	2462.212	2471.713
Mean weight		2501.224	2492.249	**2357.734**	2467.814	2485.579	2446.587	2476.249	2475.517	2476.579	2475.517
Coefficient variation		0.11869	0.31711	**0.1015**	0.4112	0.3518	0.11942	0.2173	0.32021	0.3224	0.11207

The bold items in the table indicate the optimal values for the best weight, the best mean and the best coefficient of variation

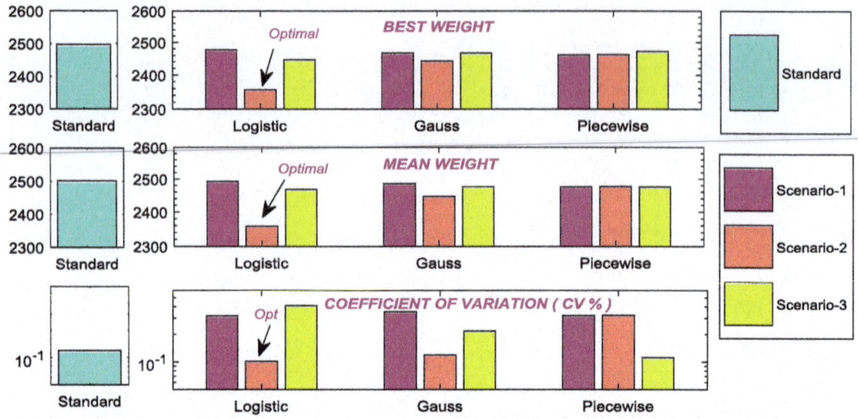

Fig. 7.6 Optimization results in standard mode and selection of chaos map for the 47-bar power transmission tower

The modulus of elasticity of the members is 203893.6 MPa and their yield stress is 253.1 MPa. The displacement limitation is set at 8 cm for all nodes and in all direction. A load case group is considered that includes 5 kN along both the X and Y axes and – 30 kN along the Z axis.

This load is applied to all free nodes of the tower. The slenderness ratio is limited to 300 for tension member and 200 for compression member and maximum stress of truss elements are imposed according to the provisions of AISC ASD [8]. The truss members are grouped into 32 symmetrical groups. The minimum cross-sectional area for all member is 20 cm^2 and maximum cross-sectional area is 1000 cm^2. In order to ensure the performance of chaotic maps and algorithms, as well as increasing the accuracy and sensitivity of calculations, each state is independently executed 30 times and the results of the best response and the average amount of responses are presented in statistical Table 7.7 and shown in Fig. 7.10. Also, the coefficient of changes in responses, which indicates the stability and robustness of the responses, has been calculated and another criterion for the efficiency of chaotic functions and algorithms has been achieved. The enhanced results for Chaotic Water Evaporation Optimization (CWEO), by Gauss chaos map and Scenario 1, a weight of 15.6306 m^3 has the optimal response.

7.6 Truss Weight Optimization with Multiple Frequency Constraints

To access the dynamic behavior of the structures and determine the position of the dampers, we will need to control the natural frequencies of the structures. Also, by creating a multi-frequency constraint, the destructive phenomenon of resonance is

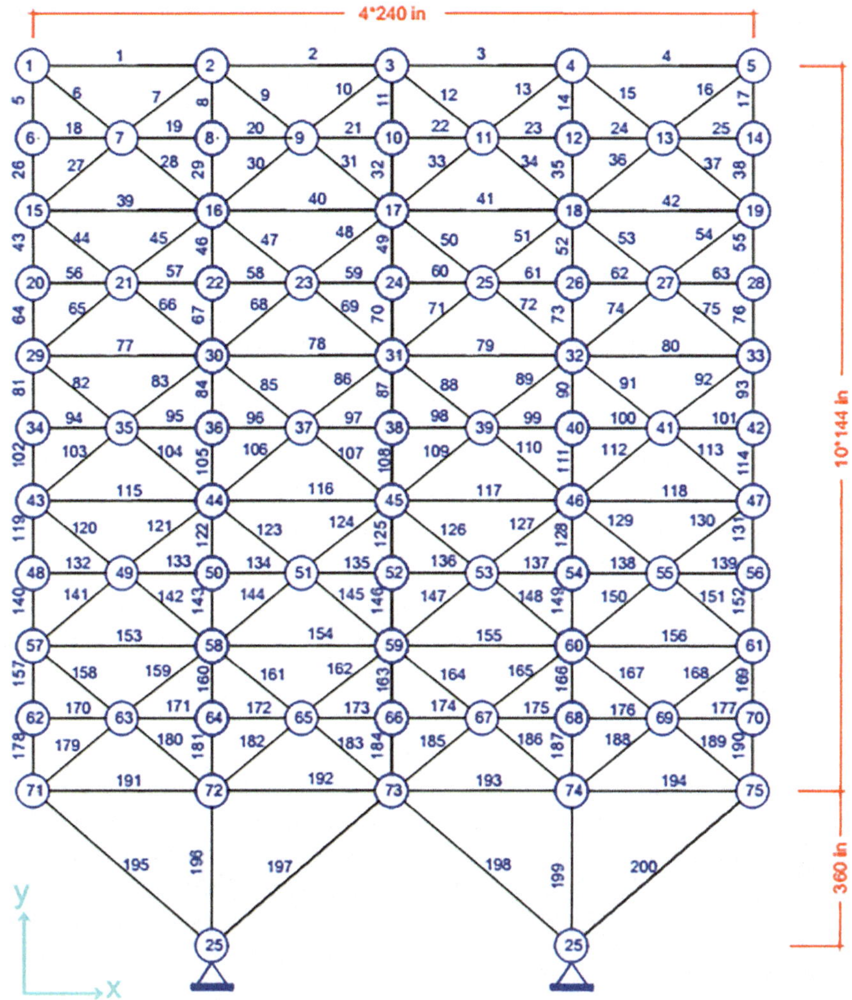

Fig. 7.7 Schematic of a 200-bar planar truss structure

avoided. The structure of the objective function in the multi-frequency constraint is strongly non-linear and non-convex and has complex implicit variables, and the optimization using meta-heuristic algorithms in the standard mode stops at the local optimum in most cases. In order to save algorithms from the trap of local optima and prevent premature convergence, chaotic dynamic series are embedded in these algorithms and replace random search distributions related to single layer strategy and drop strategy. Chaotic series, by creating a balance between single-layer and drop strategies, converges local optima to global optima and prevents optimization stagnation. In these investigations, the logistic and Gaussian chaos functions have

Table 7.5 Classification of members in the 200-bar planar truss structure

Group	Element group	Group	Element group
A_1	1, 2, 3, 4	A_{16}	82, 83, 85, 86, 88, 89, 91, 92, 103, 104, 106, 107, 109, 110, 112, 113
A_2	5, 8, 11, 14, 17	A_{17}	115, 116, 117, 118
A_3	19, 20, 21, 22, 23, 24	A_{18}	119, 122, 125, 128, 131
A_4	18, 25, 56, 63, 94, 101, 132, 139, 170, 177	A_{19}	133, 134, 135, 136, 137, 138
A_5	26, 29, 32, 35, 38	A_{20}	140, 143, 146, 149, 152
A_6	6, 7, 9, 10, 12, 13, 15, 16, 27, 28, 30, 31, 33, 34, 36, 37	A_{21}	120, 121, 123, 124, 126, 127, 129, 130, 141, 142, 144, 145, 147, 148, 150, 151
A_7	39, 40, 41, 42	A_{22}	153, 154, 155, 156
A_8	43, 46, 49, 52, 55	A_{23}	157, 160, 163, 166, 169
A_9	57, 58, 59, 60, 61, 62	A_{24}	171, 172, 173, 174, 175, 176
A_{10}	64, 67, 70, 73, 76	A_{25}	178, 181, 184, 187, 190
A_{11}	44, 45, 47, 48, 50, 51, 53, 54, 65, 66, 68, 69, 71, 72, 74, 75	A_{26}	158, 159, 161, 162, 164, 165, 167, 168, 179, 180, 182, 183, 185, 186, 188, 189
A_{12}	77, 78, 79, 80	A_{27}	191, 192, 193, 194
A_{13}	81, 84, 87, 90, 93	A_{28}	195, 197, 198, 200
A_{14}	95, 96, 97, 98, 99, 100	A_{29}	196, 199
A_{15}	102, 105, 108, 111, 114		

provided suitable conditions for improving the optimization results [6, 7]. By embedding these functions in the stages of exploration, exploitation or both parts simultaneously, three scenarios of Chaotic Water Evaporation Optimization (CWEO), are formed. Also, in order to increase diversity in the statistical space, each model is run 20 times independently, and the statistical population corresponding to the best value of cross-sections, the best average and coefficient of variation is determined. By examining the effectiveness of chaotic models, we will have a significant improvement in the optimization results and we can determine the chaos function compatible with the algorithm that creates the balance between the exploration and exploitation stages with the best efficiency. We can also identify the strengths and weaknesses of the algorithm in the stages of exploration or exploitation.

7.6.1 Coincidently Size and Topology Optimization of Skeletal Structures

In coincidently size and topology optimization problem two goals should be achieved: (1) optimal shape or topology of a ground structure should be found, (2) search for the optimal cross sections of the optimized shape for topology must be performed

Table 7.6 Optimal design comparison for 200– bar planar truss structure

Number group	Standard	Logistic map			Gauss map			Piecewise map		
	WEO	CWEO 21	CWEO 22	CWEO 23	CWEO 31	CWEO 32	CWEO 33	CWEO 41	CWEO 42	CWEO 43
A_1	0.125	0.349579	0.1	0.1	0.13495	0.1	0.1	0.1	0.1	0.12501
A_2	0.947608	1.08156	0.929501	1	0.94253	0.94223	0.931	1.1856	1.111	1.037
A_3	0.100022	0.118312	0.180559	0.1	0.1	0.12	0.131	0.1	0.100126	0.1
A_4	0.100291	0.106921	0.1	0.1	0.107787	0.1	0.1	0.1	0.101174	0.1
A_5	1.9434	1.924572	1.92879	1.93904	2.15316	1.93471	1.93176	1.9771	2.06565	1.95712
A_6	0.290113	0.306374	0.297564	0.277051	0.341639	0.284246	0.285111	0.2626	0.261756	0.292452
A_7	0.100006	0.1	0.1	0.181828	0.110157	0.170437	0.155598	0.101916	0.101915	0.102787
A_8	3.11108	3.001	3.07331	3.38808	3.3445	3.05	3.12034	3.11281	3.09636	3.11606
A_9	0.101486	0.355794	0.126717	0.1	0.1	0.1	0.12289	0.1112	0.100021	0.114044
A_{10}	4.10642	4.12017	4.07105	4.25575	4.34351	4.10495	4.15663	4.218	4.0749	4.11523
A_{11}	0.374785	0.519231	0.396611	0.406474	0.371177	0.412043	0.41484	0.3851	0.386746	0.418017
A_{12}	0.1	0.290499	0.1	0.203967	0.316316	0.228011	0.1	0.1	0.1	0.101253
A_{13}	5.26628	5.3302	5.24237	5.36794	5.63255	5.30516	5.32516	5.27731	5.22757	5.38941
A_{14}	0.138721	0.187759	0.260669	0.130388	0.1	0.1	0.154372	0.11161	0.218436	0.10456
A_{15}	6.72191	6.6126	6.24003	6.40981	6.34871	6.49669	6.38766	6.27653	6.33401	6.35181
A_{16}	0.543573	0.697209	0.57563	0.578952	0.61432	0.679083	0.587906	0.50951	0.652771	0.503556
A_{17}	0.230617	0.261	0.624413	0.158451	0.23024	0.255865	0.292944	0.45194	0.389337	0.131501
A_{18}	7.68847	7.91941	7.8025	8.01153	7.8417	7.78477	7.92123	7.936	7.63094	7.81326
A_{19}	0.103305	0.463307	0.103779	0.179584	0.315284	0.110828	0.180756	0.1	0.427138	0.16093
A_{20}	9.15305	8.9204	8.78434	8.79475	8.84589	9.10074	8.92114	8.98369	9.23529	8.89309
A_{21}	0.811887	0.920628	0.912574	0.861189	0.839714	0.868055	0.881135	0.830161	0.943708	0.710509

(continued)

Table 7.6 (continued)

Number group	Standard	Logistic map			Gauss map			Piecewise map		
	WEO	CWEO 21	CWEO 22	CWEO 23	CWEO 31	CWEO 32	CWEO 33	CWEO 41	CWEO 42	CWEO 43
A_{22}	1.89321	0.215475	0.161796	0.756125	0.1	0.646152	0.369037	0.436457	0.430512	0.868204
A_{23}	10.4418	10.9414	10.8183	11.2916	10.7287	10.7036	10.6746	10.6031	10.6589	10.5152
A_{24}	0.1	0.100969	0.1	0.1	0.259945	0.11905	0.265195	0.406568	0.138092	0.127983
A_{25}	11.5903	11.9567	11.8198	12.257	11.7379	11.7051	11.6823	11.592	11.6482	11.5034
A_{26}	1.96301	0.98164	0.931919	1.2159	0.954852	1.17268	1.21305	1.36786	1.219003	1.42645
A_{27}	6.67883	6.46148	6.81666	6.41141	6.8573	6.86943	6.87059	6.78598	7.11536	6.77846
A_{28}	11.5863	11.5193	11.702	11.41433	11.8801	11.5855	11.6091	11.64193	11.8214	11.8432
A_{29}	14.6743	13.8757	13.6043	13.9732	13.5966	13.8022	13.755	13.805	13.6993	13.7501
Best weight	27,093.58	26,342.38	26,331.60	**25,855.19**	26,175.44	26,234.29	26,150.11	26,194.99	26,440.22	26,173.25
Mean weight	27,309.54	26,567.72	26,777.25	26,449.22	**26,229.37**	26,597.52	26,423.81	26,469.67	26,619.29	26,281.52
Coefficient variation	0.96797	0.59468	1.1477	1.5117	**0.18447**	0.90513	1.2407	1.1258	0.73865	0.72058

The bold items in the table indicate the optimal values for the best weight, the best mean and the best coefficient of variation

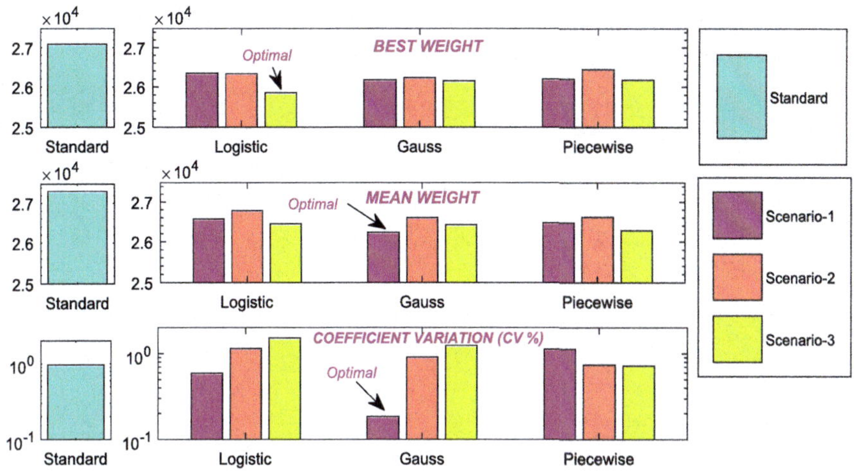

Fig. 7.8 Optimization results in standard and chaos map for the 200-bar planar truss structure

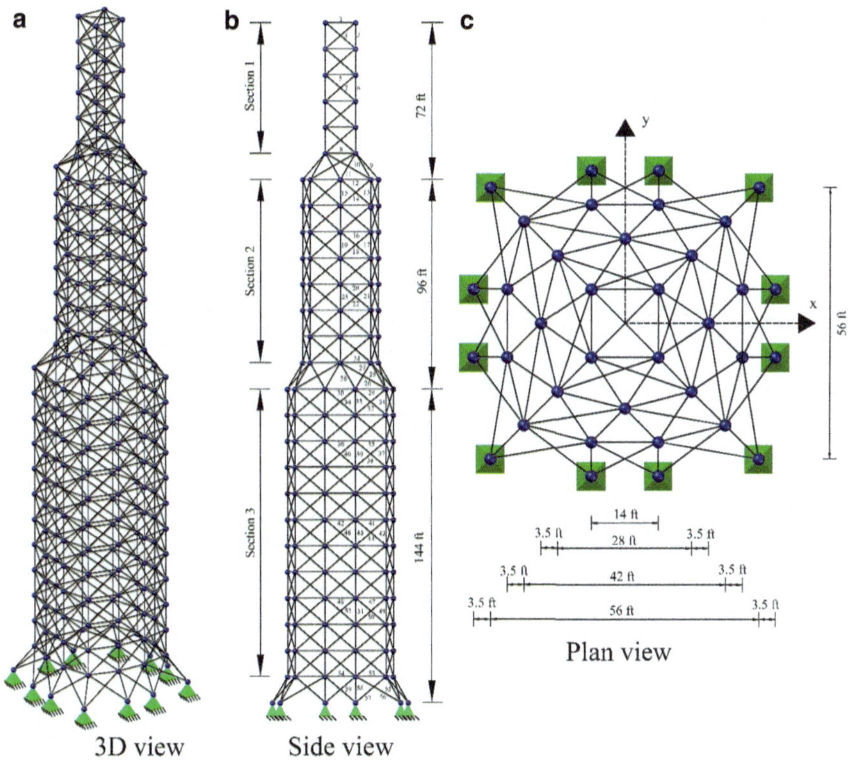

Fig. 7.9 Schematic of a 582-bar tower truss structure **a** 3D view, **b** side view, **c** top view

Table 7.7 Optimal design comparison for the 582-bar tower truss structure

Element group	Standard	Logistic			Gauss			Piecewise		
	WEO	CWEO 21	CWEO 22	CWEO 23	CWEO 31	CWEO 32	CWEO 33	CWEO 41	CWEO 42	CWEO 43
1	26.04592	20.03611	20	20.00648	20	20.0435	20.06992	20	20	20.07103
2	158.9574	162.4235	155.0819	153.4226	152.5026	152.4972	148.8069	155.886	213.9836	147.1153
3	46.71973	30.99743	31.62619	31.22713	31.94976	30.94457	31.5838	31.02384	32.91625	31.69689
4	99.77196	132.1874	112.4417	133.1365	121.6458	119.9061	125.4155	131.4441	145.7497	126.0572
5	29.60529	26.78419	27.54606	26.75797	27.25899	26.82784	27.02612	26.73377	34.59075	27.37029
6	53.11935	20.00003	20	20.01462	20	20.00518	20.00232	20.00019	23.11215	20
7	149.0111	108.8454	108.5318	106.0686	103.8829	101.7885	105.5642	103.8953	59.92312	97.96661
8	32.278	25.17342	25.13499	25.14327	25.9436	25.11589	25.5104	25.10355	40.72819	25.38092
9	28.41965	20.00745	20	20	20	20.01409	20.2796	20.00141	20.17553	20.11857
10	211.0686	102.6476	95.51973	106.1898	102.0555	89.72969	86.3706	102.5957	78.96866	89.78594
11	21.86304	22.61732	23.41761	21.69427	21.92043	21.93081	22.11233	21.79785	33.59556	26.33619
12	106.1609	142.3599	133.9639	145.7047	131.6361	141.5232	137.7514	152.829	150.5796	134.9548
13	141.5389	160.0068	144.5901	161.7154	140.6819	146.792	146.9143	161.9093	153.5396	154.672
14	108.3364	110.1462	103.2186	103.0274	89.51033	94.20685	96.41151	104.2468	67.90298	93.96321
15	118.8218	171.4053	167.3833	180.3886	166.0496	161.8645	165.2511	171.6516	313.8715	160.9479
16	31.61797	30.92305	32.99407	30.56171	31.16372	31.23694	31.03362	31.77422	33.88517	31.45599
17	179.6852	130.1768	120.456	136.1053	124.4427	126.4039	118.6331	127.7147	62.19039	128.4665
18	29.78498	23.73456	24.3025	23.8141	24.46899	24.47699	25.95758	23.88909	81.45154	25.69838
19	49.3269	20.36726	20	20.02314	20.18464	20.14972	20.03726	20.04075	51.82813	20.32288
20	79.19034	86.07434	89.55137	87.30337	78.33173	91.55356	92.13027	92.02213	225.5266	92.69829
21	26.28452	21.92162	23.01883	21.68981	23.21287	21.28459	21.35662	21.47953	34.09961	22.71545

(continued)

Table 7.7 (continued)

Element group	Standard	Logistic			Gauss			Piecewise		
	WEO	CWEO 21	CWEO 22	CWEO 23	CWEO 31	CWEO 32	CWEO 33	CWEO 41	CWEO 42	CWEO 43
22	40.59107	20.04152	20	20	20	20.21434	20	20.0231	20.49073	20.22072
23	119.5355	52.12037	51.13546	56.49925	57.4423	53.55554	49.78407	59.04454	54.02458	44.50298
24	20	20.01184	20	20.54682	20	20.08994	20	20.00013	20	20.01111
25	43.22921	20.00211	20	20.01685	20	20.10497	20.19032	20.00533	39.09334	20
26	63.76366	20.63265	20.2464	24.13628	30.51513	23.08494	20.36993	24.98993	29.84791	21.65275
27	24.99226	20	20	20.08465	20	20.00296	20.01045	20.00425	23.89143	20.73703
28	31.81752	20.03476	20.00747	20.0192	20.02903	20.12358	20.79748	20.38184	28.09535	21.20496
29	106.2915	20	20.00005	20.98138	20	20.63435	20	20.00179	76.71842	20.911
30	22.39151	20.04141	20.12119	20	20	20.01046	20.06305	20.0004	20	20.74335
31	37.74474	20.21431	20	20.0094	22.87928	20.30463	20.00125	20	20	20
32	85.55574	20.00044	20	20.36494	20.31257	20.15782	20	20.02213	20	20
Best (m^3)	19.588	16.2479	15.7648	16.2457	**15.6306**	19.6247	15.6733	16.2363	19.7227	15.7893
Mean (m^3)	20.198	16.2588	15.8138	16.2521	15.8012	19.6543	**15.714**	16.2433	20.6036	15.8494
CV (%)	1.7236	0.146122	0.4423	**0.049521**	1.0293	0.20056	0.18951	0.06832	2.049	0.3105

The bold items in the table indicate the optimal values for the best weight, the best mean and the best coefficient of variation

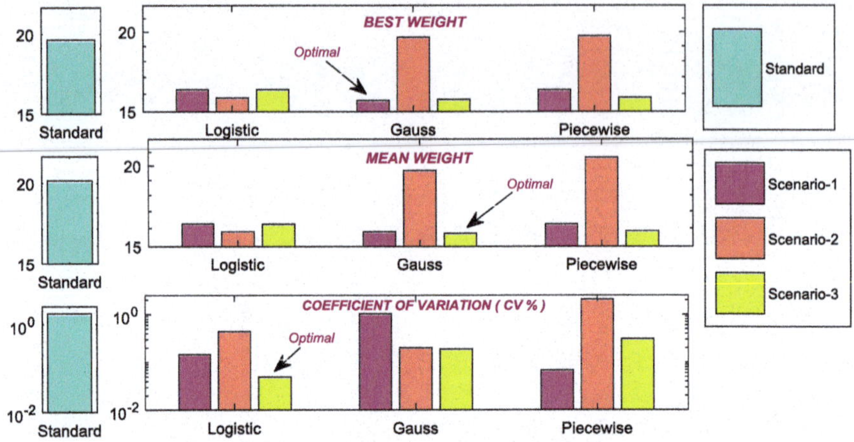

Fig. 7.10 Optimization results in standard and chaos map for the 582-bar tower truss structure

[10]. Therefore, the problem starts with the ground structure, which is composed of all possible nodes and members. Then, the node layout and the cross-sectional areas are found such that the cost of the structure is minimized. This optimization problem can formally be stated as follows:

$$Minimize \quad f(X) = \sum_{i=1}^{nm} \rho_i \cdot A_i \cdot L_i + \sum_{j=1}^{nn} b_j \tag{7.14}$$

$$\begin{aligned} & subject\ to: \\ & \begin{cases} \sigma_{i\min} \leq \sigma_{il} \leq \sigma_{i\max} \\ -\sigma_i \leq -\sigma_i^E & i = 1, 2, ..., nm \quad l = 1, 2, ..., k \\ A_i \geq 0 \end{cases} \\ & \begin{cases} \delta_{k\min} \leq \delta_{kl} \leq \delta_{k\max} \ j = 1, 2, ..., nn \quad k = 1, 2, ..., kc \\ b_j \geq 0 \end{cases} \\ & \begin{cases} \omega_m \leq \omega_m^* \ for\ some\ frequencies\ m \\ \omega_n \geq \omega_n^* \ for\ some\ frequencies\ n \end{cases} \end{aligned} \tag{7.15}$$

where f(x) is the structural cost; nm and nn denote the number of members and nodes of the ground structure, respectively; ρ_i, A_i and Li are density, cross sectional area and length of the ith member, respectively; b_j is the cost of the jth node; kc and lc are the number of displacement constraints and loading conditions, respectively; σ_{il} is the stress of the ith member under lth loading condition and σ_{imin} and σ_{imax} are its lower and upper bounds, respectively; δ_{kl} is the displacement of the kth degree of freedom under the lth loading condition, δ_{kmin} and δ_{kmax} are the corresponding lower and upper limits respectively; σ_i^E is the stress at which the ith member buckles, i.e. Euler buckling stress; ωm is the mth natural frequency of the structure and ω_m^* is its upper bound. ω_n is the nth natural frequency of the structure and ω_n^* is its lower bound.

7.6.2 Numerical Examples of Optimal Truss Design

To determine the effectiveness of Chaotic Water Evaporation Optimization (CWEO), well-known examples of truss optimization with layout and size optimization are selected as benchmark problems. To compare the improvement values, the best chaotic mode is compared with the standard model. By creating a competition between chaotic modes and triple scenarios, a strong challenge between the six independently chaotic responses is formed. In order to form the statistical population, each model has been implemented 20 times and the best weight, the best average and the best coefficient of variation have been presented. Due to the variety of investigations, the optimization results will be very close to the globall optimum. The examples provided are as follows:

- The 52-bar dome-like truss
- The 72-bar spatial truss
- The 120-bar spatial dome

7.6.2.1 A 52-Bar Dome-Like Truss

The 52-bar dome-like truss as shown in Fig. 7.11 is a well-known benchmark problem for optimizing the weight and shape of trusses with frequency constraints. This truss considers both optimizations of the section size and geometric coordinates of the nodes and the geometric shape of the structure is determined during the optimization process.

The decision variables related to the section size are classified into 8 groups according to the symmetry in the geometric shape. Geometric coordinates of all symmetric free nodes can be changed by 2 m from the initial position along the coordinate axes. In this case, the number of decision variables related to the shape of the structure and the geometric coordinates of the nodes is limited to 5 variables, and the sum of the variables, including shape and size, will be 13 variables. In all free nodes, a non-structural concentrated mass of 50 kg has affected all free nodes. The mechanical characteristics of the structure are: density of materials 7800 kg/m^3, modulus of elasticity 210,000 MPa, frequency limitations of the structure in the first mode are less than 15.916 Hz and in the second mode are greater than 28.648 Hz. For the cross section of the members, the lower limit is 1 cm^2 and the upper limit is 10 cm^2. In order to ensure the performance of turbulent functions and algorithms, as well as to increase the accuracy and sensitivity of calculations, each of the modes has been performed independently 30 times and the results related to the best response and the average value of responses are presented in Statistical Table 7.8. Also, the coefficient of variation of responses, which is a measure of the robustness

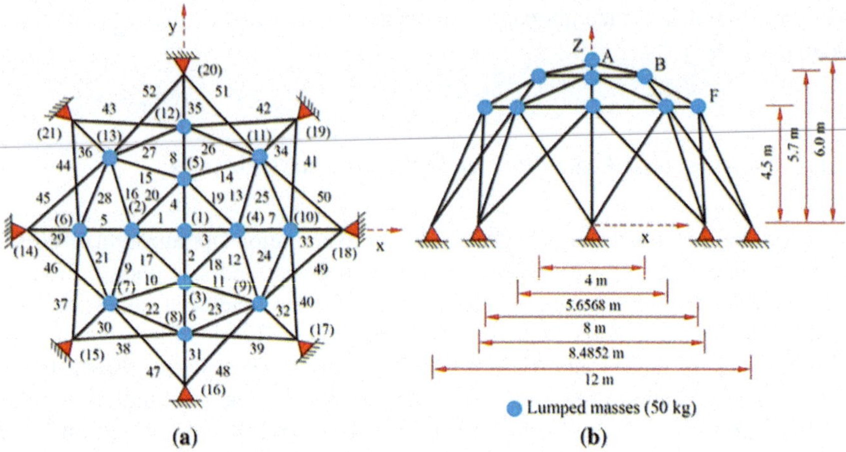

Fig. 7.11 Schematic of A 52-bar dome-like truss

and robustness of responses, has been calculated and used to compare the efficiency of turbulence functions and algorithms. For quick access to optimization information, a bar chart of each component is shown in Fig. 7.12. Examining the optimization results for different combinations of algorithms with turbulence functions and comparing it with the standard mode, shows a significant and significant improvement in reducing the weight of the 52-bar dome-like truss. The enhanced results for Chaotic Water Evaporation Optimization (CWEO), by Gauss chaos map and Scenario 3, a weight of 193.701 kg has the optimal response.

7.6.2.2 A 72-Bar Spatial Truss

The 72-bar spatial truss as shown in Fig. 7.13 is a well-known benchmark problem with weight-limit optimization. This truss only considers "optimizing the size of the sections" and the geometric shape of the structure is constant during the optimization process. The decision variables related to the section size are classified into 16 groups according to the symmetry in the geometric shape. In nodes 1 to 4, the concentrated non-structural mass of 2270 kg has affected the truss. The mechanical characteristics of the structure are: material density 2770 kg/m^3, modulus of elasticity 69,800 MPa, frequency limits of the structure in the first mode are equal to 4 Hz and in the second mode are greater than 6 Hz. For the cross section of the members, the lower limit is 0.645 cm^2 and the upper limit is 20 cm^2. In order to ensure the performance of chaotic functions and algorithms, as well as to increase the accuracy and sensitivity of calculations, each of the modes has been performed independently 20 times and the results related to the best response and the average value of responses are presented in Table 7.9. Also, the coefficient of variation of responses, which is a measure of the robustness and robustness of responses, has been calculated and used to compare the

Table 7.8 Optimal design comparison for the 52-bar dome-like truss

Decision variable	Standard	Logistic map			Gauss map		
	WEO	CWEO 21	CWEO 22	CWEO 23	CWEO 31	CWEO 32	CWEO 33
Z_A (m)	6.1396	5.9033	5.8012	5.7688	6.2319	6.0822	5.8469
X_B (m)	2.1118	2.143	2.1043	2.0429	2.0983	2.0932	2.1042
Z_B (m)	3.888	3.7436	3.7116	3.7	3.9104	3.8742	3.7401
X_F (m)	3.992	3.9351	3.8789	3.8155	4.0196	3.9438	3.9068
Z_F (m)	2.5007	2.5	2.5001	2.5041	2.5025	2.5494	2.5004
A_1 (cm^2)	1.0072	1	1.0003	1	1	1.0001	1
A_2 (cm^2)	1.2213	1.2602	1.2752	1.3053	1.2918	1.2248	1.2645
A_3 (cm^2)	1.2441	1.264	1.2564	1.2679	1.285	1.2347	1.2733
A_4 (cm^2)	1.2119	1.4756	1.5324	1.5039	1.362	1.4046	1.5107
A_5 (cm^2)	1.3431	1.4051	1.3798	1.357	1.4869	1.4356	1.4145
A_6 (cm^2)	1.002	1.0001	1.0002	1.0003	1	1	1.0006
A_7 (cm^2)	1.6297	1.4658	1.5576	1.7243	1.442	1.4748	1.5281
A_8 (cm^2)	1.4357	1.4748	1.4174	1.3053	1.4317	1.4952	1.4063
Best weight (kg)	194.911	193.762	193.757	194.206	194.974	194.924	**193.701**
Mean weight (kg)	200.138	**195.383**	199.064	196.563	199.353	201.493	200.338
Coefficient variation	2.296	**1.3182**	5.8882	2.161	1.8123	1.9536	2.4699
ω_1 (HZ)	12.9156	10.986	10.8687	10.6088	10.6823	10.7561	10.8193
ω_2 (HZ)	28.6485	28.6479	28.6477	28.6479	28.6479	28.6457	28.648

The bold items in the table indicate the optimal values for the best weight, the best mean and the best coefficient of variation

efficiency of turbulence functions and algorithms. For quick access to optimization information, a bar chart of each component is shown in Fig. 7.14. Examining the optimization results for different combinations of algorithms with turbulence functions and comparing it with the standard mode, shows a significant and significant improvement in reducing the weight of the 72-bar spatial truss.

The enhanced results for Chaotic Water Evaporation Optimization (CWEO), by Gauss chaos map and Scenario 3, a weight of 317.565 kg has the optimal response.

7.6.2.3 A 120-Bar Spatial Dome

The 120-bar spatial dome as shown in Fig. 7.15 is a well-known benchmark problem with weight-limit optimization. This truss only considers "optimizing the size of the sections" and the geometric shape of the structure is constant during the optimization

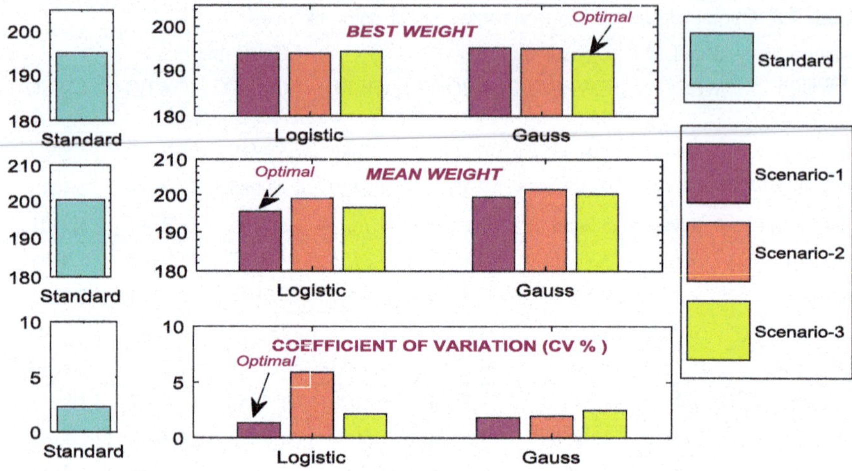

Fig. 7.12 Optimization results in standard and chaos map for the 52-bar dome-like truss

process. The decision variables related to the size of the members' sections and according to the symmetry in the geometric shape of the dome along the X and Y axes, are classified into 7 groups. Non-structural concentrated mass in all free nodes affects the structure. Their values are 3 kg in node 1500 kg in nodes 2 to 13 and 100 kg in other nodes.

The mechanical characteristics of the structure are: material density 7971.81 kg/m^3, modulus of elasticity 210,000 MPa, frequency limits of the structure in the first and second modes are greater than 9 and 11 Hz, respectively. For the cross section of the members, the range of the lower limit is 1 cm^2 and the upper limit is 129.3 cm^2. In order to ensure the performance of chaos map and algorithms, as well as to increase the accuracy and sensitivity of calculations, each of the modes has been performed independently 20 times and the results related to the best response and the average value of responses are presented in Statistical Table 7.10. Also, the coefficient of change of responses, which is a measure of the robustness and robustness of responses, has been calculated and used to compare the efficiency of turbulence functions and algorithms. For quick access to optimization information, the bar chart of each component is shown in Fig. 7.16. Examining the optimization results for different combinations of algorithms with turbulence functions and comparing it with the standard mode, shows a significant and significant improvement in reducing the weight of the 120-bar spatial dome. The enhanced results for Chaotic Water Evaporation Optimization (CWEO), by Gauss chaos map and Scenario 3, a weight of 8711.427 kg has the optimal response.

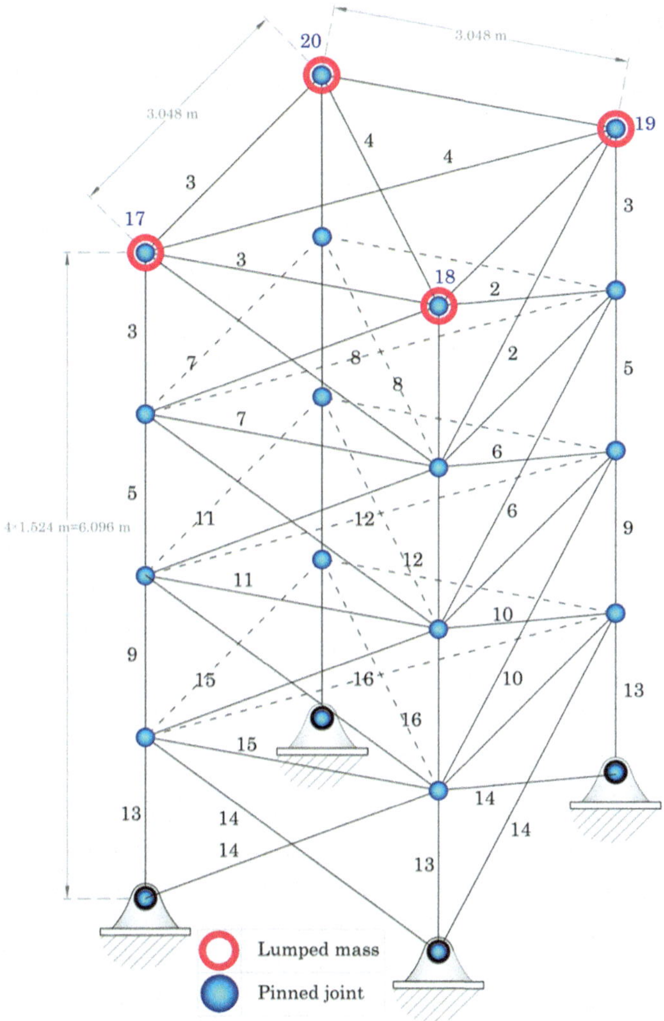

Fig. 7.13 Schematic of A 72-bar spatial truss

7.7 Concluding Remarks

In the water surface evaporation algorithm, it is inspired by the surface evaporation of tiny water molecules on a solid surface with different humidity states. In it, the dynamic simulation of the water molecule has been used. Based on this simulation, the desired surface changes from the state of repulsion to the state of water absorption, but the rate of evaporation in this sudden change does not decrease uniformly, but increases first and starts to decrease when it reaches a maximum volume. to do Also, when the humidity of the surface of the desired layer is not enough, the water molecule

Table 7.9 Optimal design comparison for the 72-bar spatial truss

Element group	Standard	Logistic map			Gauss map		
	WEO	CWEO 21	CWEO 22	CWEO 23	CWEO 31	CWEO 32	CWEO 33
A_1–A_4	3.23536	3.60799	3.61636	3.89	3.66798	3.76115	3.71363
A_5–A_{12}	7.8868	7.87735	7.90144	7.79504	8.00153	8.33608	7.58722
A_{13}–A_{16}	0.515206	0.508995	0.507706	0.521468	0.500517	0.511347	0.50231
A_{17}–A_{18}	0.607419	0.585736	0.601538	0.567576	0.547919	0.574614	0.518918
A_{19}–A_{22}	8.35094	8.29339	8.20327	8.85678	8.44578	7.79623	8.82235
A_{23}–A_{30}	8.31455	8.01917	8.08286	7.4089	7.90225	7.67853	7.55894
A_{31}–A_{34}	0.535043	0.517981	0.500778	0.507534	0.501465	0.501559	0.502233
A_{35}–A_{36}	0.52337	0.51572	0.502479	0.513692	0.504873	0.501178	0.508856
A_{37}–A_{40}	12.7995	14.7697	12.2636	14.5611	12.8894	12.8763	14.0599
A_{41}–A_{48}	7.9955	7.92004	8.31412	8.55562	7.82723	7.81921	7.88826
A_{49}–A_{52}	0.51659	0.502368	0.500236	0.500709	0.502268	0.507105	0.501254
A_{53}–A_{54}	0.512778	0.527739	0.501138	0.500043	0.502245	0.510405	0.500313
A_{55}–A_{58}	17.7751	16.2011	18.0261	15.5653	18.0523	18.0605	17.204
A_{59}–A_{66}	8.14123	8.02757	8.06601	8.35732	7.63134	8.15409	7.67987
A_{67}–A_{70}	0.524878	0.500173	0.508485	0.514102	0.500583	0.502597	0.510803
A_{71}–A_{72}	0.506898	0.535357	0.505047	0.504873	0.5008	0.50079	0.503601
Best weight (kg)	327.597	324.891	327.361	326.814	321.22	325.13	**317.565**
Mean weight (kg)	328.325	326.178	327.477	327.9	**323.972**	326.721	324.582
Coefficient variation	0.2092	0.2906	**0.0377**	0.2596	0.6972	0.313	1.4333
ω_1 (HZ)	4.001	4.0005	4	4.0009	4	4	4
ω_3 (HZ)	6.001	6.010	6	6.001	5.998	6.002	6.001

The bold items in the table indicate the optimal values for the best weight, the best mean and the best coefficient of variation

becomes a spherical droplet. In cases where the humidity of the surface of the desired layer is sufficient, the water molecule spreads and spreads as a single layer. Two strategies, single-layer and droplet, constitute the exploration and exploitation steps of the algorithm. The imbalance between layered and drip strategies is related to the random distribution used in them. By embedding logistic chaos, Gaussian and Pace-Weiss functions, the algorithm can be converged from local tele-optimums to global optima. Chaotic functions prevent premature maturation of the algorithm by creating chaotic mutations and significant improvement in optimization results is achieved. By creating a competition between 9 chaotic models, the best response is selected for

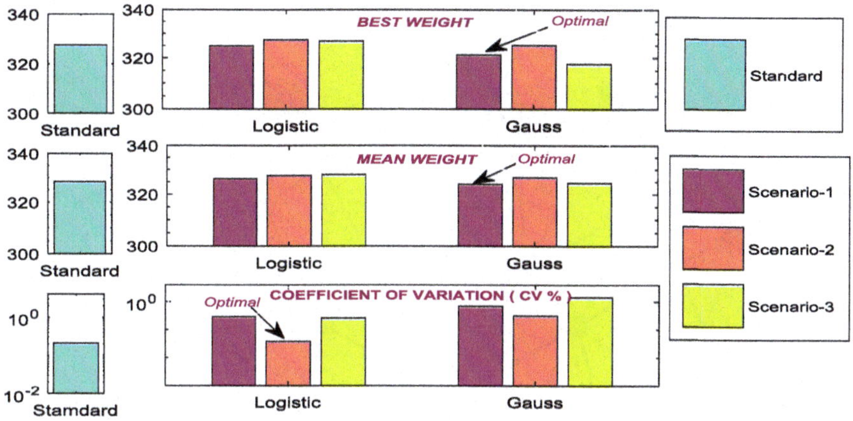

Fig. 7.14 Optimization results in standard and chaos map for the 72-bar spatial truss

comparison with the standard state. In non-linear, non-convex, complex problems, implicit modes, multi-element problems, simultaneous optimization of cross-section and geometric shape, and multi-frequency constraints, chaotic methods have high computational speed.

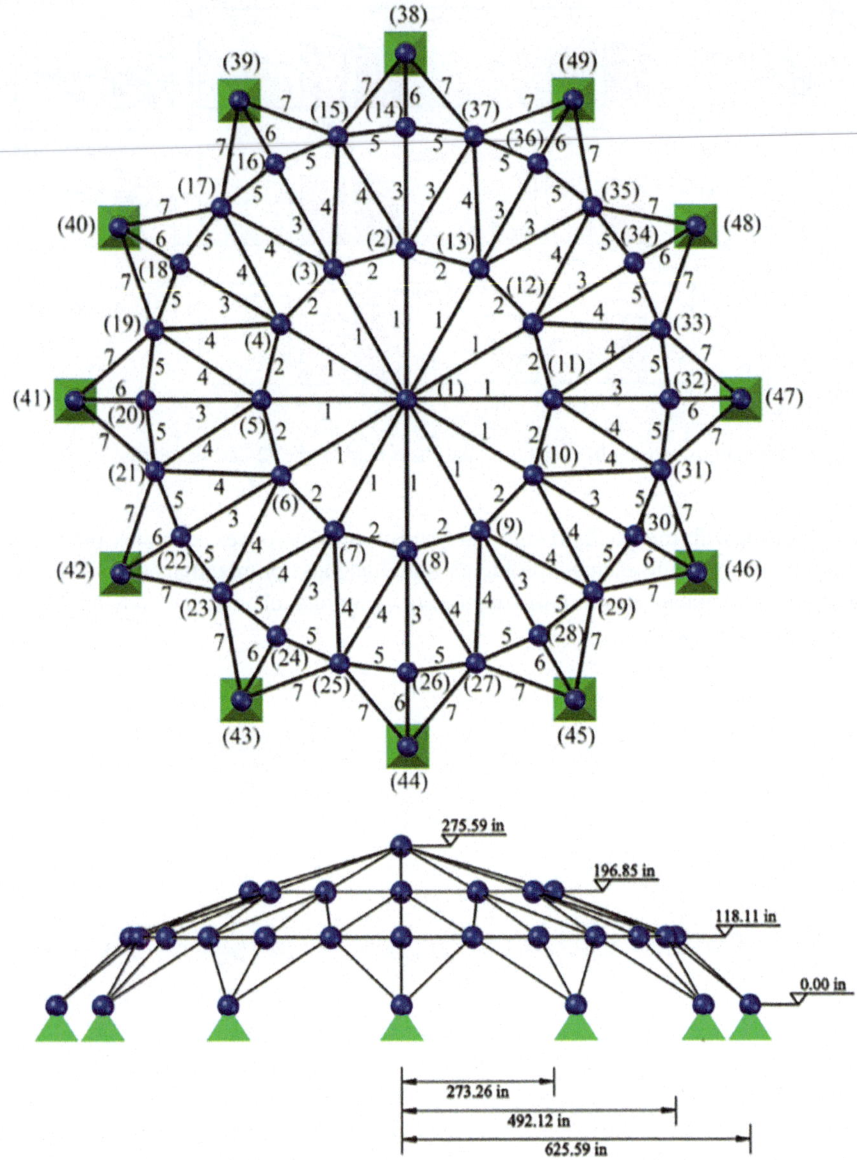

Fig. 7.15 Schematic of A 120-bar spatial dome

Table 7.10 Optimal design comparison for the 120-bar spatial dome

Decision variable	Standard	Logistic map			Gauss map		
	WEO	CWEO 21	CWEO 22	CWEO 23	CWEO 31	CWEO 32	CWEO 33
A_1	19.8311	19.5513	19.6504	19.4458	19.4614	19.2352	19.842
A_2	39.735	40.1339	40.0115	40.1934	41.1316	41.4495	39.5564
A_3	10.4857	10.3224	10.3842	11.1294	10.3218	10.2058	10.7291
A_4	20.5461	21.3523	20.7506	21.5847	20.6587	20.995	21.226
A_5	9.73952	9.9711	9.92735	9.86914	9.68173	9.95915	9.61127
A_6	11.5297	12.3439	11.8761	11.4091	11.9898	12.4936	11.5686
A_7	15.7487	14.5213	15.3069	14.3898	15.3053	14.7089	14.9451
Best weight (kg)	8720.389	8712.838	8712.904	8715.883	8714.447	8715.201	**8711.427**
Mean weight (kg)	8745.468	8725.251	8727.114	8723.127	8741.696	8748.133	**8718.489**
Coefficient variation	0.2454	0.1145	0.1778	**0.0586**	0.2539	0.3088	0.0763
ω_1 (HZ)	9.0012	9	9.0021	9	9.0001	9	9
ω_2 (HZ)	11	11	11.0014	11	11	11.0012	11

The bold items in the table indicate the optimal values for the best weight, the best mean and the best coefficient of variation

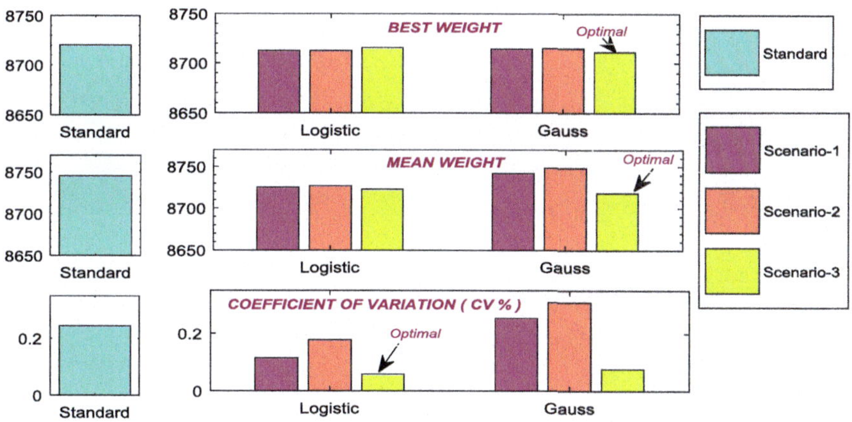

Fig. 7.16 Optimization results in standard and chaos map for the 120-bar spatial dome

References

1. Kaveh, A.: Advances in Meta-Heuristic Algorithms for Optimal Design of Structures, 2nd ed., 3rd edn. Springer, Cham, Switzerland (2021). https://doi.org/10.1007/978-3-319-05549-7

2. Kaveh, A., Bakhshpoori, T.: Water evaporation optimization: a novel physically inspired optimization algorithm. Comput. Struct. **167**, 69–85 (2016). https://doi.org/10.1016/j.compstruc.2016.01.008
3. Kaveh, A., Yousefpoor, H.: Chaos-embedded meta-heuristic algorithms for optimal design of truss structures. Scientia Iranica A **29**(6), 2868–2885 (2022). https://doi.org/10.24200/sci.2022.59812.6441
4. Kaveh, A., Yosefpoor, H.: Chaotically enhanced meta-heuristic algorithms for optimal design of truss structures with frequency constraints. Periodica Polytechnica Civil Eng. **66**(3), 900–921 (2022). https://doi.org/10.3311/PPci.20220
5. Kaveh, A., Zarfam, P., Aziminejad, A., Yosefpoor, H.: Comparison of four chaotic meta-heuristic algorithms for optimal design of large-scale truss structures. Iranian J. Sci. Technol. Trans. Civil Eng. https://doi.org/10.1007/s40996-022-00908-8
6. Ott, E.: Chaos in Dynamical Systems. Cambridge University Press, Cambridge (2002)
7. Bucolo, M., Caponetto, R., Fortuna, L., Frasca, M., Rizzo, A.: Does chaos work better than noise? IEEE Circuit Syst. Mag. **3**, 4–19 (2002)
8. American Institute of Steel Construction (AISC): Manual steel construction allowable stress design, 9th end. American Institute of Steel Construction, Chicago (1989)
9. Peitgen, H.-O., Jürgens, H., Saupe, D.: Chaos and Fractals: New Frontiers of Science. Springer, Berlin (2006)
10. Kaveh, A., Sheikholeslami, R., Talatahari, S., Keshvari-Ilkhichi, M.: Chaotic swarming of particles: a new method for size optimization of truss structures. Adv. Eng. Softw. **67**, 136–47 (2014)

Chapter 8
Chaotic Artificial Bees Colony

8.1 Introduction

Swarm intelligence is one of the techniques of artificial intelligence, which is based on the study of collective behavior in self-organized and decentralized systems. In the algorithms that are classified in the group of swarm intelligence, there is a direct or indirect relationship between the different answers of the algorithm. In other words, the local solutions related to each of the iteration steps, which are themselves low-intelligence and simple creatures, cooperate to reach the global optimal solution. While in the early algorithms like genetics, there was no connection between the answers. For example, in the crossover operator, none of the chromosomes know the best chromosome in the population and they do not know the value of the objective function for the elite chromosome. Therefore, the components that participate in swarm intelligence have excellent features [1]. These components are: population, cooperation, communication, information exchange, information flow and self-organization. Figure 8.1 shows how the main components of swarm intelligence are related.

The Artificial Bees Colony (ABC) is in the branch of swarm intelligence algorithms, which is inspired by the swarm behavior of bees to foraging resources and collect nectar for the hive. In this algorithm, each proposed answer is considered as a food source and the quality of their nectar expresses the objective function of that answer. At each stage of the iterative process, the results are refined to access new food sources with better nectar. The strategies of honey bees are classified into three groups, which includes employed or recruited, onlooker and scout bees. Employed bees modify food sources based on a specific process and the merit hive is not important for them, but onlooker bees select and modify them according to the merit of the hives. As a result, hives with high nectar quality have more chances to be selected. In this algorithm, the steps of exploitation and exploration have been assigned to two groups of employed and onlooker bees, respectively. In this meta-heuristic algorithm, in order to improve the optimization results and create interaction between

A. Kaveh and H. Yousefpoor, *Chaotic Meta-heuristic Algorithms for Optimal Design of Structures*, Studies in Computational Intelligence 1129,
https://doi.org/10.1007/978-3-031-48918-1_8

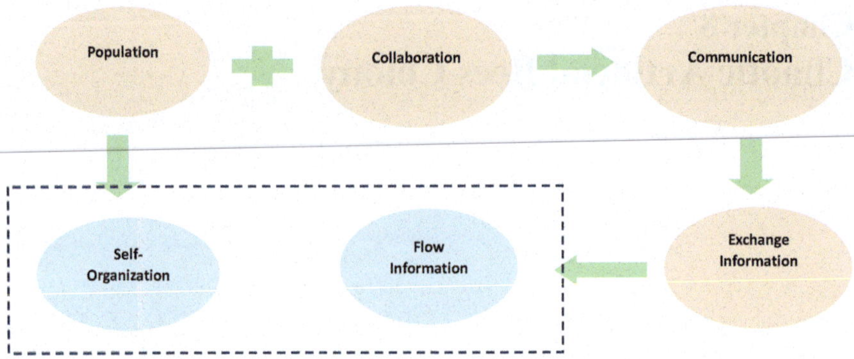

Fig. 8.1 Main components of swarm intelligence and related between

the two stages of exploration and exploitation, the random distribution related to the strategy of employed and onlooker bees is replaced by chaotic accelerators. Chaotic functions, by creating chaotic jumps, incline local optima towards global optima, and as a result, the algorithm is freed from the trap of local optima. In chaotic algorithms, premature convergence of the algorithm is improved and we can get better results with a small number of iterations. In this study, logistic, Gaussian, and Piecewise chaos functions are considered for the optimization of structures with static loads, and well-known structural examples are used to evaluate their efficiency. For optimization with multi-frequency constraints, logistic and Gaussian chaos functions have resulted in significant improvement and we did not need to check the third chaos function. In the rest of this chapter, after introducing the Artificial Bees Colony (ABC), the basic steps of the algorithm will be presented in the standard mode, and then the Chaotic Artificial Bees colony (CABC) will be formed with three scenarios. At the end, with famous examples of optimization of structures with static loading and multi-frequency constraints, the effectiveness of chaos functions will be challenged.

8.2 Standard Artificial Bees Colony (ABC)

This algorithm was presented in 2005 by Karaboga [2]. In this algorithm, the swarm behavior of bees for foraging is imitated. In terms of classification, Artificial Bees Colony (ABC) is in the branch of swarm intelligence. In this algorithm, the colony of bees operates in three different groups, each with specific responsibilities. These three groups are: employed, onlooker and scout bees. Employed bees try to find better food sources with a certain process, while onlooker bees search randomly based on the merit of food sites. Each food source or hive is known as a proposed solution and its quality is checked based on the objective function. The use of the roulette wheel plays a constructive role in this random selection of food sites (hives). The design of the algorithm is such that the possibility of searching weak sites is also considered. Also, if in the search process, sites are found that do not achieve acceptable progress after numerous reviews, we will leave those sources and think of new solutions. Therefore, if there is no result in a food site after a certain number of checks, scout bees come into action and choose new random positions. This algorithm in standard mode includes the following steps.

8.3 Basic Steps in Artificial Bees Colony

Step 1 Selection of algorithm parameters including: The number of primary bees, which is the same as the number of food sources nHB, the maximum repetition of the objective function, "maxNFEs" the maximum repetition that will abandon it if the food source position does not improve and replace it with a new random selection A, and finally, the probability percentage coefficient for leaving the food source position mr.

Step 2 Introducing the primary population of bees: In the search space, the primary answer or the primary population of bees is randomly formed. Because each bee finds a food site, we will actually have a food package equal to the number of the primary population of bees. Also, the answers of this step are evaluated by the objective function.

Step 3 Formation new food sources based on employed bee strategy: In this stage, first the calculation steps are formed by applying the random coefficient and then the new position of the food resources is determined according to Eq. (8.1):

$$\begin{aligned} Stepsize &= \boldsymbol{rand}_{ij} \otimes (HB - HB[permute(i)(j)]) \\ newHB &= HB + Stepsize \end{aligned} \tag{8.1}$$

The answers obtained from this stage are evaluated and replaced if they are improved. Otherwise, the counter of abandonment the food source and being replaced by scout bees will be added.

Step 4 Formation of new food sources based on the strategy of onlooker bees: In this stage, the food source is first selected based on merit and probabilistically (using the probabilistic choice of discrete distribution by roulette wheel). And then the calculation steps of formation and the new position of food sources are determined according to the category of Eq. (8.2):

$$\begin{aligned} &P_i = \frac{PFit_i}{\sum_{j=1}^{nHB} PFit_j} \\ &Stepsize = \boldsymbol{rand}_{ij} \otimes (HB - HB[permute(i)(j)]) \\ &if \;\; rand < mr \quad \Rightarrow \quad newHB = HB_{rws} + Stepsize \\ &if \;\; rand > mr \quad \Rightarrow \quad newHB = HB_{rws} \end{aligned} \tag{8.2}$$

The answers obtained from this stage are evaluated and replaced if they are improved. Otherwise, the counter of abandonment the food source and being replaced by scout bees will be added.

Step 5 Formation of food sources based on the strategy of scout bees: At this stage, those food sources that have not improved after a certain number of repetitions are discarded, and new random responses are determined and evaluated in the vicinity of the hives.

Step 6 Termination conditions are checked and if necessary, the operation from step 3 is repeated.

8.4 Chaos Enhanced Artificial Bees Colony (CABC)

In this algorithm, the formation of new food sources is based on two important strategies [3–5]. In the strategy of employed bees, a specific trend is taken to find food resources, while in the strategy of onlooker bees, the merits of food sites are considered. These two have the same role of exploration and exploitation process. Therefore, by replacing the chaos maps in the random selections related to these steps, we will witness a significant improvement in the performance of the algorithm. The suggested scenarios for this replacement are as follows:

Scenario 1. Replacing the chaos map in the strategy stage of employed bees: In this case, the first chaos map **CHM1** is replaced in Eqs. (8.1) and (8.3) is as follows:

$$\begin{aligned} Stepsize &= \boldsymbol{CHM1}_{ij} \otimes (HB - HB[permute(i)(j)]) \\ newHB &= HB + Stepsize \end{aligned} \tag{8.3}$$

Scenario 2. Replacing the chaos map in the onlooker bee strategy stage: In this case, the second **CHM2** chaos map is applied in Eq. (8.2) and the set of Eq. (8.4) is as follows:

$$\begin{aligned} & P_i = \frac{PFit_i}{\sum_{j=1}^{nHB} PFit_j} \\ & Stepsize = \boldsymbol{CHM2}_{ij} \otimes (HB - HB[permute(i)(j)]) \\ & if \;\; rand < mr \;\; \Rightarrow \;\; newHB = HB_{rws} + Stepsize \\ & if \;\; rand > mr \;\; \Rightarrow \;\; newHB = HB_{rws} \end{aligned} \tag{8.4}$$

Scenario 3. Replacing the chaos maps in both stages simultaneously: In this case, two chaos maps are simultaneously replaced in Eqs. (8.1) and (8.2).

8.5 Truss Weight Optimization with Static Constraints

In order to improve the optimization results in Artificial Bees Colony (ABC), the Logistic, Gossin and Piecewise chaos functions are selected. These functions are embedded in the Artificial Bee Colony (CABC) and then chaotic models are formed, and finally the competition between chaotic functions is done to reach the best optimal weight. To search for the best performance of the chaos algorithm, three scenarios are formed. In these scenarios, the chaos functions replace the distributions related to exploration, exploitation or both phases simultaneously. In each example, 9 chaotic models are formed and compared with the standard state. To form the statistical space and increase diversity in the research, each model is independently evaluated 30 times. Also, to analyze the statistical results, the best weight, the best average and the coefficient of variation are provided. In a number of studies, the application of chaos functions has increased the coefficient of variation. This performance indicates sudden jumps and increasing the difference between the best and worst answers. Due to the wide scope of investigations, there is a strong competition between chaos functions and as a result, the possibility of approaching the global optimum increases. This optimum cannot be far from the crosshairs of chaotic mutations. In each investigation, the chaos function compatible with the algorithm and the scenario corresponding to the best improvement are selected. In the Logistic chaos map, it most likely converges from local minima to general minima. Therefore, this function is suitable for improving the exploration conditions of algorithms [6]. The Gaussian chaos map is very likely to be within the local minimums and is suitable for improving exploitation conditions [7]. Finally, Piecewise map simultaneously improves both conditions [8]. Therefore, by selecting these chaos maps, the weakness of algorithms of any kind will be repaired and global optimum will not have the opportunity to escape the chaos functions. We can also find out the strengths and weaknesses of the algorithm in the stages of exploration or exploitation.

8.5.1 Formulation for Optimal Design of Skeletal Structures

In the optimization of skeletal structures, the main components include objective function, design constraints, and bound of decision variables. In optimization relationships, the weight of the structure is selected as the objective function and all design constraints must be satisfied with the least possible weight. The general form related to these problems is defined as Eq. (8.5):

$$\begin{aligned} &Find \quad \{X\} = [x_1, x_2, \ldots, x_{ng}] \\ &\qquad x_i \in D_i \\ &to \text{ minimize } W(\{X\}) = \textstyle\sum_{i=1}^{nm} \rho_i \cdot x_i \cdot L_i \\ &subject\ to: g_j(\{X\}) \le 0 \quad j = 1, 2, \ldots, n \end{aligned} \tag{8.5}$$

where {X} is the set of decision variables; ng is the dimension of decision variables (number of member groups in structure); D_i is the allowable set of values for the decision variable x_i; W({X}) presents the weight of structure; nm is the number of element; ρ_i denotes the material density of element i; L_i and x_i are the length and the cross-sectional of element i, respectively; $g_j(\{X\})$ denotes design constraints; and n is the number of the constraints.

D_i can be considered either as a continuous set or a discrete one. In the continuous problems, the decision variable can vary continuously in the optimization process

$$D_i = \{x_i \,|\, x_i \in [x_{i,\min}, x_{i,\max}]\} \tag{8.6}$$

where $x_{i,min}$ and $x_{i,max}$ is the lower and upper bond for decision variable i, respectively. If the decision variable is considered as a discrete one, selected from a set of part as:

$$D_i = \{d_{i,1}, d_{i,2}, \ldots, d_{i,r(i)}\} \tag{8.7}$$

where r(i) is the number of available discrete values for ith decision variable. In order to handle the constraints, a well-known penalty function is utilized. In this method, the aim of the optimization is redefined by introducing the cost function as:

$$\begin{aligned} f_{\cos t}(\{X\}) &= f_{penalty}(\{X\}) \times W(\{X\}) \\ f_{penalty}(\{X\}) &= (1 + \varepsilon_1 \cdot \nu)^{\varepsilon_2} \\ \nu &= \sum_{i=1}^{n} \max[o, g_j(\{X\})] \end{aligned} \tag{8.8}$$

where n represents the number of evaluated constraints for each individual design. The constant ε_1 and ε_2 are selected considering the exploration and exploitation rate of the search space. Here, ε_1 is set to unity, ε_2 is selected in a way that it decreases the penalties and reduces the cross-sectional areas. Thus, in the first steps of the search process, ε_2 is set to 1.5 and ultimately increased to 3.

8.5.2 Introduction of Selected Chaos Map

By forming sentences of chaos functions, the dynamic series of chaos is determined. By embedding chaotic dynamic series in the artificial bee colony, the optimization results will converge towards global optima, and these series will have a good potential to prevent meta-heuristic algorithms from getting trapped in local optima. With chaotic meta-heuristic algorithms, the non-linear and non-convex behavior of the objective function in the optimization of truss structures can be easily controlled and adjusted. Among the advantages of the series formed by the mediation of chaos functions are non-repetition and non-convergence towards a certain limit. The standard mode of Artificial Bees Colony (ABC), will have premature convergence when dealing with non-linear and non-convex problems. The reason for that is the random distribution related to the main phases of the algorithm, including the phase of recruited bees, the phase of onlooker bees and the phase of scout bees. By substituting chaotic series in place of random distributions, the local optimum jumps to the global optimum. Therefore, by choosing these functions, the weakness of algorithms of any kind is improved. The relationships of these chaos functions have been introduced in the previous chapter.

8.5.3 Numerical Examples of Optimal Truss Design

By comparing the performance of chaotic models and choosing the most appropriate chaos function, the competition between 9 chaotic responses is formed. Also, in order to know the quality and quantity of answers, the results of the best chaotic model are compared with the standard mode. Due to the variety of reviews, the optimal responses will be largely close to the global optimality. The examples provided are as follows:

- The 52-bar planar truss;
- The 72-bar spatial truss;
- The 120-bar dome shaped truss.

8.5.3.1 A 52-Bar Planar Truss

The topology and nodal numbering of the 52-bar planer truss are shown in Fig. 8.2. The material density is 7860 kg/m^3 and the modulus of elasticity is 207 GPa. The members are subjected to stress limitation 180 MPa in both tension and compression. For loading, the intensity of the loads in the X and Y directions is 100 and 200 kN, respectively, and both loads are applied simultaneously. Also, truss members are classified into 12 groups according to geometric symmetry. The cross-section areas of the members were chosen from the 64 discrete values of the AISC code. In order to ensure the performance of chaos map and algorithms, as well as to increase the accuracy and sensitivity of calculations, each of the modes has been performed independently 30 times and the results related to the best response and the average value of responses are presented in Table 8.1. The graph of these results is also shown in Fig. 8.3. The coefficient of variation of the responses, which indicates the stability and robustness of the responses, has been calculated and another criterion for the efficiency of chaos map and algorithms has been obtained. After reviewing the optimization results for different combinations of algorithms with chaos map and comparing it with the initial and standard mode, a significant improvement in reducing the weight of the 52-bar planar truss results. The enhanced results for

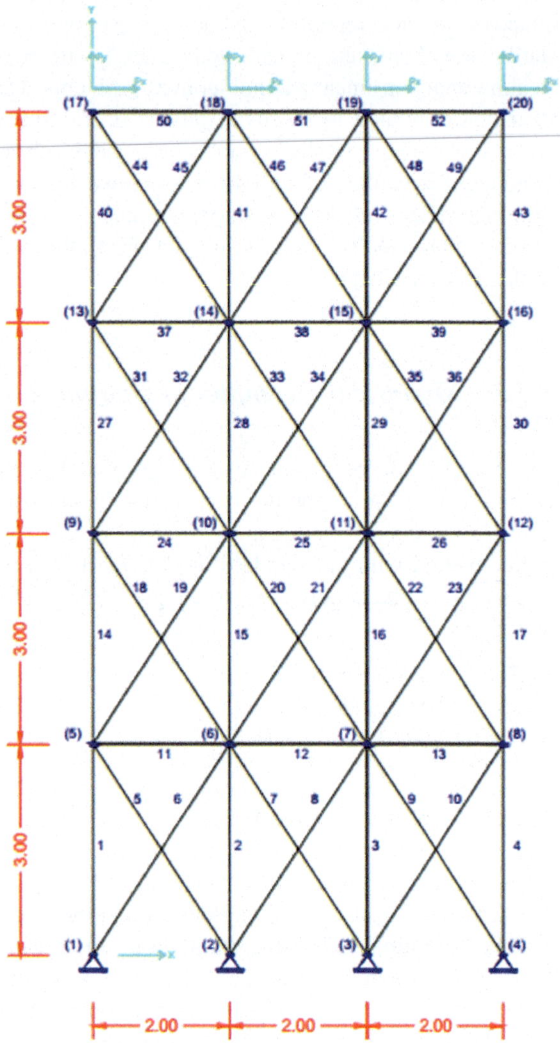

Fig. 8.2 Schematic of a 52-bar planar truss

Chaotic Artificial Bees Colony (CABC), by Piecewise chaos map and scenario 2, a weight of 1899.897 kg has the optimal response.

8.5.3.2 A 72-Bar Spatial Truss

The topology and nodal numbering of the 72-bar spatial truss are shown in Fig. 8.4. The material density is 0.1 lb/in^3 and the modulus of elasticity is 10,000 ksi. Permissible members' stress of 25 ksi is assumed. The maximum displacement for each

Table 8.1 Optimal design comparison for the 52-bar planar truss

Number group	Element group	Standard	Logistic map			Gauss map			Piecewise map		
		ABC	CABC 21	CABC 22	CABC 23	CABC 31	CABC 32	CABC 33	CABC 41	CABC 42	CABC 43
1	A_1–A_4	4658.055	4658.055	4658.055	4658.055	4658.055	4658.055	4658.055	4658.055	4658.055	4658.055
2	A_5–A_{10}	1161.288	1161.288	1161.288	1161.288	1161.288	1161.288	1161.288	1161.288	1161.288	1161.288
3	A_{11}–A_{13}	506.451	494.193	494.193	506.451	494.193	494.193	494.193	494.193	506.451	494.193
4	A_{14}–A_{17}	3303.219	3303.219	3303.219	3303.219	3303.219	3303.219	3303.219	3303.219	3303.219	3303.219
5	A_{18}–A_{23}	1008.385	939.998	1008.385	939.998	1008.385	939.998	939.998	939.998	939.998	1008.385
6	A_{24}–A_{26}	506.451	494.193	494.193	506.451	494.193	641.289	494.193	494.193	506.451	506.451
7	A_{27}–A_{30}	2238.705	2238.705	2290.318	2290.318	2290.318	2238.705	2238.705	2238.705	2238.705	2290.318
8	A_{31}–A_{36}	1008.385	1008.385	1008.385	1008.385	1008.385	1008.385	1008.385	1008.385	1008.385	1008.385
9	A_{37}–A_{39}	494.193	388.386	388.386	388.386	388.386	494.193	506.451	494.193	388.386	388.386
10	A_{40}–A_{43}	1283.868	1283.868	1283.868	1283.868	1283.868	1283.868	1283.868	1283.868	1283.868	1283.868
11	A_{44}–A_{49}	1161.288	1161.288	1161.288	1161.288	1161.288	1161.288	1161.288	1161.288	1161.288	1161.288
12	A_{50}–A_{52}	494.193	792.256	494.193	498.193	506.451	498.193	494.193	494.193	506.451	494.193
Best	Weight	1915.39	1912.22	1914.659	1904.376	1915.238	1909.731	1903.183	1902.605	**1899.897**	1915.237
Mean	Weight	1923.171	1917.384	1920.842	1911.273	1923.171	1914.304	1911.135	1908.879	**1901.327**	1921.953
	CV	0.2352	0.16104	0.18062	0.18526	0.2152	0.16177	0.19104	0.17823	**0.10942**	0.17664

The bold item in the table indicate the optimal values for the best weight, the best mean and the best coefficient of variation

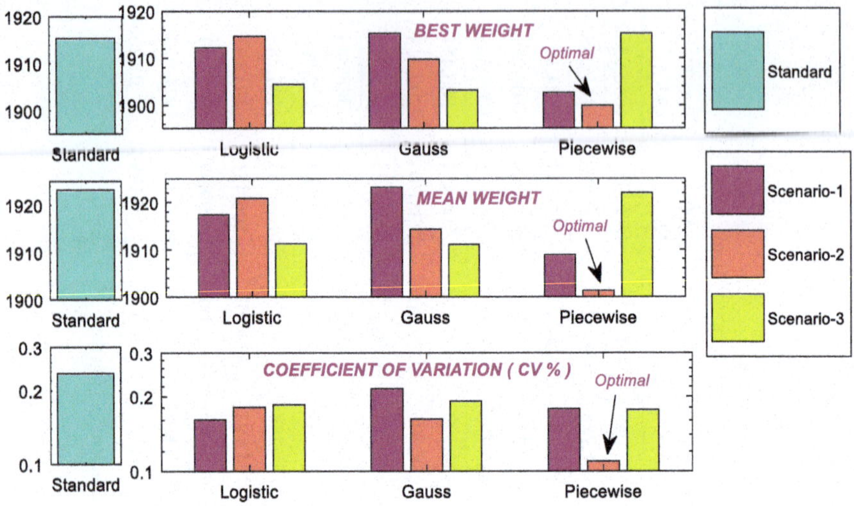

Fig. 8.3 Optimization results in standard and chaos map for the 52-bar planar truss

of the upper nodes is limited at 0.25 in in all directions. The two loading conditions are applied according to Table 8.2. Also, truss members are classified into 16 groups according to considering the geometric symmetry. The cross section of the members, which are the same as the decision variables, is considered in the range of 0.1–4 in^2. In order to ensure the performance of chaos map and algorithms, as well as to increase the accuracy and sensitivity of calculations, each of the modes has been performed independently 30 times and the results related to the best response and the average value of responses are presented in Table 8.3. A graph of these results is also shown in Fig. 8.5. The coefficient of variation of the responses, which indicates the stability and robustness of the responses, has been calculated and another criterion for the efficiency of chaos map and algorithms has been obtained. Examining the optimization results for different combinations of algorithms with chaos map and comparing it with the initial and standard modes, shows a significant improvement in reducing the weight of the 72-bar spatial truss. The enhanced results for Chaotic Artificial Bees Colony (CABC), by Logistic chaos map and scenario 2, a weight of 379.0452 lb has the optimal response.

8.5.3.3 A 120-Bar Dome Shaped Truss

The topology and nodal numbering of the 120-bar dome shaped truss shown in Fig. 8.6. The material density is 0.288 lb/in^3 and the modulus of elasticity is 30,450 lb/in^2. Permissible members' tensile and compressive stresses are proposed in accordance with AISC ASD code. The maximum deformation for each node in all directions is set at 0.1969 in. All non-abutment nodes are affected by vertical loading, with intensities of – 13.49 kips in node one and – 6.744 kips in spheres 2–14 and 2.248 kips in the other nodes, respectively.

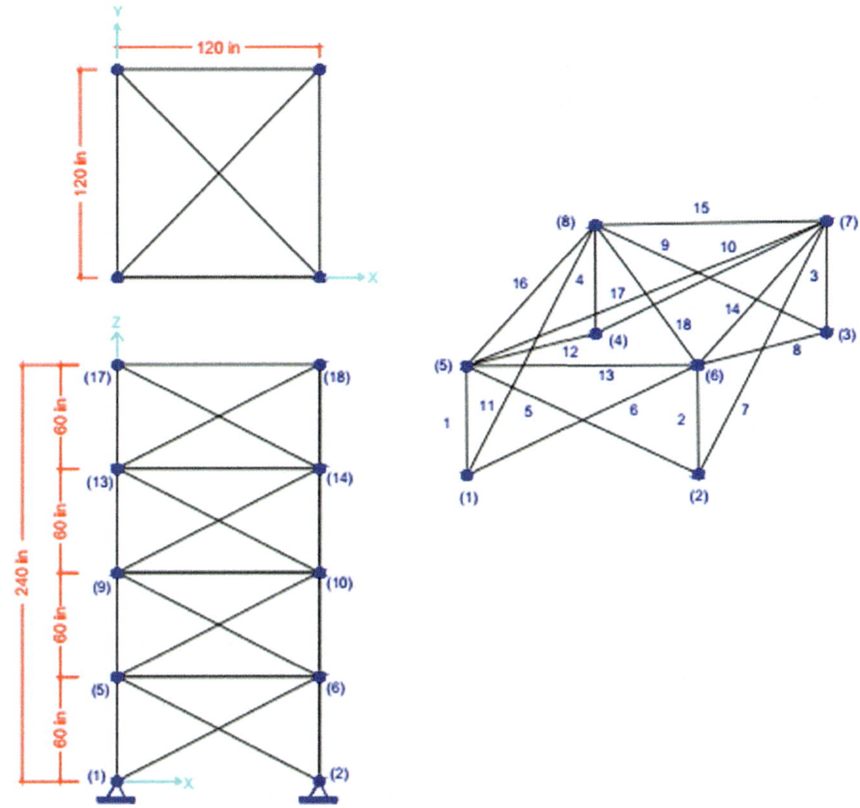

Fig. 8.4 Schematic of a 72-bar spatial truss and typical story

Table 8.2 Loading conditions for the 72- bar spatial truss

Node	Load case 1			Load case 2		
	F_x (kips)	F_y (kips)	F_z (kips)	F_x (kips)	F_y (kips)	F_z (kips)
17	5	5	− 5	0	0	− 5
18	0	0	0	0	0	− 5
19	0	0	0	0	0	− 5
20	0	0	0	0	0	− 5

Also, truss members are classified into seven groups according to their shape and geometric symmetry. The minimum cross-sectional area of the members, which is the lower limit of the decision variables, is 0.775 in^2, and the maximum cross-sectional area is 20 in^2. The allowable stress in tension and pressure is determined based on the following equations.

Table 8.3 Optimal design comparison for the 72-bar spatial truss

Number group	Element group	Standard	Logistic map			Gauss map			Piecewise map		
		ABC	CABC 21	CABC 22	CABC 23	CABC 31	CABC 32	CABC 33	CABC 41	CABC 42	CABC 43
1	A_1–A_4	1.9256	1.8687	1.8933	1.8711	1.8662	1.9124	1.8962	1.877	1.8661	1.8973
2	A_5–A_{12}	0.51796	0.49825	0.52105	0.51072	0.51658	0.51064	0.51938	0.51199	0.51308	0.51288
3	A_{13}–A_{16}	0.1	0.10042	0.10005	0.10013	0.1	0.10008	0.10005	0.1	0.10008	0.10002
4	A_{17}–A_{18}	0.10016	0.1	0.1	0.10001	0.10005	0.1	0.10006	0.10019	0.10023	0.10004
5	A_{19}–A_{22}	1.2912	1.3177	1.2736	1.2747	1.2653	1.2785	1.2593	1.2756	1.2549	1.2746
6	A_{23}–A_{30}	0.5158	0.51898	0.50159	0.51863	0.51432	0.50856	0.51179	0.51261	0.50891	0.51159
7	A_{31}–A_{34}	0.1	0.10029	0.1	0.1	0.10057	0.1	0.1	0.10009	0.10028	0.10004
8	A_{35}–A_{36}	0.10025	0.10068	0.1	0.10054	0.1	0.10002	0.10017	0.10002	0.1	0.1
9	A_{37}–A_{40}	0.49931	0.53861	0.55002	0.5246	0.5166	0.51699	0.51696	0.52409	0.53044	0.52467
10	A_{41}–A_{48}	0.5005	0.50755	0.51388	0.51732	0.51699	0.51541	0.51481	0.51669	0.5163	0.51633
11	A_{49}–A_{52}	0.10082	0.10117	0.1	0.1	0.1	0.10008	0.1	0.10005	0.1	0.1
12	A_{53}–A_{54}	0.1	0.1	0.1	0.10052	0.10151	0.10092	0.10042	0.10041	0.10083	0.10043
13	A_{55}–A_{58}	0.15792	0.15674	0.15673	0.15672	0.15673	0.15668	0.15677	0.15657	0.15607	0.15655
14	A_{59}–A_{66}	0.54639	0.53582	0.53788	0.54254	0.53993	0.54116	0.53974	0.54163	0.55125	0.54365
15	A_{67}–A_{70}	0.41524	0.41497	0.41661	0.40535	0.4135	0.41008	0.41396	0.41273	0.41354	0.40652
16	A_{71}–A_{72}	0.54712	0.59359	0.55526	0.56276	0.56662	0.57203	0.56565	0.57675	0.56951	0.56752
Best	Weight	379.4327	379.0893	**379.0452**	379.27	379.1449	379.264	379.439	379.557	379.4677	379.578
Mean	Weight	379.7669	379.6248	**379.3893**	379.5782	379.5252	379.4397	379.579	379.602	379.5971	379.603
	CV	0.063808	0.10128	0.074592	0.048879	0.05629	0.036727	0.02283	**0.01090**	0.019144	0.02707

The bold items in the table indicate the optimal values for the best weight, the best mean and the best coefficient of variation

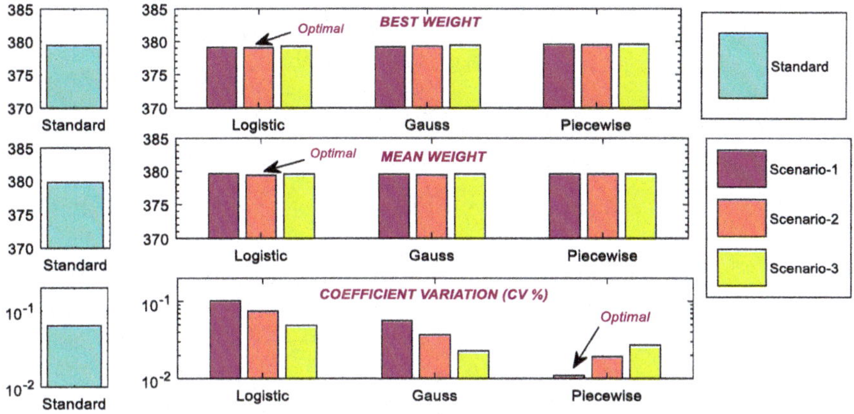

Fig. 8.5 Optimization results in standard mode and chaos map for the 72- bar spatial truss

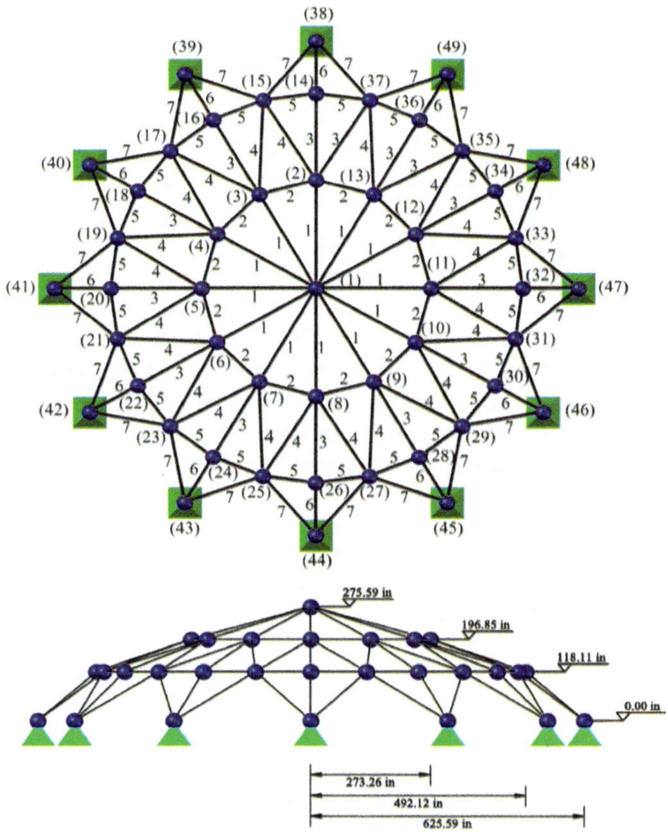

Fig. 8.6 Schematic of a 120-bar dome shaped truss

$$\sigma_i = \begin{cases} 0.6F_y & for\ \sigma_i \geq 0 \\ \sigma_i^- & for\ \sigma_i < 0 \end{cases} \tag{8.9}$$

For compressive stresses, we have the following equations:

$$\sigma_i^- = \begin{cases} (1 - \frac{\lambda^2}{2C^2})F_y/FS & for\ \lambda < C \\ \frac{12\pi^2 E}{23\lambda^2} & for\ \lambda \geq C \end{cases} \tag{8.10}$$

In this equation we have:

$$FS = (\frac{5}{3} + \frac{3\lambda}{8C} - \frac{\lambda^3}{8C^3});\ C = \sqrt{(\frac{2\pi^2 E}{F_y})};\ \lambda = \frac{kl}{r};$$
$$r = aA^b;\ a = 0.4993;\ b = 0.6777 \tag{8.11}$$

The components used include E modulus of elasticity, Fy the yield stress of steel, C the ratio of the limit slenderness which separates the elastic and inelastic buckling region in comparison with the existing slandering λ. Also, k is the effective length factor of the l and r is the radius of rotation of the limb. In order to ensure the performance of chaos map and algorithms, as well as to increase the accuracy and sensitivity of calculations, each of the modes has been performed independently 30 times and the results related to the best response and the average value of responses are presented in Table 8.4. A graph of these results is shown in Fig. 8.7.

Also, the coefficient of variation of the responses, which indicates the stability and robustness of the responses, has been calculated and another criterion for the efficiency of chaos map and algorithms has been obtained. The minimum and maximum cross-sectional area of all members is 0.775 in^2 and 20 in^2 respectively. By examining the optimization results for different combinations of algorithms chaos map and comparing it with standard mode, a significant improvement in reducing the weight of the 120-bar dome shaped truss is achieved. The enhanced results for Chaotic Artificial Bees Colony (CABC), by Logistic chaos map and scenario 2, a weight of 33,147.79 lb has the optimal response.

Table 8.4 Optimal design comparison for the 120-bar dome shaped truss

Number group	Standard	Logistic map			Gauss map			Piecewise map		
	ABC	CABC 21	CABC 22	CABC 23	CABC 31	CABC 32	CABC 33	CABC 41	CABC 42	CABC 43
1	3.02449	3.02635	3.02638	3.02357	3.02709	3.02671	3.0272	3.02589	3.02473	3.02472
2	14.0917	14.7363	14.4572	14.6334	14.5699	14.5957	14.4108	14.497	14.5973	14.3771
3	5.30673	5.24264	5.36363	4.98742	5.17581	5.22001	5.23009	5.1342	5.18075	5.15007
4	3.03762	3.00327	2.97999	3.05158	3.03971	3.01466	3.06689	3.08586	3.09454	3.1093
5	8.84705	8.41424	8.53467	8.59447	8.55875	8.51008	8.58735	8.54775	8.5143	8.62785
6	3.37363	3.38702	3.29277	3.48504	3.34317	3.39635	3.31954	3.41318	3.3094	3.33719
7	2.51659	2.51127	2.52417	2.50012	2.50054	2.50428	2.49668	2.49542	2.49578	2.49583
Best weight	33,336.87	33,158.04	**33,147.79**	33,186.54	33,171.32	33,170.02	33,189.27	33,206.52	33,213.9	33,225.28
Mean weight	33,381.86	33,220.32	**33,190.3**	33,221.69	33,213.94	33,214.93	33,216.55	33,219.61	33,226.55	33,244.77
CV	0.12707	0.14627	0.093615	0.094903	0.087668	0.11602	0.063459	0.037097	**0.030296**	0.036228

The bold items in the table indicate the optimal values for the best weight, the best mean and the best coefficient of variation

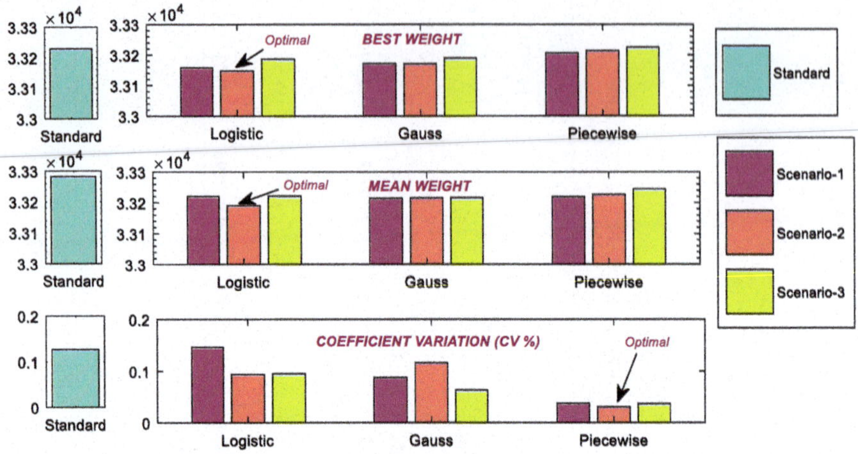

Fig. 8.7 Optimization results in standard mode and chaos map for the 120-bar dome shaped truss

8.6 Truss Weight Optimization with Multiple Frequency Constraints

To prevent the effects of destructive phenomena, such as earthquakes and wind, the natural frequency of the structure is compared with the frequency of the input load. Therefore, the natural frequency of structures should be constrained in a certain range. The objective function in the frequency constraint is associated with implicit and complex modes, and the optimization using meta-heuristic algorithms in the standard mode stops at the local optimum in most cases. In the Chaotic Artificial Bees Colony (CABC), by embedding chaotic dynamic series, include Logistic and Gaussian map, a significant improvement in the optimization results is achieved [6, 7]. These series can replace the random distributions used in the strategy of recruited bees, onlooker bees and scout bees. By balancing these strategies, chaos series converges the local optimum to the global optimum and avoids optimization stagnation. In each investigation, 6 chaotic models have been formed and compared with the standard state. To create statistical information and increase diversity in research, each model is independently evaluated 30 times. Finally, to analyze the statistical results, the best weight, the best average and the coefficient of variation are reported. Also, useful information is obtained about the weakness and strength of the algorithm and how to improve the results.

8.6.1 Formulation of the Structural Optimization with Frequency Constraints

In optimization problems for the cross section and geometric shape of trusses that are associated with frequency constraints, the goal is to minimize the weight of the structure so that the constraints for a number of natural frequencies for vibration modes are satisfied [9]. The cross section of the members along with the coordinates of some nodes are introduced as decision variables. These variables are selected continuously. The upper and lower bounds are also specified for variables in some cases. These optimization problems are defined in mathematical form according Eq. (8.12):

$$\begin{array}{ll} Find & X = \{A, S\},\quad A = \{A_1, A_2, \ldots, A_{na}\}, S = \{S_1, S_2, \ldots, S_{ns}\} \\ to\, Minimize: & W(A, S) = \sum_{i=1}^{nm} \rho_i \times L_i(S) \times A_i \\ subjected\ to: & \begin{cases} \tilde{\omega}_i^L \le \omega_i \le \tilde{\omega}_i^U & i = 1, 2, \ldots, n\omega \\ A_j^L \le A_j \le A_j^U & j = 1, 2, \ldots, na \\ S_k^L \le S_k \le S_k^U & k = 1, 2, \ldots, ns \end{cases} \end{array} \tag{8.12}$$

In this relation, X is the vector of decision variables, A is the variable of the cross-section of the members, na is the symbol to represent the number of variables of the cross-section, A_i is the value of the cross-section of the ith variable, S is the variable of shape and arrangement, ns is the number of shape variables with the same coordinates Nodes are, S_i is the numerical value of the ith variable, W expresses the weight of the truss, nm indicates the total number of members, ρ_i the specific gravity of the material belonging to the ith member of the truss, L_i the length corresponding to the ith member which can be expressed through the variables Determine, ω_i expression of the ith natural frequency of the truss, ω_i^L and ω_i^u, respectively, represent the lower limit and the upper limit of the fixed frequency of the base, $n\omega$ indicates the total number of frequency restrictions, Aj^L and Aj^u, respectively, express the lower limit and the upper limit of the A_j cross section, and $S_k{}^L$ and $S_k{}^U$ to Arrange the lower limit and limit of kth variable S_k. Given that meta-heuristic algorithms are used for unbounded problems; we use the penalty function in modeling to convert the bound to unbound state. In this method, if no violation has been committed, the amount of the fine will be zero, otherwise, if there is a violation, the amount of the penalty function will be obtained from Eqs. (8.13)–(8.16):

$$V_i = \begin{cases} 0 & if\ \tilde{\omega}_i^L \le \omega_i \le \tilde{\omega}_i^U \\ 1 - \frac{\omega_i}{\tilde{\omega}_i^L}\ if\ \tilde{\omega}_i^L > \omega_i \\ 1 - \frac{\tilde{\omega}_i^U}{\omega_i}\ if\ \omega_i > \tilde{\omega}_i^U \end{cases} \tag{8.13}$$

$$\gamma = \sum_{i=1}^{n\omega} V_i \tag{8.14}$$

$$F_{penalty} = (1 + \varepsilon_1 \times \gamma)^{\varepsilon_2} \tag{8.15}$$

$$to\ Minimize\ \ Mer(A, S) = W(A, S) \times F_{penalty} \tag{8.16}$$

In these relations γ represents the set of violations and ε_1 and ε_2 are selected based on the search and extraction ratio. In this study, ε_1 units and ε_2 are selected with incremental linear changes in the range of 1.3–3 at the end of the iteration. Finally, "*Mer*" is the merit function or the objective function after the penalty is imposed.

8.6.2 Numerical Examples of Optimal Truss Design

For the challenge in results of Chaotic Artificial Bees Colony (CABC), well-known examples of truss optimization are chosen, with layout and size optimization as benchmark problems. To compare the improvement values, the best chaotic mode is compared with the standard model. Due to the variety of reviews, the optimal responses will be largely close to the global optimality. The examples provided are as follows:

- The planar 10-bar truss;
- The simply supported 37-bar planar truss;
- 200-bar planar truss structure.

8.6.2.1 A Planar 10-Bar Truss

The planar 10-bar truss as depicted in Fig. 8.8 is a well-known benchmark problem in the field of weight optimization of the structures with frequency constraints. This truss only considers "optimizing the size of the sections" and the geometric shape of the structure is constant during the optimization process. The non-structural concentrated mass of 454 kg affected the free nodes. The mechanical properties of the structure are: The density of materials is 2770 kg/m^3, the modulus of elasticity is 69,800 MPa, the frequency limits of the structure in the first, second and third modes are greater than 7, 15 and 20 Hz, respectively. The cross-sectional area of each member is selected as an independent decision variable. Therefore, ten design variables with a lower limit of 0.645 cm^2 and an upper limit of 50 cm^2 are considered. In order to ensure the performance of chaotic functions and algorithms, as well as to increase the accuracy and sensitivity of calculations, each of the modes has been performed independently 20 times and the results related to the best response and the average value of responses are presented in Statistical Table 8.5. Also, the coefficient of variation of responses, which is a measure of the robustness of responses, has been calculated and used to compare the efficiency of chaos maps and standard algorithms. For quick access to optimization information, a bar chart of each component is

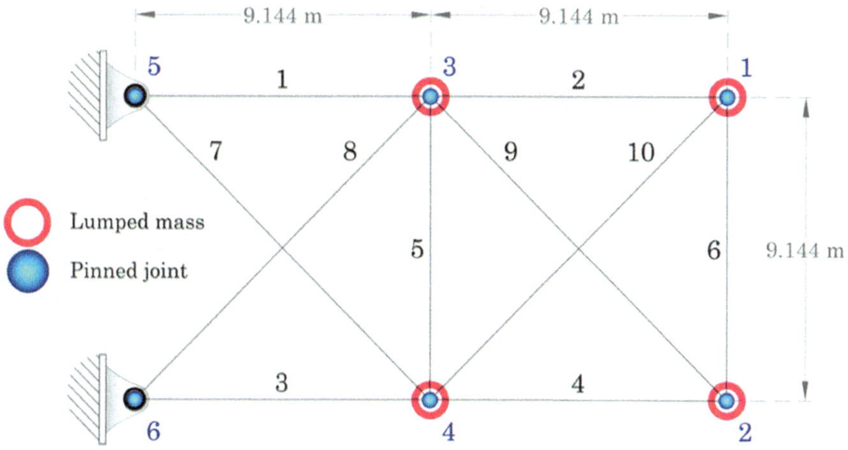

Fig. 8.8 Schematic of a planar 10-bar truss

shown in Fig. 8.9. Examining the optimization results for different combinations of algorithms with chaos maps and comparing it with the standard mode, shows a significant improvement in reducing the weight of the planar 10-bar truss.

The enhanced results for Chaotic Artificial Bees Colony (CABC), by Logistic chaos map and scenario 2, a weight of 523.054 kg has the optimal response.

8.6.2.2 A Simply Supported 37-Bar Planar Truss

A 37-member planar truss with a simple abutment as shown in Fig. 8.10 is a well-known benchmark problem for optimizing the weight and shape of a truss with a frequency limit. This truss considers both optimizations of the section size and geometric coordinates of the nodes and the geometric shape of the structure is determined during the optimization process. The lower edge member is considered to be a rod member with a rectangular cross-section and an area of 40 cm^2. The other members are modeled as rod members. These decision variables are grouped according to symmetry in the geometric shape. The y-axis coordinates form the shape variables in the upper year, taking into account the symmetry. In the lower axis, a non-structural concentrated mass of 10 kg affects all free nodes. The mechanical properties of the structure are: The density of the materials is 7800 kg/m^3, the modulus of elasticity is 21,000 MPa, the frequency limits of the structure in the first, second and third modes are greater than 20, 40 and 60 Hz, respectively. For the cross section of the members, the lower limit is 1 cm^2 and the upper limit is 10 cm^2. The total size and shape decision variables are 19 variables, of which 14 variables are related to the size and cross-sectional area of the members and 5 variables are related to the shape of the structure and the coordinates of the nodes. There are also three frequency limits. In order to ensure the performance of turbulent functions and

Table 8.5 Optimal design comparison for the planar 10-bar truss

Number group	Standard	Logistic map			Gauss map		
	ABC	CABC 21	CABC 22	CABC 23	CABC 31	CABC 32	CABC 33
A_1	34.3926	33.3516	34.4607	36.6934	34.4308	34.1884	33.932
A_2	16.833	14.8657	14.5129	15.9159	15.8868	15.2129	14.9608
A_3	35.8411	35.8384	33.1225	33.7065	34.9843	35.7227	35.3021
A_4	16.634	15.758	16.4164	15.0082	13.9592	15.1839	13.7466
A_5	0.645	0.73371	0.645	0.645	0.670056	0.757492	0.677549
A_6	5.05619	4.5599	4.94455	4.73826	4.63367	4.62183	4.74318
A_7	22.5799	21.6513	22.7812	23.3608	25.3139	22.6526	25.0941
A_8	21.7853	24.793	24.1704	23.0754	21.9318	24.2534	23.022
A_9	12.5682	13.1335	12.3842	12.7185	11.7162	12.5213	12.1332
A_{10}	12.8138	12.2699	13.0742	11.5034	13.149	12.1755	12.7344
Best weight (kg)	526.939	523.587	**523.054**	523.378	523.156	524.178	523.236
Mean weight (kg)	529.252	526.096	525.043	525.815	**524.005**	524.935	524.209
Coefficient variation	0.4253	0.4989	0.2508	0.3971	0.161	**0.1019**	0.1091
ω_1 (HZ)	6.98132	6.98027	6.95798	6.97455	6.98127	6.98599	6.98262
ω_2 (HZ)	16.3246	16.2394	16.1654	16.2545	16.172	16.3406	16.1777
ω_3 (HZ)	19.9979	19.9868	20.0928	20.0065	20.0109	20.0004	20.0042

The bold items in the table indicate the optimal values for the best weight, the best mean and the best coefficient of variation

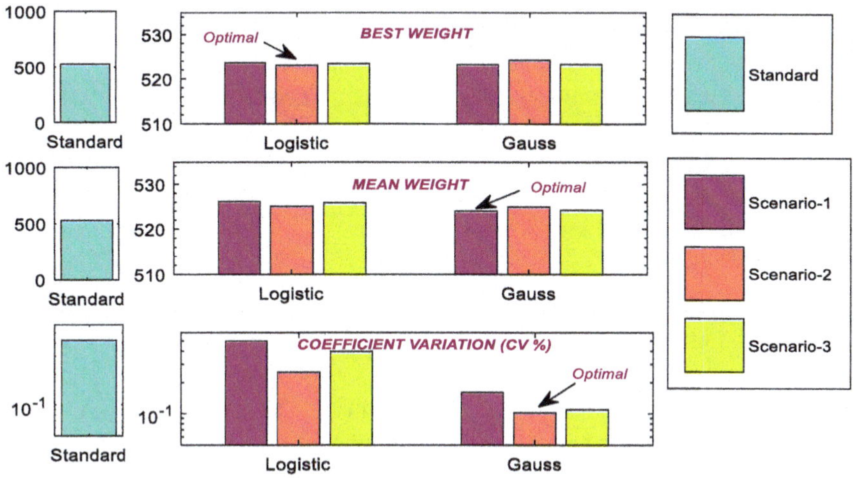

Fig. 8.9 Optimization results in standard mode and chaos map for the planar 10-bar truss

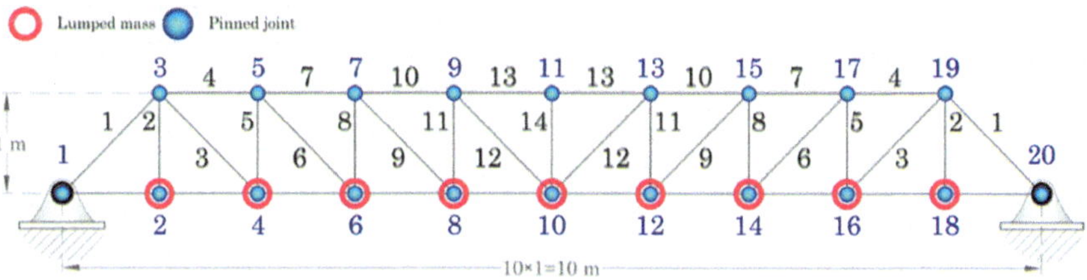

Fig. 8.10 Schematic of a simply supported 37-bar planar truss

algorithms, as well as to increase the accuracy and sensitivity of calculations, each of the independent modes has been performed 20 times and the results related to the best response and the average value of responses. Is presented in statistical Table 8.6.

Also, the coefficient of variation of responses, which is a measure of the robustness and robustness of responses, has been calculated and used to compare the efficiency of turbulence functions and algorithms. For quick access to optimization information, a bar chart of each component is shown in Fig. 8.11. Examining the optimization results for different combinations of algorithms with turbulence functions and comparing it with the standard mode, shows a significant improvement in reducing the weight of A simply supported 37-bar planar truss. The enhanced results for Chaotic Artificial Bees Colony (CABC), by Logistic chaos map and scenario 2, a weight of 358.542 kg has the optimal response.

Table 8.6 Optimal design comparison for the simply supported 37-bar planar truss

Decision variable	Standard	Logistic map			Gauss map		
	ABC	CABC 21	CABC 22	CABC 23	CABC 31	CABC 32	CABC 33
Y_3, Y_{19} (m)	0.82953	0.83961	0.9225	0.87519	0.90304	0.87308	0.87519
Y_5, Y_{17} (m)	1.1901	1.2416	1.2354	1.2127	1.2774	1.2547	1.2505
Y_7, Y_{15} (m)	1.4068	1.4052	1.4013	1.3647	1.4696	1.4238	1.4447
Y_9, Y_{13} (m)	1.5111	1.5706	1.527	1.4752	1.5955	1.548	1.5596
Y_{11} (m)	1.5911	1.5865	1.5758	1.5474	1.6532	1.6188	1.6071
A_1, A_{27}	3.1993	3.6465	2.8729	2.6393	2.9543	3.2324	3.1037
A_2, A_{26}	1	1	1.0179	1	1.0325	1.0667	1.0726
A_3, A_{24}	1.093	1.007	1	1.0013	1.0082	1	1
A_4, A_{25}	2.8973	2.512	2.9496	3.1193	2.7716	2.7912	2.882
A_5, A_{23}	1	1.0881	1.1047	1.1588	1.1891	1.0995	1.1082
A_6, A_{21}	1.3123	1.2374	1.2011	1.1974	1.2653	1.2372	1.2989
A_7, A_{22}	3.3666	2.4746	2.7576	2.8613	2.7132	2.7608	2.686
A_8, A_{20}	1.2621	1.4182	1.1823	1.688	1.3598	1.2764	1.3605
A_9, A_{18}	1.4843	1.4142	1.4744	1.3687	1.4376	1.4577	1.437
A_{10}, A_{19}	2.9163	2.5512	2.6853	2.7725	2.5531	2.8363	2.7648
A_{11}, A_{17}	1.1861	1.2396	1.1828	1	1.2109	1.1216	1.1498
A_{12}, A_{15}	1.4432	1.3788	1.2808	1.3395	1.3486	1.3361	1.3432
A_{13}, A_{16}	2.6136	2.7482	2.6166	3.1787	2.5049	2.6952	2.5933
A_{14}	1.0588	1	1.0175	1	1.0101	1.034	1.0027
Best weight (kg)	360.832	359.624	**358.542**	359.452	359.953	359.815	359.782
Mean weight (kg)	361.71	360.413	**360.082**	360.107	360.147	360.17	360.173
Coefficient variation	0.3122	0.1602	0.2436	0.1993	0.1467	0.1059	**0.0663**
ω_1 (HZ)	20.001	20	20	20.011	20.005	20	20.024
ω_2 (HZ)	40	40	40.002	40.010	40.0303	40	40.012
ω_3 (HZ)	60.017	60	60	60.021	60.052	60	60.025

The bold items in the table indicate the optimal values for the best weight, the best mean and the best coefficient of variation

8.6.2.3 A 200-Bar Planar Truss Structure

The 200-bar planar truss structure as shown in Fig. 8.12 is a well-known benchmark problem for weight optimization with frequency limits. This truss only considers "optimizing the size of the sections" and the geometric shape of the structure is constant during the optimization process. The decision variables related to the

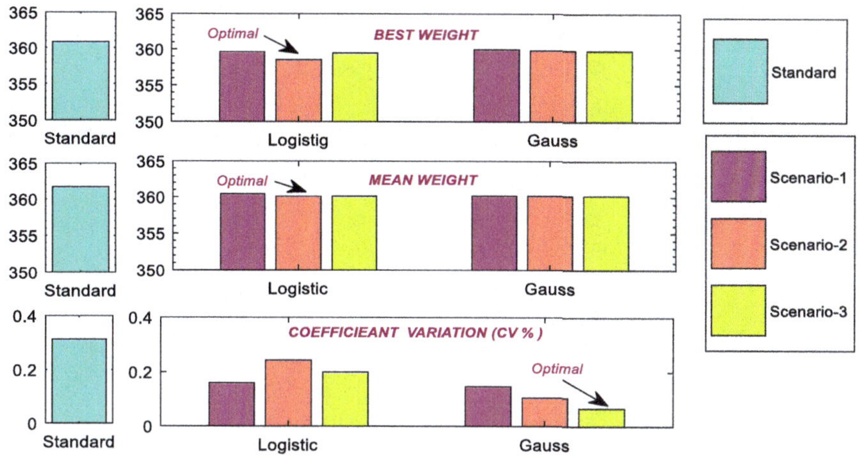

Fig. 8.11 Optimization results in standard mode and chaos map for the simply supported 37-bar planar truss

sections size of the members are classified into 29 groups according Table 8.7. Non-structural concentrated mass in nodes 1–5 and in the amount of 100 kg affect the structure. The mechanical characteristics of the structure are: material density 7860 kg/m^3, modulus of elasticity 210,000 MPa, frequency limits of the structure in the first, second and third modes are greater than 5, 10 and 15 Hz, respectively. For the cross section of the members, the lower limit range is 0.1 cm^2 and the upper limit is 25 cm^2. In order to ensure the performance of turbulent functions and algorithms, as well as to increase the accuracy and sensitivity of calculations, each of the modes has been performed independently 20 times and the results related to the best response and the average value of responses are presented in Statistical Table 8.8. Also, the coefficient of change of responses, which is a measure of the robustness and robustness of responses, has been calculated and used to compare the efficiency of turbulence functions and algorithms. For quick access to optimization information, a bar chart of each component is shown in Fig. 8.13.

Examining the optimization results for different combinations of algorithms with turbulence functions and comparing it with the standard mode, shows a significant and significant improvement in reducing the weight of the 200-bar planar truss structure. The enhanced results for Chaotic Artificial Bees Colony (CABC), by Logistic chaos map and scenario 2, a weight of 2125.214 kg has the optimal response.

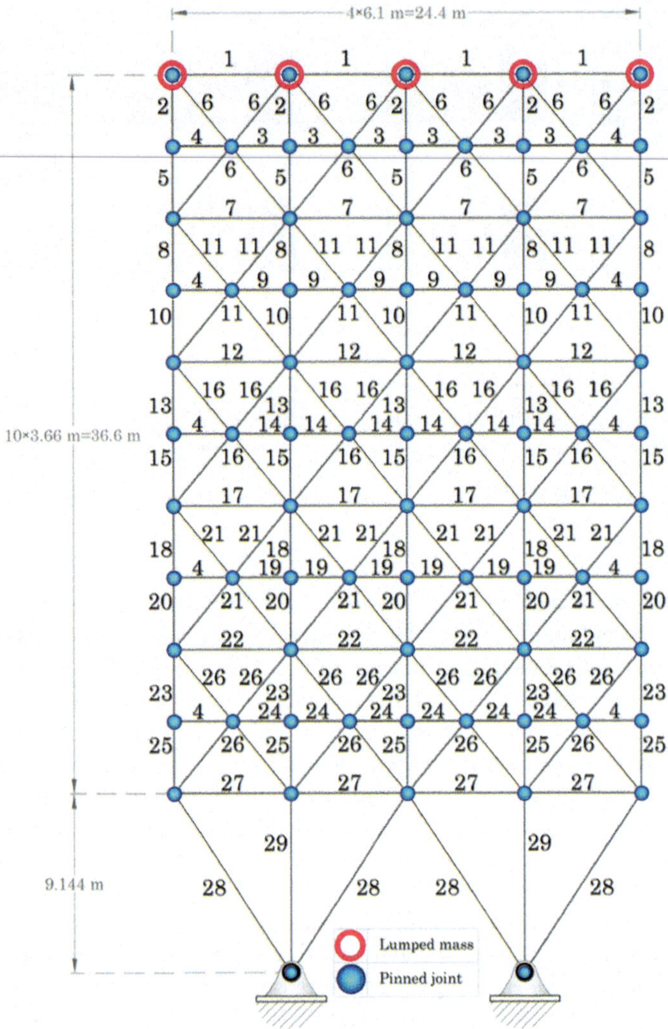

Fig. 8.12 Schematic of a 200-bar planar truss structure

8.7 Concluding Remarks

In this algorithm, it is inspired by the group behavior of bees to search for food. In fact, each proposed answer represents a food source and its quality is checked based on the objective function. The location of these food sources is improved by the bees in the iterative process of searching. Bees are placed in colonies in three groups with different responsibilities. These three groups include: recruited bees, onlooker bees and scout bees. Recruited bees try to find better food sources with a certain process, while onlooker bees search randomly based on the merit of food sites. The strategy

Table 8.7 Classification of members in the 200-bar planar truss structure

Group	Element group	Group	Element group
A_1	1, 2, 3, 4	A_{16}	82, 83, 85, 86, 88, 89, 91, 92, 103, 104, 106, 107, 109, 110, 112, 113
A_2	5, 8, 11, 14, 17	A_{17}	115, 116, 117, 118
A_3	19, 20, 21, 22, 23, 24	A_{18}	119, 122, 125, 128, 131
A_4	18, 25, 56, 63, 94, 101, 132, 139, 170, 177	A_{19}	133, 134, 135, 136, 137, 138
A_5	26, 29, 32, 35, 38	A_{20}	140, 143, 146, 149, 152
A_6	6, 7, 9, 10, 12, 13, 15, 16, 27, 28, 30, 31, 33, 34, 36, 37	A_{21}	120, 121, 123, 124, 126, 127, 129, 130, 141, 142, 144, 145, 147, 148, 150, 151
A_7	39, 40, 41, 42	A_{22}	153, 154, 155, 156
A_8	43, 46, 49, 52, 55	A_{23}	157, 160, 163, 166, 169
A_9	57, 58, 59, 60, 61, 62	A_{24}	171, 172, 173, 174, 175, 176
A_{10}	64, 67, 70, 73, 76	A_{25}	178, 181, 184, 187, 190
A_{11}	44, 45, 47, 48, 50, 51, 53, 54, 65, 66, 68, 69, 71, 72, 74, 75	A_{26}	158, 159, 161, 162, 164, 165, 167, 168, 179, 180, 182, 183, 185, 186, 188, 189
A_{12}	77, 78, 79, 80	A_{27}	191, 192, 193, 194
A_{13}	81, 84, 87, 90, 93	A_{28}	195, 197, 198, 200
A_{14}	95, 96, 97, 98, 99, 100	A_{29}	196, 199
A_{15}	102, 105, 108, 111, 114		

of recruited and onlooker bees is considered to perform exploration and exploitation steps. The existence of random distributions in these strategies is, on the one hand, to provide diversity in the search space, but it causes the optimization operation to stagnate and stop at the local optimal position. Therefore, chaos dynamic functions have a high potential to save the artificial bee colony algorithm. By embedding logistic, Gaussian, and Pace-Weiss chaos functions, local optima converge to global optima, and premature convergence is eliminated. By forming chaotic models, 9 different chaotic modes compete and the best response is compared with the standard mode. For problems with multi-frequency limitation, size and layout optimization, the standard mode of the artificial bee colony faces a wide stagnation, which is resolved in chaotic modes. By examining the chaotic state, the weaknesses and strengths of algorithms in dealing with optimization problems are determined.

Table 8.8 Optimal design comparison for the 200-bar planar truss structure

Number group	Standard	Logistic map			Gauss map		
	ABC	CABC 21	CABC 22	CABC 23	CABC 31	CABC 32	CABC 33
A_1	0.2635	0.328489	0.319768	0.302415	0.337738	0.296739	0.291457
A_2	0.518562	0.44407	0.492775	0.458246	0.486987	0.445526	0.66047
A_3	0.1	0.134728	0.128467	0.1	0.1	0.1	0.102233
A_4	0.110684	0.102673	0.1	0.1	0.1	0.1	0.148897
A_5	0.718629	0.549849	0.515959	0.548701	0.493796	0.566181	0.592174
A_6	0.820769	0.810343	0.912779	0.813143	0.821976	0.812191	0.825876
A_7	0.1	0.1	0.122802	0.1	0.100082	0.101664	0.108664
A_8	1.54501	1.52295	1.36426	1.43463	1.38102	1.4936	1.49153
A_9	0.123271	0.120967	0.1	0.102645	0.1	0.104589	0.102012
A_{10}	1.55616	1.43362	1.57586	1.56386	1.55248	1.64111	1.52397
A_{11}	1.09365	1.22419	1.16284	1.17889	1.1723	1.17852	1.2046
A_{12}	0.402115	0.108261	0.1	0.165523	0.175181	0.1	0.1
A_{13}	3.10977	2.80711	2.76204	2.97917	2.9764	3.04783	2.40077
A_{14}	0.1	0.223349	0.112442	0.1	0.102453	0.115514	0.186685
A_{15}	3.6571	3.42746	3.49766	3.29294	3.01091	3.17959	3.87405
A_{16}	1.78293	1.57793	1.56433	1.58496	1.56361	1.59409	1.49537
A_{17}	0.186635	0.149383	0.369972	0.21657	0.295435	0.304129	0.415404
A_{18}	4.42889	5.37992	4.99193	5.09198	5.17394	5.27336	4.96251
A_{19}	0.361134	0.1	0.157209	0.102444	0.1	0.13417	0.115867
A_{20}	5.82063	5.44708	5.30275	5.54898	5.54386	5.59457	4.67933

(continued)

Table 8.8 (continued)

Number group	Standard	Logistic map			Gauss map		
	ABC	CABC 21	CABC 22	CABC 23	CABC 31	CABC 32	CABC 33
A_{21}	1.96555	2.00029	2.1364	2.10788	2.13472	2.08368	2.31367
A_{22}	0.837621	0.51339	0.538075	0.68766	0.54547	0.816327	0.975726
A_{23}	7.21331	7.26212	7.1213	7.72273	7.58061	7.46041	7.91111
A_{24}	0.1	0.171203	0.480098	0.120448	0.224181	0.1	0.227007
A_{25}	8.00951	8.08962	7.48155	7.97364	8.08797	7.93347	9.49588
A_{26}	3.2264	2.8294	2.72518	2.74418	2.78697	2.85998	2.59252
A_{27}	9.78171	9.99703	10.7284	10.2809	10.4064	10.058	9.92082
A_{28}	20.0785	21.1901	20.4264	21.1739	21.0686	20.9215	20.7199
A_{29}	12.9077	10.0335	10.8459	11.004	11.1322	11.159	11.1406
Best weigh	2173.027	2132.403	**2125.214**	2155.989	2156.027	2154.701	2161.785
Mean weight	2196.935	2160.781	**2147.07**	2157.174	2157.878	2158.673	2168.495
Coefficient variation	0.7442	0.8161	0.7832	0.0311	0.0656	0.1055	0.3122
ω_1 (HZ)	5.0128	5.001	5	5.01	5.0042	5	5.0022
ω_2 (HZ)	12.209	12.5305	12.5042	12.2708	12.5271	12.2399	12.4603
ω_3 (HZ)	15.3319	15.1269	15.1578	15.0777	15.1122	15.094	15.3385

The bold items in the table indicate the optimal values for the best weight, the best mean and the best coefficient of variation

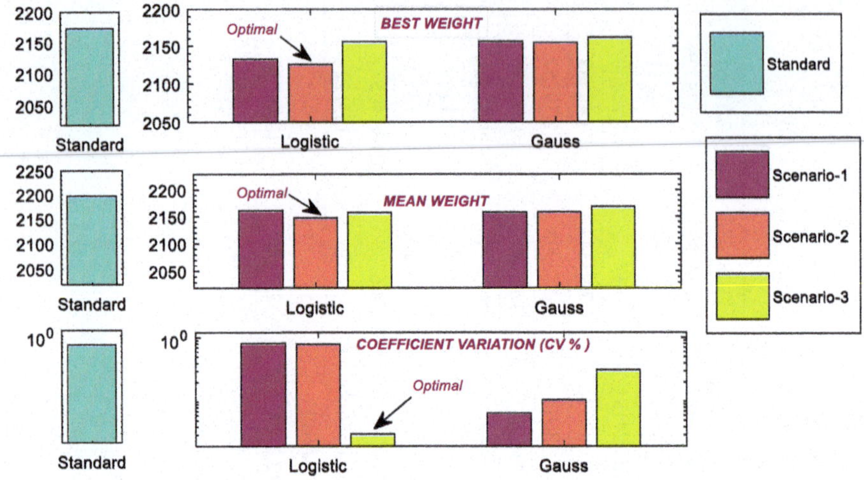

Fig. 8.13 Optimization results in standard mode and chaos map for the 200-bar planar truss structure

References

1. Kaveh, A.: Advances in Metaheuristic Algorithms for Optimal Design of Structures, 2nd edn. Springer, Cham, Switzerland (2021). https://doi.org/10.1007/978-3-319-05549-7
2. Karaboga, D.: An idea based on honey bee swarm for numerical optimization. Erciyes University, Kayseri, Turkey, Rep. TR06 (2005)
3. Kaveh, A., Yousefpoor, H.: Chaos-embedded meta-heuristic algorithms for optimal design of truss structures. Sci. Iranica A **29**(6), 2868–2885 (2022). https://doi.org/10.24200/sci.2022.59812.6441
4. Kaveh, A., Yosefpoor, H.: Chaotically enhanced meta-heuristic algorithms for optimal design of truss structures with frequency constraints. Periodica Polytech. Civ. Eng. **66**(3), 900–921 (2022). https://doi.org/10.3311/PPci.20220
5. Kaveh, A., Zarfam, P., Aziminejad, A., Yosefpoor, H.: Comparison of four chaotic meta-heuristic algorithms for optimal design of large-scale truss structures. Iran. J. Sci. Technol. Trans. Civ. Eng. https://doi.org/10.1007/s40996-022-00908-8
6. Ott, E.: Chaos in Dynamical Systems. Cambridge University Press, Cambridge (2002)
7. Bucolo, M., Caponetto, R., Fortuna, L., Frasca, M., Rizzo, A.: Does chaos work better than noise? IEEE Circuit Syst. Mag. (3), 4–19 (2002)
8. Peitgen, H.-O., Jürgens, H., Saupe, D.: "Chaos and Fractals": New Frontiers of Science. Springer, Berlin (2006)
9. Kaveh, A., Sheikholeslami, R., Talatahari, S., Keshvari-Ilkhichi, M.: Chaotic swarming of particles: a new method for size optimization of truss structures. Adv. Eng. Softw. **67**, 136–47 (2014)

Chapter 9
Chaotic Imperialist Competitive Algorithm

9.1 Introduction

The optimization algorithms that we have reviewed so far have been inspired by natural processes and they have not paid attention to other human evolutionary manifestations. In this chapter, the algorithm with a different inspiration is proposed for optimization, which imitates not a natural phenomenon, but a socio-political phenomenon. In this algorithm, it has specially looked at the process of colonization as a stage of human socio-political evolution and by mathematical modeling of this historical phenomenon, it has been used as the source of inspiration for a powerful and efficient algorithm in the fields of scientific optimization [1]. The Imperialist Competitive Algorithm (ICA), like other evolutionary algorithms, starts with a number of random initial populations that are made up of countries. Some of the best countries are chosen as imperialists and the rest of the countries are also known as their colonies. The Imperialist countries of each empire, depending on their power, pull the colonized countries towards them. In the competition between the empires, any empire that does not succeed, the number of its colonial countries will decrease and it will gradually be removed from the competition scene. Therefore, in the course of the Imperialist competition, the power of the big empires gradually increased and the weak empires decrease, and eventually fell. In this regard, the empires will have to develop the colonial countries to increase their power. As a result of this policy, countries move towards the absolute optimality of the objective function. The presented model for the assimilation policy of emperors is according to Fig. 9.1a. In this model, the movement is made in the direction of the connecting line between the colonies and the emperor, and the colonial country is dragged to its new position. The assimilation coefficient in this movement is introduced by β and is associated with a random distribution. This random distribution is proposed in the standard version of the algorithm of a uniform distribution. Although the empires countries have been seriously following the assimilation policy, but the events are not progressing completely according to their applied policy and there will be deviations

A. Kaveh and H. Yousefpoor, *Chaotic Meta-heuristic Algorithms for Optimal Design of Structures*, Studies in Computational Intelligence 1129,
https://doi.org/10.1007/978-3-031-48918-1_9

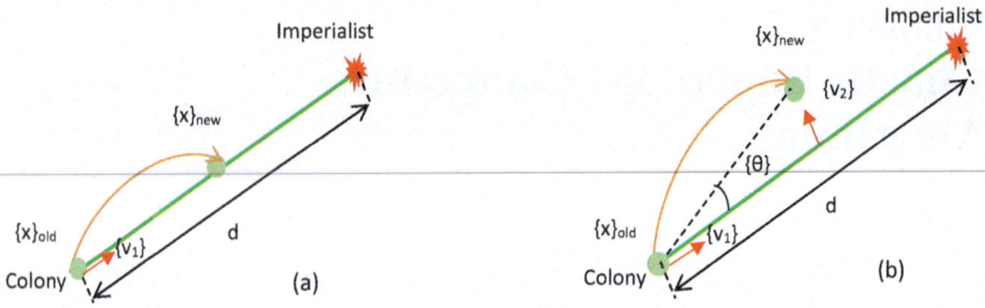

Fig. 9.1 Movement of colonies to its new location **a** based on assimilation policy **b** in a deviated direction

in the work. Accordingly, the previous movement will be accompanied by an angular deviation of θ. This deviation constitutes another step of the algorithm and is defined by adding the angle θ with a random distribution. Figure 9.1b shows the amount of deviation in the assimilation path towards the empire.

In order to determine the new location of the colony, after applying the items related to the assimilation policy and the amount of deviation, it will be according to Eq. (9.1).

$$\begin{aligned} \{x\}_{new} &= \{x\}_{old} + \beta \times d \times \{\boldsymbol{rand}\} \otimes \{v_1\} + U(-1, 1) \times \tan(\theta) \times d \times \{v_2\} \\ &\{v_1\} \cdot \{v_2\} = 0, \ \|\{v_2\}\| = 1 \end{aligned} \tag{9.1}$$

These two steps play the main role in the exploration and exploitation steps. The assimilation policy in imperialist countries and internal competition between empires will play a key role in finding local optima and form the role of algorithm extraction. On the other hand, deviating from the path of assimilation and in some cases external competition between emperors will cause unknown spaces of decision variables to be examined and will play the role of exploration for the algorithm [2]. In the standard version, it is proposed to select the range of β in the interval [0,2], which in this proposal, the value of β = 1 puts the colony exactly in the position of the imperialist, and numbers greater than 1 cause that there is no limit for the colonies and can Go beyond the set boundaries. Therefore, some colonies may reach a better position than the imperialists in the search for answers. A look at the ups and downs of the rotation of power in countries clearly shows that countries that were once at the peak of political-military power fell after a while, and on the other hand, countries that once had no power took over. The suggestion of an upper bound of 2 for β is so that the choice is accessible to the relative state, and the statistical mean of this choice lies in an infinite number of moves right on the emperor. From a statistical point of view, the mean, median and expectation to the assimilation policy will be imperialist. In order to improve the results and create interaction between the two exploration and exploitation stages, chaotic functions replace the random distribution of assimilation and deviation steps. These functions transfer the local optima to global optima by creating chaotic jumps and save the algorithm from the trap of local optima.

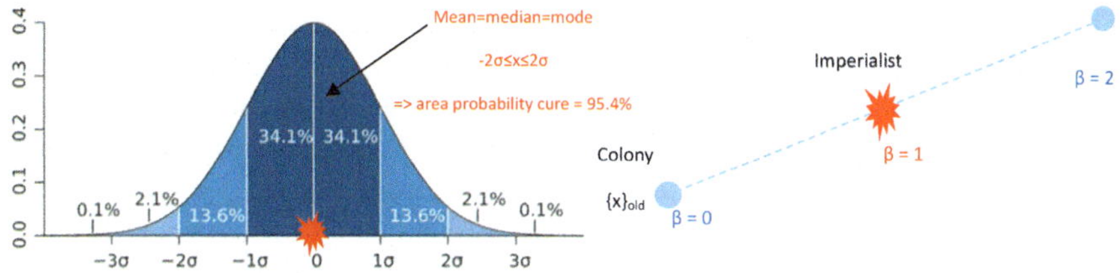

Fig. 9.2 The emperor's position in the place of the mean, median and expectation of the normal distribution

In chaotic Imperialist Competitive Algorithm (CICA), premature convergence is eliminated and we can obtain improved results in a small number of iterations. Also, for problems with high complexity (NP-Hard), non-linear and non-convex, the algorithm should continue processing until reaching the sub optima without stopping at local optima. In this chapter, Logistic, Gaussian, and Piecewise chaos functions have been chosen to form chaotic meta-heuristic algorithms. Also, for the purpose of evaluation, optimization of structures with static loads and well-known structural problems have been considered. For optimization with multi-frequency constraints, logistic and Gaussian chaos functions have resulted in significant improvement and we did not need to check the third chaos function (Fig. 9.2).

In the rest of this chapter, after introducing the colonial competition algorithm, the basic steps of the algorithm will be presented in the standard mode, and then the chaotic mode of the colonial competition algorithm will be formed with three scenarios. In the end, the effectiveness of chaos functions will be challenged with famous optimization examples of static constraints and multi-frequency constraints.

9.2 Standard Imperialist Competitive Algorithm (ICA)

The source of inspiration for this algorithm is the political, cultural and social sphere and examines the formation process of empires and colonial countries in the world of politics. This algorithm was proposed by Atashpoz et al. [3]. In this algorithm, first, an initial population of countries is formed, and among them, a number of the best countries are introduced as emperors. In the development of the superior power, the emperors think about capturing colonies and increasing the territory of their empire. Therefore, they apply the policy of attracting colonial countries. The countries of each empire are competing with each other to bring their conditions to the emperor, but the main competition is between the emperors themselves. Therefore, just like a sports team, at the same time that team members are trying to reach the captain and have an intra-group competition among themselves, but the main competition is the out-group competition and between teams. Therefore, we have two types of competition, including intra-group and extra-group. At this stage, the weak colonial countries move towards the emperor in order to improve their level of progress.

This stage is known as the assimilation policy stage and the coefficient chosen for movement is between zero and two ($0 \leq \beta \leq 2$). In the next step, in order to search for unknown places in space, the Revolution step is executed and a random change in the answers is obtained. At this stage, various answers are created to get access to better answers. By applying this step, if the results were successful, it is accepted. Now, in the competition within the group, if a country is found that is better than the existing emperor, it will replace the emperor and the previous country will be considered as a colony. In inter-group competition, emperors compete, and in the comparison index, the total value of the objective function for the emperor and a percentage of the average value of the objective function for the colonies are considered. This percentage of colonies' participation in the original version is suggested as ten percent ($\xi = 0.1$). In this competition, the weakest empire loses one of its colonies and the strongest empire takes over it. In this process, if the empire has no clones, that empire will fall. In the iterative process of steps, if only one emperor is left, the operation is stopped and that empire is introduced as the final answer of this period.

9.3 Basic Steps in Imperialist Competitive Algorithm

Step 1 Formation of the initial population of the countries: In the search space, the initial population of countries is formed randomly. A few of the best countries are designated as empires, and others are designated as colonies.
Step 2 Applying the policy of assimilation colonialists: The difference vector is formed along each colony and its corresponding emperor, and by applying the absorption coefficient β, a random number with a certain distribution function and revolution phase, the new position of the colonies is determined. The categories of Eq. (9.2) are related to the implementation of the policy of assimilation colonialists:

$$\begin{aligned} & d = eC_{ij} - EC_{ij} \\ & Assimilation = \beta \times d \otimes \boldsymbol{rand}(size(EC\{i\}\{j,:\})) \\ & Revolution = -\gamma + 2 \times \gamma \otimes \boldsymbol{rand}(size(EC\{i\}\{j,:\})) \\ & i = 1, 2, \ldots, nE;\ j = 1, 2, \ldots, nCOL(i) \end{aligned} \tag{9.2}$$

In these relations, d is the vector of the distance between each colony and its corresponding emperor, β is the assimilation coefficient that is chosen in the range of 0–2 and here 2 is meant, EC and eC are respectively the position of the colony and the emperor corresponding to that colony, γ is the deviation angle. Which is proposed in the original version of 0.1 radians, Assimilation and Revolution respectively, are the values of absorption and mutation, newEC is the new position of the colony, *i* is the number of the emperor and finally *j* is the number of colonies related to the same emperor.

Step 3 intra-group competition: At this stage, if a colony is better than its emperor, their places are exchanged.
Step 4 inter-group competition: The TC index for each empire is calculated according to Eq. (9.3) and based on it, a colony is removed from the weakest empire and transferred to the empire with the highest probability of acquisition.

$$TC_i = ePFit_i + \xi \times mean(EPFit\{i\}\{j, :\})$$
$$i = 1, 2, \ldots, nE; j = 1, 2, \ldots, nCOL(i) \tag{9.3}$$

In this regard, ξ is the percentage of colony participation, i is the number of the emperor, and j is the number of colonies related to the same emperor.
Step 5 Collapse of the weak emperor: If we are in a situation where the weakest empire does not have any colonies, that empire is collapse and removed from the list of empires.
Step 6 Termination conditions: If there is only one empire left, the operation is stopped and the winner of the competition is announced, otherwise the operation is repeated from step 2.

9.4 Chaos-Embedded Imperialist Competitive Algorithm (CICA)

In this algorithm, to determine the new position of each colony in its respective empire, two important strategies of assimilation and revolution are performed [4–6]. These strategies play the role of exploration and exploitation steps in the algorithm. By replacing chaos maps in their related random selections, we will see a significant improvement in algorithm optimization results. The suggested scenarios for this replacement are as follows:

Scenario 1 Replacing the chaos map in the assimilation strategy stage: In this scenario, the first **CHM1** chaos map replaces the random selection of the assimilation stage in Eq. (9.2) and the result is obtained according to Eq. (9.4):

$$Assimilation = \beta \times d \otimes \boldsymbol{CHM1}(size(EC\{i\}\{j, :\})) \tag{9.4}$$

Scenario 2 Replacing the chaos map in the revolution strategy phase: In this case, the second **CHM2** chaos map replaces the random selection of the revolution phase in Eq. (9.3) and the result will be according to Eq. (9.5):

$$Revolution = -\gamma + 2 \times \gamma \otimes \boldsymbol{CHM2}(size(EC\{i\}\{j, :\})) \tag{9.5}$$

Scenario 3 Replacement of chaos maps in both stages of assimilation and revolution simultaneously: In this case, both chaos maps are simultaneously replaced by the random distributions of the algorithm in Eqs. (9.2) and (9.3).

9.5 Truss Weight Optimization with Static Constraints

By embedding chaos functions in meta-heuristic algorithms, a significant improvement in optimization results is obtained. For this purpose, we use the Logistic, Gossin and Piecewise chaos functions in the Imperialist Competitive Algorithm (CICA). With the formation of chaotic models, the competition between the chaotic functions to achieve the best optimal weight is carried out. To search for the best efficiency of each of the chaos functions in meta-heuristic algorithms, three scenarios are formed. In these scenarios, chaos functions replace random distributions corresponding to exploration, exploitation, or both phases simultaneously. For each structural model, 9 chaotic models are formed and compared with the standard state. Also, in order to form a statistical space and increase diversity in the research, each model is independently evaluated 30 times. To analyze these statistical results, the best weight, the best mean and the coefficient of variation are presented. In a number of models, the use of chaos functions has increased the coefficient of variation. This behavior shows sudden mutagenesis and increasing difference between best and worst response. Due to the wide scope of investigations, there is a strong competition between chaos functions and as a result, the possibility of approaching the global optimum increases. This optimum cannot be far from the crosshairs of chaotic mutations. In each investigation, the chaos function compatible with the algorithm and the scenario corresponding to the best improvement are selected. In the Logistic chaos map, it most likely converges from local minima to general minima [7]. Therefore, this function is suitable for improving the exploration conditions of algorithms. The Gaussian chaos map is very likely to be within the local minimums and is suitable for improving exploitation conditions [8]. Finally, Piecewise map simultaneously improves both conditions [9]. Therefore, by selecting these chaos maps, the weakness of algorithms of any kind will be repaired and global optimum will not have the opportunity to escape the chaos functions. We can also find out the strengths and weaknesses of the algorithm in the stages of exploration or exploitation.

9.5.1 Formulation of the Structural Optimization Problems

Optimization problems for truss structures include objective function, design constraints, and bound of decision variables. In optimization relationships, the weight of the structure is selected as the objective function and all design constraints must be satisfied with the least possible weight. The general form related to these problems is defined as Eq. (9.6):

$$\begin{aligned} &Find && A = \{A_1, A_2, A_3, \ldots, A_n\} \\ &to\ Minimize && W(A) = \textstyle\sum_{i=1}^{n} \gamma_i \times A_i \times L_i \\ &Subjected\ to && g_j(A) \le 0\ ;\ \ j = 1, 2, 3, \ldots, m \\ & && h_k(A) \le 0\ ;\ \ k = 1, 2, 3, \ldots, p \\ & && \{A_L\} \le \{A\} \le \{A_U\} \end{aligned} \tag{9.6}$$

In this category of Eq. (9.1), A is the cross-section vector of the members, W is the total weight of the structure, n is the number of members of the structure, g_i and h_i are design constraints that can include stress, slender constraints and node displacement. Also, A_l and A_u are the lower and upper bounds of the decision variables. Despite the selected algorithms are used to optimize unbounded problems. In modeling, the penalty function method is used to convert the bound function to the unbound function. In this method, if there is no violation, the amount of the fine will be zero. Otherwise, and if there is a violation, the value of the penalty function is obtained from Eqs. (9.7)–(9.11):

$$\sigma_i \le \sigma^{\max} \Rightarrow V_i = \max(0, \frac{\sigma_i}{\sigma^{\max}} - 1);\quad i = 1, 2, 3, \ldots, n \tag{9.7}$$

$$\delta_j \le \delta^{\max} \Rightarrow V_j = \max(\frac{\delta_j}{\delta^{\max}} - 1);\quad j = 1, 2, 3, \ldots, n \tag{9.8}$$

$$\lambda_k \le \lambda^{\max} \Rightarrow V_k = \max(0, \frac{\lambda_k}{\lambda^{\max}} - 1);\quad k = 1, 2, 3, \ldots, p \tag{9.9}$$

$$F_{penalty}(A) = 1 + \gamma_p \times \left\{ \sum_{i=1}^{m} V_i + \sum_{j=1}^{n} V_j + \sum_{k=1}^{p} V_k \right\} \tag{9.10}$$

$$to\ Minimize\quad Mer(A) = W(A) \times F_{penalty}(A) \tag{9.11}$$

Equations (9.7), (9.8) and (9.9) are related to stress, displacement and slenderness, respectively. The penalty function is presented in Eq. (9.10) and the objective function is formed after the penalty (merit function) in Eq. (9.11).

9.5.2 Introduction of Selected Chaos Map

Chaos dynamic series are extracted by sequentially forming sentences of chaos functions. Then, the formed series are embedded in the Imperialist Competitive Algorithm (CICA). The optimization results with this algorithm show that the local optima converge to the global optimum and these series will have a good potential to prevent meta-heuristic algorithms from getting trapped in the local optima. With chaotic meta-heuristic algorithms, the non-linear and non-convex behavior of the objective function in the optimization of truss structures can be easily controlled and adjusted. Among the advantages of chaotic series is non-repetition and non-convergence towards a certain limit. In the standard case of Imperialist Competitive Algorithm (ICA), the movement of the colonized countries towards the empire (assimilation phase) and the act of sudden revolution take place with a uniform random distribution. Therefore, in problems with many variables and complex conditions, optimization with standard mode will lead to premature convergence. Chaotic series converge local optima to global optima with sudden jumps.

9.5.3 Numerical Examples of Optimal Truss Design

To evaluate the Chaotic Imperialist Competitive Algorithm (CICA), well-known examples of truss optimization are selected as benchmark problems. By comparing the performance of chaotic models and choosing the best chaos function, the competition between 9 chaos responses is formed. Also, in order to know the quality and quantity of answers, the results of the best chaotic model are compared with the standard mode. Due to the variety of reviews, the optimal responses will be largely close to the global optimality. The examples provided are as follows:

- The 25-bar spatial truss
- The 47-bar power transmission tower
- The 200-bar planar truss structure
- The 582-bar tower truss structure.

9.5.3.1 A 25-Bar Spatial Truss

The topology and nodal numbering of the 25-bar spatial truss is as shown in Fig. 9.3. The material density is 0.1 lb/in^3 and the modulus of elasticity of the members is 10,000 ksi.

The structure includes 25 members, which are divided into eight groups according to Table 9.1. Also, allowable compressive and tensile stresses have been determined for each type. This spatial truss is subjected to two Loading conditions as shown in Table 9.2. Also, the maximum displacement limitations for the nodes in all directions of the axes is $\pm$ 0.35 in. The range of choice of members' cross-section or range of decision variables is in the range of 0.01–3.4 in^2.

In order to ensure the performance of chaotic maps and algorithms, as well as to increase the accuracy and sensitivity of calculations, each of the modes has been performed independently 30 times and the results related to the best response and the average value of responses are presented in Table 9.3. The graph of these results is also shown in Fig. 9.4. The coefficient of variation of the responses, which indicates the stability and robustness of the responses, has been calculated and another criterion for the efficiency of chaos functions and algorithms has been obtained. Examining the optimization results for different combinations of algorithms with chaos map

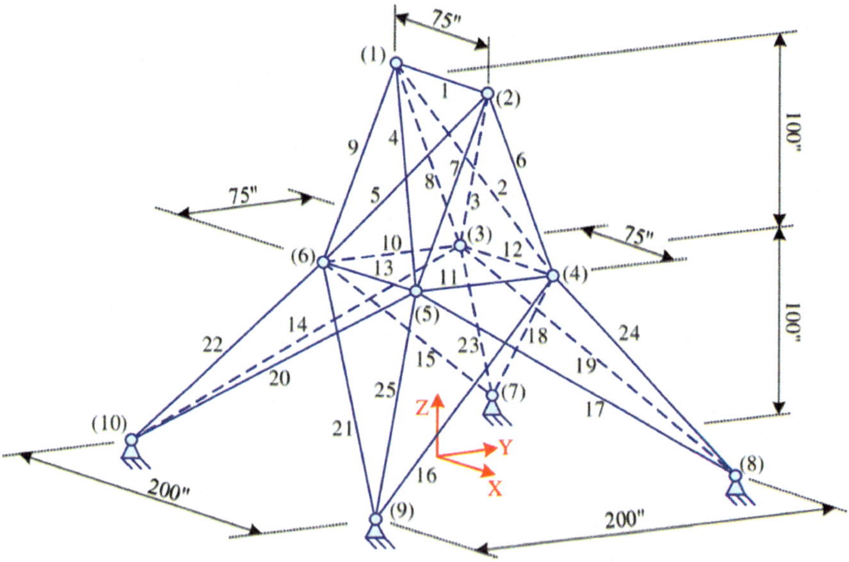

Fig. 9.3 Schematic of a twenty-five-bar spatial truss

Table 9.1 Grouping of members and stresses limitation for the 25-member spatial truss

Number group	Element group	Compressive stress limitation (ksi)	Tensile stress limitation (ksi)
1	A_1	35.092	40
2	A_2–A_5	11.590	40
3	A_6–A_9	17.305	40
4	A_{10}–A_{11}	35.092	40
5	A_{12}–A_{13}	35.092	40
6	A_{14}–A_{17}	6.759	40
7	A_{18}–A_{21}	6.959	40
8	A_{22}–A_{25}	11.082	40

Table 9.2 Loading condition for the 25-member spatial truss

Node point	Load case 1			Load case 2		
	F_x (kips)	F_y (kips)	F_z (kips)	F_x (kips)	F_y (kips)	F_z (kips)
1	0	20	− 5	1	10	− 5
2	0	− 20	− 5	0	10	− 5
3	0	0	0	0.5	0	0
4	0	0	0	0.5	0	0

and comparing it with standard modes, shows a significant improvement in reducing the weight of the 25-bar spatial truss. The enhanced results for Chaotic Imperialist Competitive Algorithm (CICA), by Logistic chaos map and scenario 2, a weight of 544.2739 lb has the optimal response.

9.5.3.2 A 47-Bar Power Transmission Tower

The topology and nodal numbering of the power transmission tower that consists of 47 members and 22 nodes, as shown in Fig. 9.5. The density of material is 0.3 lb/in^3 and the modulus of elasticity of the members is 30,000 ksi. Both stress and buckling limits must be satisfied for all member. The allowable stress of tensile members is 20 ksi and for compressive members 15 ksi is recommended. Euler buckling compressive stress for members is also calculated from the following equation.

$$\sigma_i^e = \frac{-KEA_i}{L_i^2}; \quad i = 1, 2, \ldots, 47 \tag{9.12}$$

Where K is a constant coefficient that must be selected according to the geometric type of the section, E is the modulus of elasticity of the material, A_i is cross-sectional area of member and L_i is the length of member. Here we consider k to be 3.96. Structural nodes are affected by a combination of triple loading modes. The first group includes loads of 6 kip in the positive direction of the X-axis and 14 kip in the negative direction of the Y-axis, which are acting in nodes 17 and 22. The second group consists of loads of 6 kg in the positive direction of the X-axis and 14 kg in the negative direction of the Y-axis, which acting only at node 17.

The third group includes loads of 6 kP in the positive direction of the X-axis and 14 kP in the negative direction of the Y-axis, which acting only in node 22. The first group represents the oblique loads of both power transmission lines and the second and third groups represent the state where one of the two lines breaks. Also, truss members are classified into 27 groups according to geometric symmetry. The cross-section areas of the members, were chosen from the 64 discrete values of the AISC code. In order to ensure the performance of chaos map and algorithms, as well as to increase the accuracy and sensitivity of calculations, each of the modes has been performed independently 30 times and the results related to the best response and the average value of responses are presented in Table 9.4. The graph of these results is also shown in Fig. 9.6. The coefficient of variation of the responses, which indicates the stability and robustness of the responses, has been calculated and another criterion for the efficiency of chaos maps and algorithms has been obtained. By examining the optimization results for different combinations of algorithms with chaos map and comparing it with the standard mode, a significant improvement in reducing the weight of the 47-bar power transmission tower is achieved. The enhanced results for Chaotic Imperialist Competitive Algorithm (CICA), by Logistic chaos map and scenario 2, a weight of 2354.303 lb has the optimal response.

Table 9.3 Optimal design comparison for the 25-bar spatial truss

Number group	Element group	Standard	Logistic map			Gauss map			Piecewise map		
		ICA	CICA 21	CICA 22	CICA 23	CICA 31	CICA 32	CICA 33	CICA 41	CICA 42	CICA 43
1	A_1	0.086923	0.022927	0.01	0.01	0.01206	0.64068	0.01	0.032296	0.01	0.01056
2	A_2–A_5	2.1883	1.9577	2.0215	1.8951	1.8521	2.1928	2.0003	1.8653	1.5531	1.7967
3	A_6–A_9	2.7537	3.0162	2.9475	3.1219	3.0727	2.6857	2.9777	3.1713	3.2098	3.2109
4	A_{10}–A_{11}	0.01	0.010001	0.010066	0.01	0.01	0.01	0.01	0.01	0.01	0.01005
5	A_{12}–A_{13}	0.01	0.01	0.016165	0.025599	0.01	0.01	0.01	0.01	0.01	0.010001
6	A_{14}–A_{17}	0.7119	0.72533	0.67361	0.7902	0.68242	0.73866	0.64881	0.66277	0.6803	0.7771
7	A_{18}–A_{21}	1.6296	1.6894	1.6699	1.6912	1.7708	1.6167	1.6757	1.6987	2.0143	1.742
8	A_{22}–A_{25}	2.6986	2.6045	2.6891	2.5129	2.6195	2.6463	2.7148	2.6278	2.4505	2.4828
	Best weight	546.5506	544.5207	**544.2739**	545.8455	544.9381	545.2392	544.3579	544.7809	544.636	545.4103
	Mean weight	551.7089	546.0748	548.6032	548.097	549.0756	547.3528	**545.5663**	549.9591	547.0149	548.745
	CV	1.3349	0.39625	1.0462	1.41809	0.96075	0.31247	**0.21299**	0.87189	0.44585	0.48315

The bold items in the table indicate the optimal values for the best weight, the best mean and the best coefficient of variation

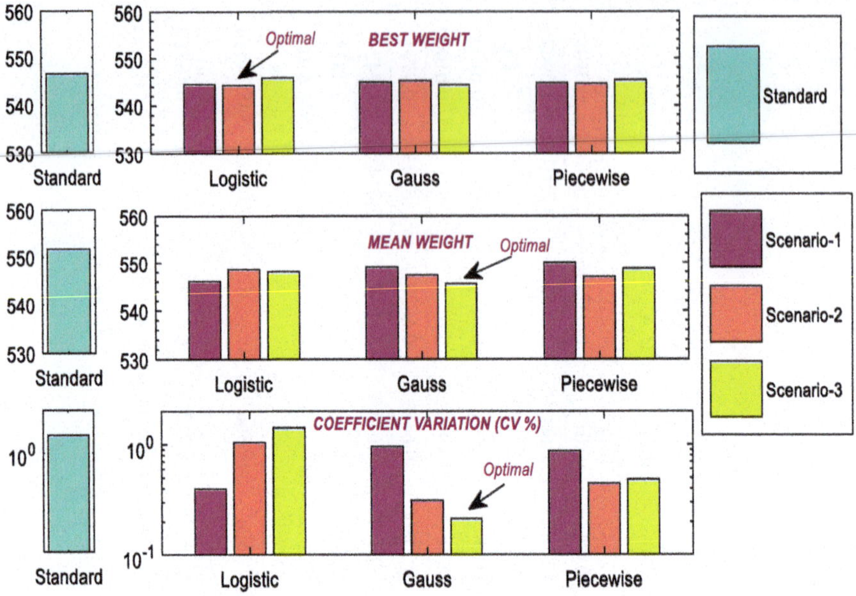

Fig. 9.4 Optimization results in standard mode and selection of chaos map for the 25-member spatial truss

9.5.3.3 A 200-Bar Planar Truss Structure

The topology and nodal numbering of the 200-bar planar truss structure are shown in Fig. 9.7. The material density is 0.283 lb/in^3 and the module of elasticity is 30,000 ksi. Allowable stress of members 10 klb has been assumed. No deformation restrictions are intended for optimization. Truss loading is done in three independent groups. The first group is the lateral load of the structure, which includes a load of 1 kip along the positive axis of Xs and is applied in nodes 1, 6, 15, 20, 29, 34, 43, 48, 57, 62 and 71. The second group covers the gravity load of the structure, which includes a load of 10 kip in the negative direction of the axis of the Ys and in nodes 1, 2, 3, 4, 5, 6, 8, 10, 12, 14, 15, 16, 17, 18, 19 and… 71, 72, 73, 74 and 75 apply. The third group applies both loading groups together. Truss members are also classified into 29 groups according Table 9.5. The minimum cross section of the members, which is the lower limit of the decision variables, is considered 0.1 in^2 and the maximum cross section is 16 in^2. In order to ensure the performance of chaotic maps and algorithms, as well as increasing the accuracy and sensitivity of calculations, each state is independently executed 30 times and the results of the best response and the average amount of responses are presented in statistical Table 9.6, and shown in Fig. 9.8. Also, the coefficient of changes in responses, which indicates the stability and robustness of the responses, has been calculated and another criterion for the efficiency of chaotic functions and algorithms has been achieved.

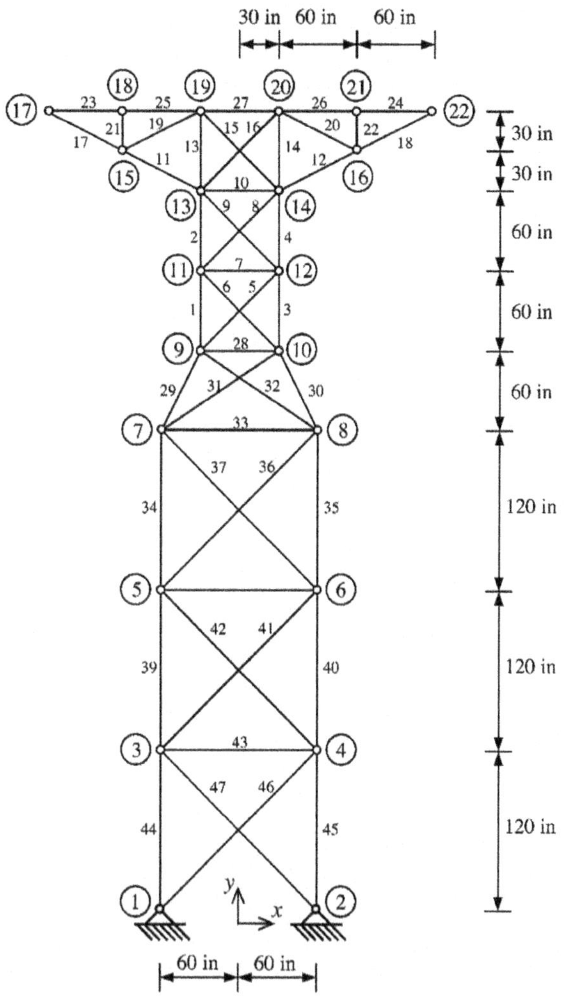

Fig. 9.5 Schematic of a 47-bar power transmission tower

The enhanced results for Chaotic Imperialist Competitive Algorithm (CICA), by Gauss chaos map and scenario 1, a weight of 25,480.081 lb has the optimal response.

9.5.3.4 A 582-Bar Tower Truss Structure

The topology and nodal numbering of the 582-bar tower truss structure are shown in Fig. 9.9. The optimization of this tower with continuous variables has been done for the first time by Kaveh et al. The modulus of elasticity of the members is 203893.6 MPa and their yield stress is 253.1 MPa. The displacement limitation is set at 8 cm for all nodes and in all direction. A load case group is considered that includes 5 kN along both the X and Y axes and – 30 kN along the Z axis. This

Table 9.4 Optimal design comparison for the 47-bar power transmission tower

Number group	Element group	Standard	Logistic map			Gauss map			Piecewise map		
		ICA	CICA 21	CICA 22	CICA 23	CICA 31	CICA 32	CICA 33	CICA 41	CICA 42	CICA 43
1	A_1–A_3	3.81	3.87	3.81	3.84	3.81	3.81	3.81	3.84	3.84	3.84
2	A_2–A_4	3.38	3.38	3.38	3.38	3.38	3.47	3.38	3.38	3.38	3.38
3	A_5–A_6	0.766	0.766	0.766	0.785	0.994	0.766	0.994	0.785	0.785	0.994
4	A_7	0.111	0.111	0.111	0.25	0.25	0.111	0.196	0.141	0.111	0.196
5	A_8–A_9	0.994	0.994	0.785	0.994	0.994	0.994	0.994	0.994	0.994	0.994
6	A_{10}	2.13	1.99	1.85	1.99	2.13	1.8	1.99	1.99	1.99	1.8
7	A_{11}–A_{12}	2.13	2.13	2.13	2.13	2.13	2.13	2.13	2.13	2.13	2.13
8	A_{13}–A_{14}	1.457	1.228	1.228	1.228	1.457	1.33	1.228	1.457	1.35	1.228
9	A_{15}–A_{16}	1.8	1.62	1.563	1.563	1.563	1.62	1.62	1.563	1.8	1.62
10	A_{17}–A_{18}	2.13	2.13	2.13	2.13	2.13	2.38	2.13	2.13	2.13	2.13
11	A_{19}–A_{20}	0.111	0.111	0.111	0.111	0.111	0.111	0.111	0.141	0.141	0.141
12	A_{21}–A_{22}	0.111	0.111	0.111	0.111	0.141	0.111	0.141	0.111	0.111	0.111
13	A_{23}–A_{24}	1.8	1.8	1.8	1.8	1.8	1.8	1.8	1.8	1.8	1.8
14	A_{25}–A_{26}	1.8	1.8	1.8	1.8	1.8	1.8	1.8	1.8	1.8	1.8
15	A_{27}	2.62	1.563	1.457	1.457	1.457	2.13	1.563	1.457	1.99	1.8
16	A_{28}	0.994	1.228	0.543	0.563	1.13	0.766	0.563	0.442	0.563	0.391
17	A_{29}–A_{30}	3.84	3.84	3.63	3.63	3.84	3.65	3.63	3.63	3.63	3.63
18	A_{31}–A_{32}	1.457	1.457	1.457	1.457	1.457	1.563	1.563	1.457	1.563	1.563
19	A_{33}	0.391	0.25	0.25	0.307	0.25	0.25	0.994	0.307	0.307	0.563
20	A_{34}–A_{35}	3.05	3.09	3.09	3.09	3.05	3.05	3.05	3.09	3.09	3.05

(continued)

Table 9.4 (continued)

Number group	Element group	Standard	Logistic map			Gauss map			Piecewise map		
		ICA	CICA 21	CICA 22	CICA 23	CICA 31	CICA 32	CICA 33	CICA 41	CICA 42	CICA 43
21	A_{36}–A_{37}	1.457	1.266	1.228	1.266	1.266	1.457	1.457	1.266	1.266	1.457
22	A_{38}	0.111	0.25	0.371	0.307	0.141	0.391	0.111	0.307	0.25	0.111
23	A_{39}–A_{40}	3.71	3.68	3.68	3.72	3.68	3.72	3.84	3.73	3.68	3.70
24	A_{41}–A_{42}	1.563	1.62	1.563	1.62	1.62	1.62	1.563	1.563	1.563	1.563
25	A_{43}	0.25	0.25	0.111	0.111	0.102	0.196	0.141	0.196	0.25	0.111
26	A_{44}–A_{45}	4.58	4.87	4.54	4.68	4.71	4.55	4.53	4.56	4.61	4.58
27	A_{46}–A_{47}	1.62	1.563	1.457	1.457	1.563	1.457	1.563	1.457	1.563	1.457
	Best weight	2470.583	2440.081	**2354.303**	2394.293	2433.109	2438.111	2453.716	2389.392	2426.403	2419.344
	Mean weight	2473.513	2461.016	**2356.687**	2406.46	2449.506	2450.827	2465.527	2395.594	2438.943	2422.641
	CV	0.05617	0.84259	**0.04744**	0.21540	0.2513	0.2271	0.2259	0.17306	0.23161	0.05347

The bold items in the table indicate the optimal values for the best weight, the best mean and the best coefficient of variation

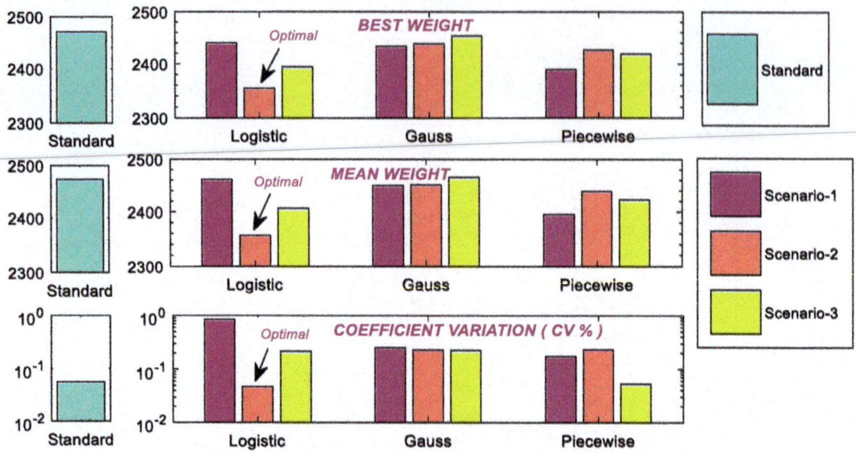

Fig. 9.6 Optimization results in standard mode and selection of chaos map for the 47-bar power transmission tower

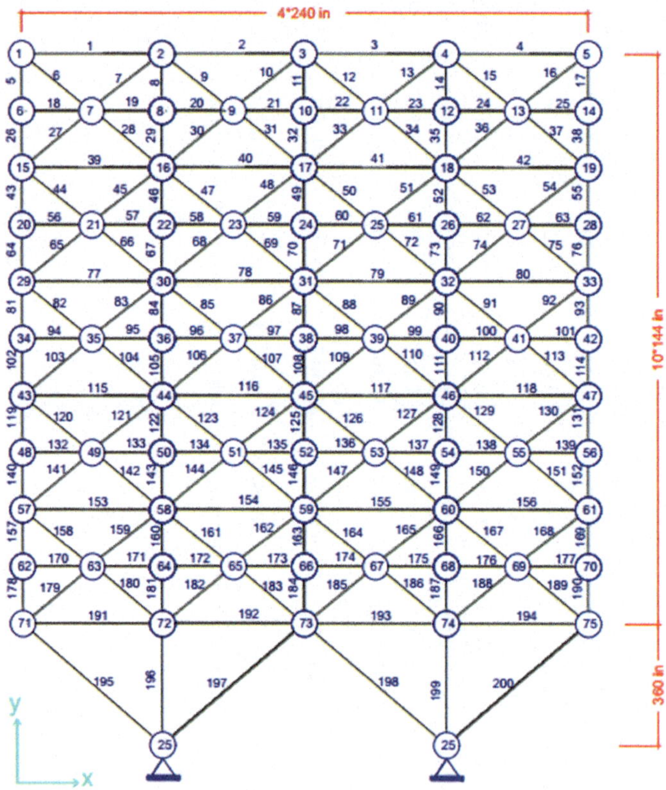

Fig. 9.7 Schematic of a 200-bar planar truss structure

Table 9.5 Classification of members in the 200-bar planar truss structure

Group	Element group	Group	Element group
A_1	1, 2, 3, 4	A_{16}	82, 83, 85, 86, 88, 89, 91, 92, 103, 104, 106, 107, 109, 110, 112, 113
A_2	5, 8, 11, 14, 17	A_{17}	115, 116, 117, 118
A_3	19, 20, 21, 22, 23, 24	A_{18}	119, 122, 125, 128, 131
A_4	18, 25, 56, 63, 94, 101, 132, 139, 170, 177	A_{19}	133, 134, 135, 136, 137, 138
A_5	26, 29, 32, 35, 38	A_{20}	140, 143, 146, 149, 152
A_6	6, 7, 9, 10, 12, 13, 15, 16, 27, 28, 30, 31, 33, 34, 36, 37	A_{21}	120, 121, 123, 124, 126, 127, 129, 130, 141, 142, 144, 145, 147, 148, 150, 151
A_7	39, 40, 41, 42	A_{22}	153, 154, 155, 156
A_8	43, 46, 49, 52, 55	A_{23}	157, 160, 163, 166, 169
A_9	57, 58, 59, 60, 61, 62	A_{24}	171, 172, 173, 174, 175, 176
A_{10}	64, 67, 70, 73, 76	A_{25}	178, 181, 184, 187, 190
A_{11}	44, 45, 47, 48, 50, 51, 53, 54, 65, 66, 68, 69, 71, 72, 74, 75	A_{26}	158, 159, 161, 162, 164, 165, 167, 168, 179, 180, 182, 183, 185, 186, 188, 189
A_{12}	77, 78, 79, 80	A_{27}	191, 192, 193, 194
A_{13}	81, 84, 87, 90, 93	A_{28}	195, 197, 198, 200
A_{14}	95, 96, 97, 98, 99, 100	A_{29}	196, 199
A_{15}	102, 105, 108, 111, 114		

load is applied to all free nodes of the tower. The slenderness ratio is limited to 300 for tension member and 200 for compression member and maximum stress of truss elements are imposed according to the provisions of AISC ASD. The truss members are grouped into 32 symmetrical groups. The minimum cross-sectional area for all member is 20 cm^2 and maximum cross-sectional area is 1000 cm^2.

In order to ensure the performance of chaotic maps and algorithms, as well as increasing the accuracy and sensitivity of calculations, each state is independently executed 30 times and the results of the best response and the average amount of responses are presented in statistical Table 9.7, and shown in Fig. 9.10. Also, the coefficient of changes in responses, which indicates the stability and robustness of the responses, has been calculated and another criterion for the efficiency of chaotic functions and algorithms has been achieved. The enhanced results for Chaotic Imperialist Competitive Algorithm (CICA), by Piecewise chaos map and scenario 2, a weight of 15.6083 m^3 has the optimal response.

Table 9.6 Optimal design comparison for 200-bar planar truss structure

Number group	Standard	Logistic map			Gauss map			Piecewise map		
	ICA	CICA 21	CICA 22	CICA 23	CICA 31	CICA 32	CICA 33	CICA 41	CICA 42	CICA 43
A_1	0.1095	0.117	0.1	0.111	0.1479	0.111	0.1	0.1285	0.110262	0.120674
A_2	1.1224	0.895543	1.015	0.9251	0.941	1	0.93154	0.9484	0.936658	0.93562
A_3	0.1019	0.163965	0.1	0.1	0.1	0.100049	0.1764	0.1281	0.100137	0.100085
A_4	0.1	0.101292	0.1	0.1	0.1	0.1	0.1	0.1	0.1	0.100409
A_5	2.1197	1.93614	1.9486	1.92869	1.9431	1.9665	1.9299	2.031	1.9375	1.94845
A_6	0.25103	0.394947	0.26284	0.2805	0.2965	0.2621	0.296764	0.2895	0.27389	0.285325
A_7	0.1	0.107374	0.1002	0.1	0.1	0.1	0.1	0.1176	0.1167	0.11465
A_8	2.9743	2.98759	2.9991	2.9695	3.1046	2.9822	3.07331	3.2795	3.05335	3.05195
A_9	0.4452	0.410583	0.32621	0.526834	0.1009	0.1	0.126194	0.1124	0.1	0.110836
A_{10}	4.4955	4.01517	3.99846	3.9686	4.1034	3.97069	4.07107	4.1254	4.06961	4.08343
A_{11}	0.505099	0.457111	0.422806	0.50937	0.4038	0.4235	0.39124	0.4152	0.368677	0.407175
A_{12}	0.110845	0.100099	0.1001	0.10222	0.1839	0.88654	0.1	0.21	0.102191	0.599616
A_{13}	5.60999	4.99759	5.22184	5.72086	5.4249	5.69502	5.25891	5.4325	5.26739	5.29848
A_{14}	0.1	1.72358	0.1	0.161	0.1011	0.1	0.259899	0.1281	0.101109	0.101527
A_{15}	6.60371	6.05768	6.2298	6.41853	6.4278	6.42303	6.24104	6.4532	6.24761	6.25051
A_{16}	0.495299	0.947128	0.501418	0.588334	0.5724	0.875	0.55611	0.5935	0.465163	0.75277
A_{17}	0.531754	0.326569	1.15339	0.271	0.135	0.15291	0.622176	0.1431	1.35663	0.518927
A_{18}	7.7991	8.01927	7.66198	7.88627	7.9602	8.00124	7.7708	8.1602	7.5917	8.09504
A_{19}	0.114199	0.103055	0.1	0.25	0.1009	0.23	0.1	0.1165	0.151206	0.129255
A_{20}	8.62169	8.91345	8.66733	8.96057	8.9599	8.98783	8.77001	9.4175	8.59443	9.09746

(continued)

Table 9.6 (continued)

Number group	Standard	Logistic map			Gauss map			Piecewise map		
	ICA	CICA 21	CICA 22	CICA 23	CICA 31	CICA 32	CICA 33	CICA 41	CICA 42	CICA 43
A_{21}	0.9395	0.898925	1.10411	1.00411	0.70387	0.986953	0.911	0.70159	1.20564	0.990607
A_{22}	1.2409	0.1485	0.243655	0.211459	0.4224	0.24	0.161509	0.6059	0.5966	0.249386
A_{23}	10.6992	10.9216	10.982	10.9295	10.8456	10.9977	10.7995	10.8379	11.1098	11.3902
A_{24}	0.1	0.295041	0.120099	0.1	0.1	0.15	0.1	0.1164	0.595874	0.341984
A_{25}	11.6846	11.9381	11.9251	11.9143	11.8448	11.9868	11.7972	11.8399	12.1091	12.3967
A_{26}	1.5399	1.00717	1.07435	0.95714	1.0359	0.992845	0.92145	1.1999	1.54211	1.27388
A_{27}	6.6889	6.67999	6.87771	6.88572	6.6863	6.71187	6.81581	6.6139	4.8863	4.99476
A_{28}	11.227	11.7711	11.5667	11.3647	10.912	11.2263	11.7013	10.924	10.99904	10.72496
A_{29}	14.1425	13.7568	13.7922	13.8655	13.8131	13.8901	13.5925	13.9521	14.6849	14.5828
Best weigh	26,720.44	26,693.60	26,205.50	26,134.56	**25,480.08**	26,369.87	25,804.03	25,929.78	26,259.27	26,103.99
Mean weight	27,175.68	27,600.31	27,458.34	**26,624.71**	26,835.21	27,641.83	27,936.88	27,431.26	26,917.74	27,274.59
Coefficient variation	1.6774	1.7346	1.4809	1.4891	1.3967	1.7181	1.8903	1.6398	**1.3751**	1.653

The bold items in the table indicate the optimal values for the best weight, the best mean and the best coefficient of variation

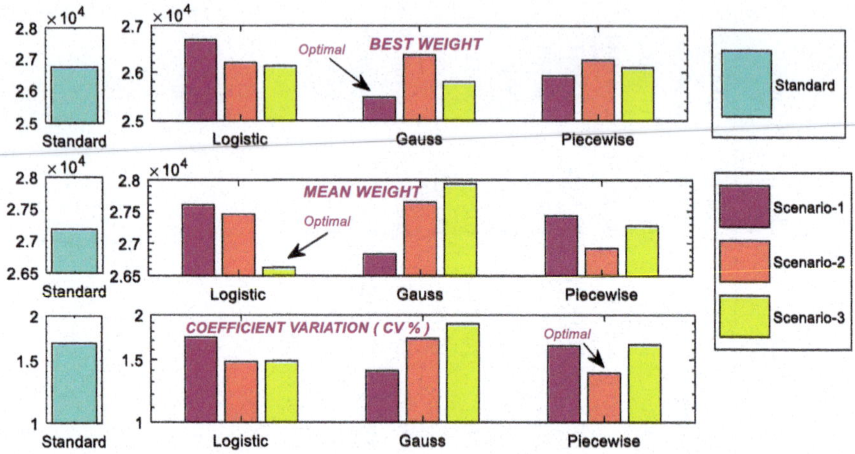

Fig. 9.8 Optimization results in standard and chaos map for the 200-bar planar truss structure

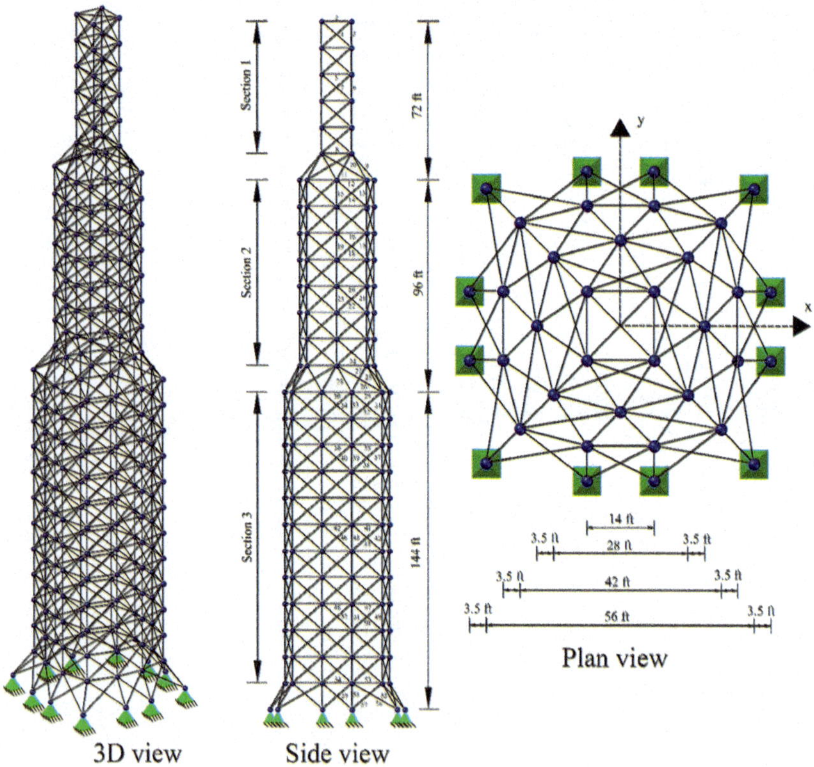

Fig. 9.9 Schematic of a 582-bar tower truss structure **a** 3D view, **b** side view, **c** top view

Table 9.7 Optimal design comparison for the 582-bar tower truss structure

Element group	Standard	Logistic map			Gauss map			Piecewise map		
	ICA	CICA 21	CICA 22	CICA 23	CICA 31	CICA 32	CICA 33	CICA 41	CICA 42	CICA 43
1	20	20.38073	20.39156	20	20	23.9789	20.02547	20	20.00244	20
2	143.7549	150.4822	135.9029	153.9418	155.6664	151.2457	91.03906	152.5026	141.9118	103.5824
3	54.23346	31.94984	34.5207	31.00371	30.89922	44.21157	38.34884	31.94976	31.04415	59.46837
4	230.2861	127.3296	117.6387	130.8899	130.0213	108.5098	137.9407	121.6458	133.2846	80.54247
5	36.54006	27.18174	27.29126	26.72705	26.70754	38.75168	34.73569	27.25899	26.66697	37.48744
6	20	20.65863	20	20.00002	20.0003	29.72342	54.94447	20	20.00017	23.79508
7	78.37087	102.8283	97.98491	107.1131	101.8026	129.4964	113.7192	103.8829	94.54256	109.8489
8	45.50036	25.52239	25.94217	25.39445	25.09662	27.18698	25.19783	25.9436	25.22956	32.46105
9	24.34342	20.16649	20.44191	20.30351	20.00214	37.30426	20.03756	20	20.00003	20.09996
10	116.38	88.96193	106.7845	95.29747	99.23998	149.0051	114.89	102.0555	95.78139	167.0256
11	44.43415	22.31875	21.8565	21.77726	21.7389	45.47919	33.30927	21.92043	21.77571	36.21004
12	161.2864	151.8453	156.9395	139.9543	144.296	60.76999	230.3872	131.6361	143.22	142.9643
13	283.1685	139.3159	152.7118	160.6952	157.1701	134.3326	129.5192	140.6819	148.2035	241.5209
14	62.17853	101.8106	77.74408	106.7071	111.7653	81.9867	86.06573	89.51033	89.55971	72.46719
15	74.81496	152.6356	147.4186	179.3485	185.5611	263.6728	80.30392	166.0496	168.3043	143.5049
16	58.70227	32.4359	36.18939	30.68122	30.35484	41.5139	69.62999	31.16372	31.00569	78.30012
17	72.66767	123.4409	133.6122	130.7583	135.3074	151.916	161.5196	124.4427	129.5225	105.5862
18	75.47715	23.93428	24.91609	23.73209	23.6295	35.63621	30.75871	24.46899	23.64643	38.29353
19	66.92678	20	20.17178	20.06163	20.00011	20	20	20.18464	20	37.67404
20	53.92177	71.08082	88.06825	101.7936	97.76061	128.6412	142.6644	78.33173	91.77708	89.18258

(continued)

Table 9.7 (continued)

Element group	Standard	Logistic map			Gauss map			Piecewise map		
	ICA	CICA 21	CICA 22	CICA 23	CICA 31	CICA 32	CICA 33	CICA 41	CICA 42	CICA 43
21	35.17136	23.5839	22.43484	21.69252	21.20534	60.95129	24.68398	23.21287	21.2666	54.61555
22	39.9163	20.85571	20.46485	20	20	20	31.33891	20	20.0779	86.42193
23	38.16209	53.5077	56.23128	58.18016	53.34051	154.0778	168.6296	57.4423	53.14451	99.24446
24	23.59572	20	20	20.00004	20	27.94148	57.92289	20	20.03029	20
25	20	20	20.64912	20.00331	20.00222	20	29.62064	20	20.01294	37.74398
26	24.74211	29.91861	23.49894	27.63556	23.47372	20	20	30.51513	23.65135	68.93575
27	21.22509	20	20	20.05688	20.00966	24.00738	26.31934	20	20	20
28	20	22.94528	20	20	20.00097	86.22331	30.08882	20.02903	20.00001	22.17979
29	20	20	20	20.17851	20.64659	20	20.1142	20	20.03032	20
30	20.83695	20.34429	20	20.01193	20.00281	35.67238	66.61977	20	20.00092	26.43832
31	53.94997	21.60968	21.61437	20.00603	20	20	20.11895	22.87928	20.08437	25.03157
32	43.29713	20.04536	21.47922	20.54306	20	44.68363	20	20.31257	20.00036	20.44394
Best (m^3)	21.0091	15.7672	15.7351	16.2481	16.2259	20.0055	19.5788	15.7904	**15.6083**	20.3924
Mean (m^3)	22.2032	15.8472	15.9966	16.254	16.2328	20.5424	20.5304	15.8523	**15.6287**	22.8241
CV (%)	1.9326	0.3854	1.398	0.128732	**0.073045**	2.1683	1.0737	1.1263	0.09567	1.4564

The bold items in the table indicate the optimal values for the best weight, the best mean and the best coefficient of variation

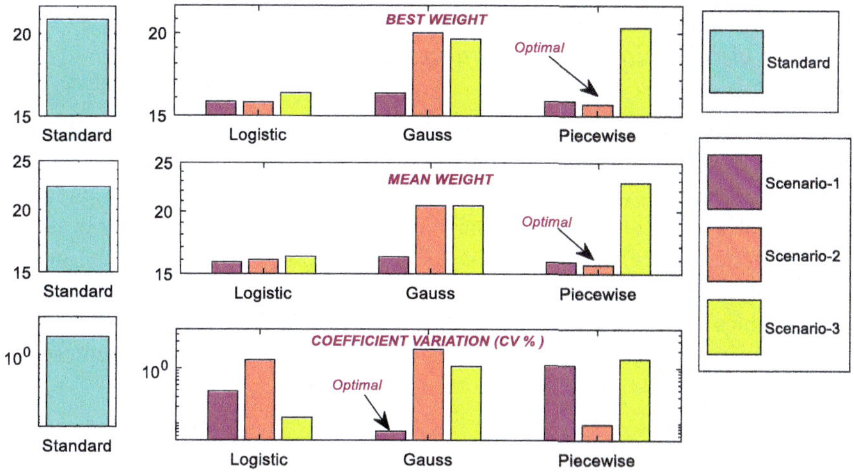

Fig. 9.10 Optimization results in standard and chaos map for the 582-bar tower truss structure

9.6 Truss Weight Optimization with Multiple Frequency Constraints

When the natural frequency of the structure is the same as the frequency of incoming loads such as earthquake and wind, the destructive phenomenon of resonance occurs. Therefore, in the optimization problem, one of the limitations will be determining the specific range for the natural frequencies of the structure. Structural problems in the frequency Constraints are associated with implicit and complex modes, and the optimization using meta-heuristic algorithms in the standard mode stops at the local optimum in most cases. In Chaotic Imperialist Competitive Algorithm (CICA), by embedding the chaotic dynamic series, include Logistic and Gaussian chaos map, a significant improvement in the optimization results is achieved. These chaotic dynamic series can replace random distributions in colonial competition strategies. These strategies include the strategy of absorbing the colonial countries by the emperors (assimilation), and the revolution strategy. By balancing these strategies, the chaos series converges the local optimum to the global optimum and avoids optimization stagnation. In each study, 6 chaotic modes are formed and compared with the standard mode. To create statistical information and increase diversity in research, each model is independently evaluated 30 times. Finally, to analyze the statistical results, the best weight, the best average and the coefficient of variation are reported. Also, useful information is obtained about the weakness and strength of the algorithm and how to improve the results.

9.6.1 Formulation of the Structural Size and Layout Optimization with Frequency Constraints

In truss design with multi-frequency constraints and optimization of size and shape, the goal is to minimize the weight of the structure while satisfying some limiting on natural frequencies of the structure [10]. In this type of optimization, the decision variables include the cross-section of the members or the geometric coordinates of some nodes. There will be no changes in the topology of the structure, and the connection conditions remain unchanged during the optimization process. The decision variables should be chosen within the corresponding range of the upper bound and the lower bound of their search space. Optimization problems can be represented by the mathematical relationship according Eq. (9.13):

$$\begin{aligned}
&Find\ \ X = [x_1, x_2, \ldots, x_n] \\
&to\ minimize\ P(X) = f(X) \times f_{penalty}(X) \\
&subjected\ to \\
&\omega_j \leq \omega_j^*\ \ for\ some\ natural\ frequencies\ j \\
&\omega_k \geq \omega_k^*\ \ for\ some\ natural\ frequencies\ k \\
&x_{i,\min} \leq x_i \leq x_{i,\max}
\end{aligned} \tag{9.13}$$

In this equation, X is the vector of decision variables, including the geometric coordinates of the nodes and the cross-sectional area of the members, n is the number of variables or the dimension of the search space, which is introduced based on the grouping of members and according to the symmetry of the structure. P(X) is the penalized objective function or merit function that must reach a minimum value. f(X) is the cost function that is considered as the weight in the optimization of structures. $f_{penalty}(X)$ is the penalty function that is used to apply constraints. In cases where some constraints related to the problem are violated, this function enters the calculation process with a suitable weight. w_j expresses the jth natural frequency of the structure and w_j^* is its upper bound. W_k expresses the kth natural frequency of the structure and w_k^* is its lower bound. X_{imin} and X_{imax} are the lower and upper bound of the decision variable x_i, respectively. The cost function is expressed by the mathematical relationship according to Eq. (9.14):

$$f(X) = \sum_{i=1}^{nm} \gamma_i \cdot L_i \cdot A_i \tag{9.14}$$

In this equation, γ_i is the specific material density of member i, L_i is the length of member I, and A_i is the cross-sectional area of member i. The penalty function is defined by the mathematical relation according to Eq. (9.15):

$$f_{penalty}(X) = (1 + \varepsilon_1 \cdot \nu)^{\varepsilon_2}, \quad \nu = \sum_{i=1}^{q} \nu_i \tag{9.15}$$

where q is the number of frequency constraints and ν_i expresses the violation value which is defined from the following equation:

$$\nu_i = \begin{cases} 0 & if\ the\ ith\ constrain\ is\ satisfied \\ \left|1 - \frac{\omega_i}{\omega_i^*}\right| & else \end{cases} \tag{9.16}$$

The parameters ε_1 and ε_2 are chosen based on creating a balance between exploration and exploitation in the decision variables space. In these investigations, the value of ε_1 is set as unity, and the parameter ε_2 starts from 1.5 and increases linearly to 6 in all test examples. These values penalize the unfeasible solutions more severely as the optimization process proceeds. As a result, in the early stages, the agents are free to explore the search space easily, but in the final iterations, the conditions of exploitation prevail and among them, the best local optimum is selected as the sub global optimum.

9.6.2 Numerical Examples of Optimal Truss Design

To validate results of the Chaotic Imperialist Competitive Algorithm (CICA), well-known examples of truss optimization with layout and size optimization are selected as benchmark problems. A significant improvement in the results is achieved by creating a competition between the perturbed states and the triple scenarios. In fact, 7 independent answers compete with each other. Due to the variety of checks, the optimal responses will be very close to the global optimal. The examples provided are as follows:

- The 52-bar dome-like truss
- The 72-bar spatial truss
- The 120-bar spatial dome.

9.6.2.1 A 52-Bar Dome-Like Truss

The 52-bar dome-like truss as shown in Fig. 9.11 is a well-known benchmark problem for optimizing the weight and shape of trusses with frequency constraints. This truss considers both optimizations of the section size and geometric coordinates of the nodes and the geometric shape of the structure is determined during the optimization process. The decision variables related to the section size are classified into 8 groups according to the symmetry in the geometric shape. Geometric coordinates of all symmetric free nodes can be changed by 2 m from the initial position along the coordinate axes.

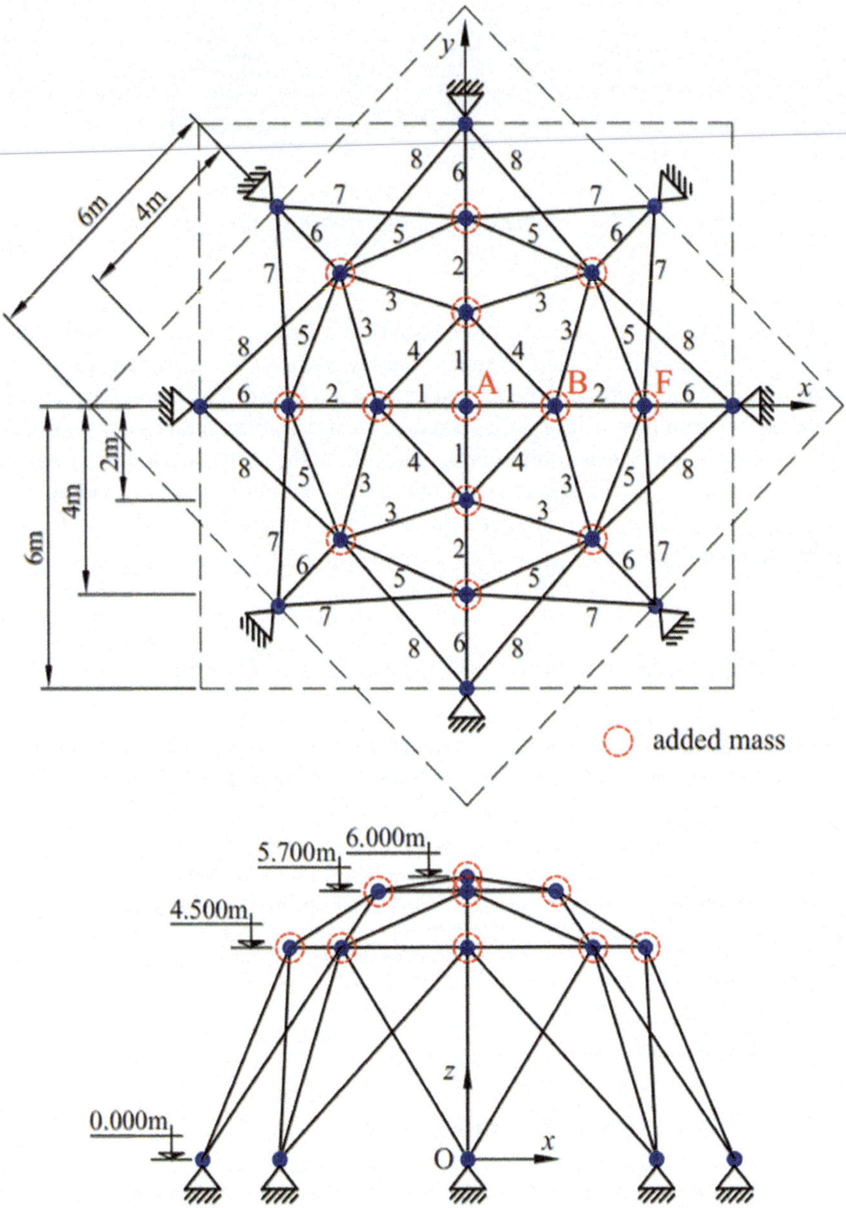

Fig. 9.11 Schematic of A 52-bar dome-like truss

In this case, the number of decision variables related to the shape of the structure and the geometric coordinates of the nodes is limited to 5 variables, and the sum of the variables, including shape and size, will be 13 variables. In all free nodes, a non-structural concentrated mass of 50 kg has affected all free nodes. The mechanical characteristics of the structure are: density of materials 7800 kg/m^3, modulus of elasticity 210,000 MPa, frequency limitations of the structure in the first mode are less than 15.916 Hz and in the second mode are greater than 28.648 Hz. For the cross section of the members, the lower limit is 1 cm^2 and the upper limit is 10 cm^2. In order to ensure the performance of turbulent functions and algorithms, as well as to increase the accuracy and sensitivity of calculations, each of the modes has been performed independently 30 times and the results related to the best response and the average value of responses are presented in Statistical Table 9.8. Also, the coefficient of variation of responses, which is a measure of the robustness and robustness of responses, has been calculated and used to compare the efficiency of turbulence functions and algorithms. For quick access to optimization information, a bar chart of each component is shown in Fig. 9.12. Examining the optimization results for different combinations of algorithms with turbulence functions and comparing it with the standard mode, shows a significant and significant improvement in reducing the weight of the 52-bar dome-like truss. The enhanced results for Chaotic Imperialist Competitive Algorithm (CICA), by Gauss chaos map and scenario 1, a weight of 192.702 kg has the optimal response.

9.6.2.2 A 72-Bar Spatial Truss

The 72-bar spatial truss as shown in Fig. 9.13 is a well-known benchmark problem with weight-limit optimization. This truss only considers "optimizing the size of the sections" and the geometric shape of the structure is constant during the optimization process. The decision variables related to the section size are classified into 16 groups according to the symmetry in the geometric shape. In nodes 1–4, the concentrated non-structural mass of 2270 kg has affected the truss. The mechanical characteristics of the structure are: material density 2770 kg/m^3, modulus of elasticity 69,800 MPa, frequency limits of the structure in the first mode are equal to 4 Hz and in the second mode are greater than 6 Hz.

For the cross section of the members, the lower limit is 0.645 cm^2 and the upper limit is 20 cm^2. In order to ensure the performance of chaotic functions and algorithms, as well as to increase the accuracy and sensitivity of calculations, each of the modes has been performed independently 30 times and the results related to the best response and the average value of responses are presented in Table 9.9. Also, the coefficient of variation of responses, which is a measure of the robustness and robustness of responses, has been calculated and used to compare the efficiency of turbulence functions and algorithms. For quick access to optimization information, a bar chart of each component is shown in Fig. 9.14. Examining the optimization results for different combinations of algorithms with turbulence functions and comparing it with the standard mode, shows a significant and significant improvement in reducing

Table 9.8 Optimal design comparison for the 52-bar dome-like truss

Decision	Standard	Logistic map			Gauss map		
Variable	ICA	CICA 21	CICA 22	CICA 23	CICA 31	CICA 32	CICA 33
Z_A (m)	5.9602	5.4103	5.0963	5.9403	6.3231	6.4206	6.5837
X_B (m)	2.3873	2.8668	1.8421	2.3296	1.94	2.3558	2.1428
Z_B (m)	3.8091	4.9216	3.7	3.7242	4.2327	3.8584	4.1243
X_F (m)	4.1292	4.5391	3.7484	3.9761	3.7456	4.1032	3.7663
Z_F (m)	2.5022	2.718	2.5	2.5076	2.9079	2.5	2.8752
A_1 (cm^2)	1	1	1	1	1.0051	1	1.0002
A_2 (cm^2)	1.0014	1	1.5206	1.0101	1	1.012	1.0756
A_3 (cm^2)	1.3035	1.3895	1.434	1.1369	1.1086	1.2138	1.0017
A_4 (cm^2)	1.43	1.18	1.5665	1.4493	1.1852	1.2822	1.5374
A_5 (cm^2)	1.5827	1.0324	1.4836	1.3614	1.4937	1.6105	1.5711
A_6 (cm^2)	1.0267	1	1	1	1	1	1
A_7 (cm^2)	1.4119	1	1.4905	1.7362	1.4152	1.821	1.3818
A_8 (cm^2)	1.3943	1.7288	1.4997	1.3335	1.5583	1	1.5873
Best weight (kg)	195.769	194.097	197.552	193.877	**192.702**	193.65	196.602
Mean weight (kg)	226.63	211.332	205.225	209.588	**199.528**	203.351	219.634
Coefficient variation	9.7778	8.244	4.37	4.9768	**2.36**	6.0891	9.9891
ω_1 (HZ)	14.8248	13.9112	14.1345	14.98185	15.94442	14.3017	12.635
ω_2 (HZ)	29.8248	29.8631	28.6692	29.349	28.88346	29.9358	28.3772

The bold items in the table indicate the optimal values for the best weight, the best mean and the best coefficient of variation

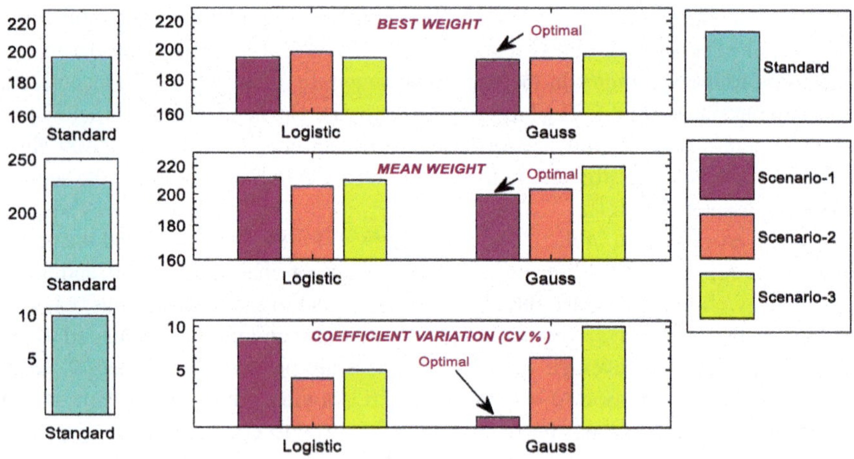

Fig. 9.12 Optimization results in standard and chaos map for the 52-bar dome-like truss

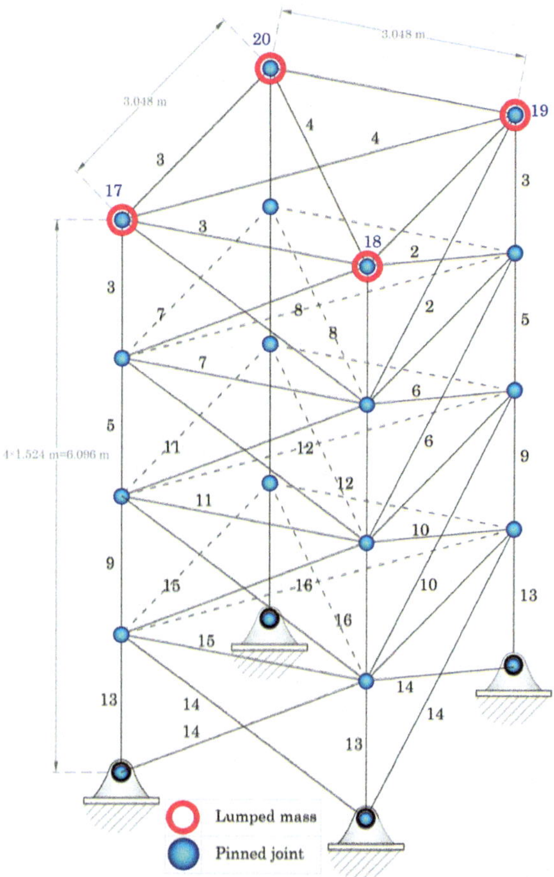

Fig. 9.13 Schematic of A 72-bar spatial truss

the weight of the 72-bar spatial truss. The enhanced results for Chaotic Imperialist Competitive Algorithm (CICA), by Gauss chaos map and scenario 1, a weight of 320.436 kg has the optimal response.

9.6.2.3 A 120-Bar Spatial Dome

The 120-bar spatial dome as shown in Fig. 9.15 is a well-known benchmark problem with weight-limit optimization. This truss only considers "optimizing the size of the sections" and the geometric shape of the structure is constant during the optimization process.

The decision variables related to the size of the members' sections and according to the symmetry in the geometric shape of the dome along the X and Y axes, are classified into 7 groups. Non-structural concentrated mass in all free nodes affects the structure. Their values are 3 kg in node 1500 kg in nodes 2–13 and 100 kg in

Table 9.9 Optimal design comparison for the 72-bar spatial truss

Element group	Standard	Logistic map			Gauss map		
	ICA	CICA 21	CICA 22	CICA 23	CICA 31	CICA 32	CICA 33
A_1–A_4	3.79196	8.35295	6.95848	6.90573	6.89973	7.03119	6.56318
A_5–A_{12}	7.59489	7.77493	7.65729	7.63633	7.62657	7.60728	7.75825
A_{13}–A_{16}	0.500009	0.5	0.5	0.5	0.5	0.50068	0.5
A_{17}–A_{18}	0.566058	0.5	0.5	0.500182	0.5	0.508065	0.5
A_{19}–A_{22}	10.9838	6.42034	7.69326	8.90582	7.69514	7.39321	9.48157
A_{23}–A_{30}	8.31701	7.84376	7.75703	7.52422	7.57292	7.7164	7.72526
A_{31}–A_{34}	0.5	0.500014	0.5	0.5	0.5	0.500029	0.5
A_{35}–A_{36}	0.522941	0.58547	0.547198	0.5	0.501813	0.510991	0.500033
A_{37}–A_{40}	13.524	13.4808	11.9527	12.789	14.4284	13.0338	13.1083
A_{41}–A_{48}	8.7967	7.76058	7.69983	7.76289	7.98533	7.66969	7.82492
A_{49}–A_{52}	0.500775	0.5	0.5	0.5	0.5	0.5	0.5
A_{53}–A_{54}	0.5	0.500001	0.5	0.5	0.5	0.5	0.5
A_{55}–A_{58}	14.9309	18.4248	18.8092	16.7262	16.552	17.9328	16.0496
A_{59}–A_{66}	7.76862	7.64927	7.75562	7.91926	7.52965	7.83978	7.62915
A_{67}–A_{70}	0.500005	0.500002	0.5	0.5	0.5	0.5	0.5
A_{71}–A_{72}	0.5	0.5	0.5	0.5	0.5	0.504936	0.5
Best weight (kg)	329.999	324.872	321.445	320.981	**320.436**	321.077	321.488
Mean weight (kg)	330.588	329.265	329.535	329.045	**326.172**	331.968	329.941
Coefficient variation	**0.1469**	0.867	1.3951	2.8363	1.5307	2.991	1.4532
ω_1 (HZ)	4	4	4	4.0001	4	4.00023	4
ω_3 (HZ)	6.08	6.0023	6.0001	6.0027	6.03	6.0100	6

The bold items in the table indicate the optimal values for the best weight, the best mean and the best coefficient of variation

other nodes. The mechanical characteristics of the structure are: material density 7971.81 kg/m^3, modulus of elasticity 210,000 MPa, frequency limits of the structure in the first and second modes are greater than 9 and 11 Hz, respectively. For the cross section of the members, the range of the lower limit is 1 cm^2 and the upper limit is 129.3 cm^2. In order to ensure the performance of chaos map and algorithms, as well as to increase the accuracy and sensitivity of calculations, each of the modes has been performed independently 20 times and the results related to the best response and the average value of responses are presented in Statistical Table 9.10. Also, the coefficient of change of responses, which is a measure of the robustness and robustness of

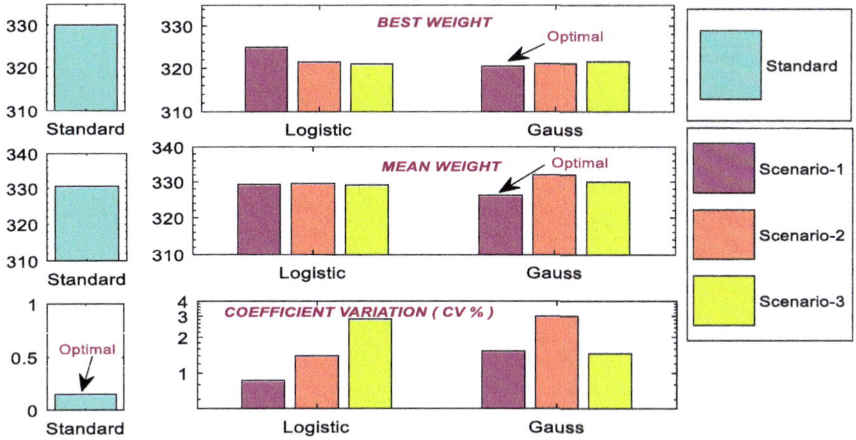

Fig. 9.14 Optimization results in standard and chaos map for the 72-bar spatial truss

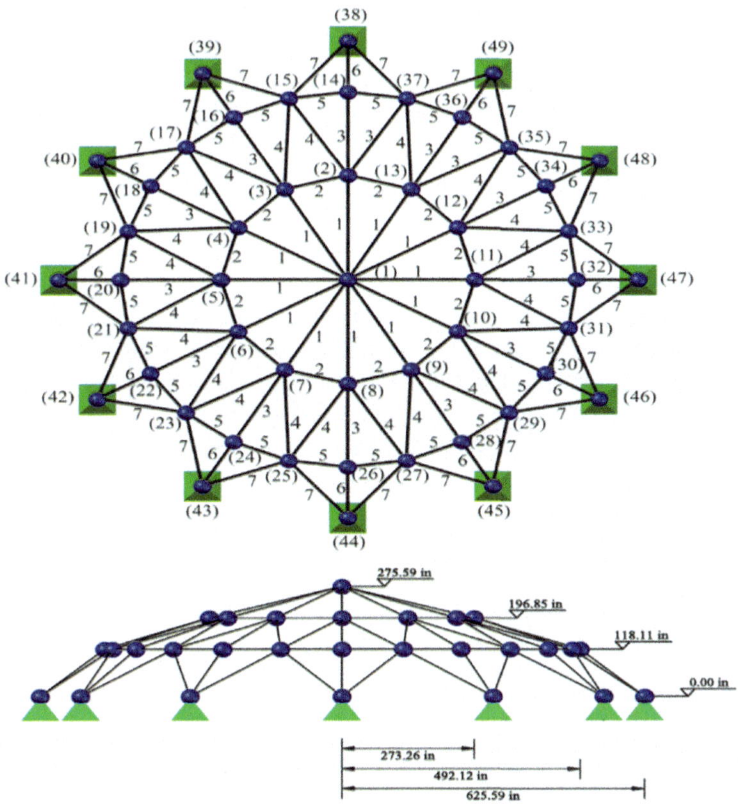

Fig. 9.15 Schematic of A 120-bar spatial dome

Table 9.10 Optimal design comparison for the 120-bar spatial dome

Decision Variable	Standard ICA	Logistic map CICA 21	CICA 22	CICA 23	Gauss map CICA 31	CICA 32	CICA 33
A_1	19.7961	19.791	19.6885	19.5558	19.6738	19.5558	19.6713
A_2	40.268	39.1279	40.7154	40.4643	39.6328	40.4643	39.5383
A_3	10.696	10.1864	10.5288	10.9318	10.7077	10.9318	10.6313
A_4	20.9889	21.3749	20.9967	21.0939	20.8326	21.0939	21.0394
A_5	9.33509	10.2202	9.37043	9.62433	9.94007	9.62433	10.116
A_6	11.7797	12.6381	11.8585	11.3525	11.5466	11.3525	11.8548
A_7	15.1826	14.5577	15.0992	15.0426	15.3143	15.0426	14.9835
Best weight (kg)	8723.153	8722.479	8719.007	8720.762	**8715.68**	8720.762	8719.504
Mean weight (kg)	8734.105	8740.481	8729.41	8729.574	8730.292	8730.199	**8724.906**
Coefficient variation	0.1146	0.1614	0.1268	**0.0651**	0.1159	0.0705	0.0679
ω_1 (HZ)	9.00427	9.00312	9.00274	9.00076	9	9.0001	9
ω_2 (HZ)	11	11.0025	11.0004	11.0015	11.0008	11.0015	11.001

The bold items in the table indicate the optimal values for the best weight, the best mean and the best coefficient of variation

responses, has been calculated and used to compare the efficiency of turbulence functions and algorithms. For quick access to optimization information, the bar chart of each component is shown in Fig. 9.16. Examining the optimization results for different combinations of algorithms with turbulence functions and comparing it with the standard mode, shows a significant and significant improvement in reducing the weight of the 120-bar spatial dome. The enhanced results for Chaotic Imperialist Competitive Algorithm (CICA), by Gauss chaos map and scenario 1, a weight of 8715.68 kg has the optimal response.

9.7 Concluding Remarks

In this algorithm, the evolution in the political, cultural and social fields is inspired. This algorithm starts with the formation of an initial population of countries. In this regard, a number of the best countries are chosen as emperors and based on their superior power, they think of acquiring colonies. Therefore, these powerful countries will think of increasing their empire by forming an empire and by applying the policy of attracting colonial countries. The countries of each empire are competing with each other to bring their conditions to the emperor, but the main competition

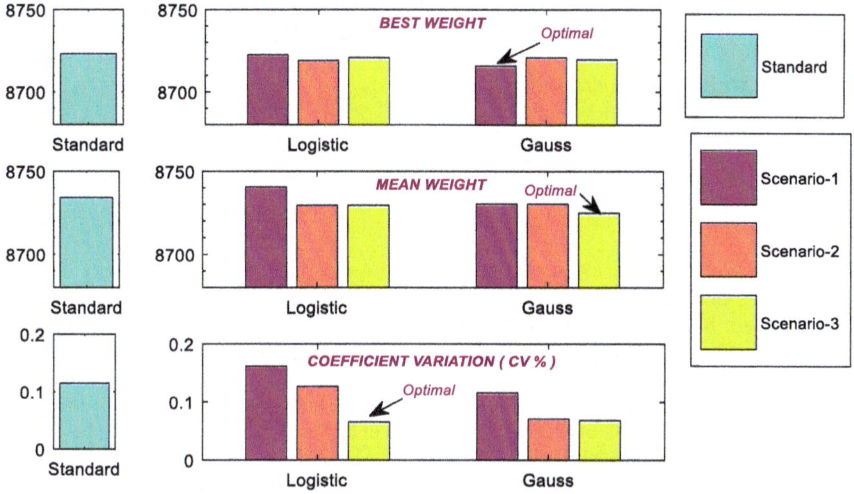

Fig. 9.16 Optimization results in standard and chaos map for the 120-bar spatial dome

is the competition between the emperors themselves. Therefore, like a sports team, at the same time that the members of the team are trying to reach the captain, and they have an intra-group competition among themselves, but the main competition is the out-group competition and between the teams. Therefore, we have two types of competition, which include in-group and out-group. At this stage, the weak colonial countries move towards the emperor in order to improve their level of development. This stage is known as the assimilation policy stage and the coefficient chosen for movement is between 0 and 2. ($0 \leq \beta \leq 2$) In the next step, in order to search for unknown places in space, the Revolution step is executed and a random change is made in the answers. At this stage, various answers are created to get access to better answers. By applying this step, if the results were successful, it is accepted. Now, in the competition within the group, if a country is found that is better than the existing emperor, it will replace the emperor and the previous one will be considered as a colony. In intergroup competition, emperors compete, which is considered in the comparison index of the total value of the objective function for the colonizer and a percentage of the average value of the objective function for the colonies. This percentage of the participation of the colonies is suggested as ten percent in the original version. ($\xi = 0.1$) In this competition, the weakest empire loses one of its colonies and the strongest empire acquires it. In this process, if a colony has no clones, that empire will fall. In the iterative process of steps, if only one emperor is left, the operation is stopped and that empire is introduced as the final answer of this round. In this algorithm, the strategy of attraction and revolution is combined with random distributions. These distributions are aimed at increasing diversity, but in most cases, they cause the algorithm to get stuck in local optima. With premature convergence, the optimization operation faces a massive stagnation. Therefore, the chaotic dynamic series converges the local optima to the overall optima by creating jumps. Also, the balance between the exploration and exploitation steps is achieved. By creating a competition between 9 chaotic modes, the best answer is selected and compared with the standard mode.

References

1. Kaveh, A.: Advances in Metaheuristic Algorithms for Optimal Design of Structures, 2nd edn. Springer, Cham, Switzerland (2021). https://doi.org/10.1007/978-3-319-05549-7
2. Talatahari, S., Kaveh, A., Sheikholeslami, R.: Chaotic imperialist competitive algorithm for optimum design of truss structures. Struct. Multidisc. Optim. **46**(3), 355–367 (2012). https://doi.org/10.1007/s00158-011-0754-4
3. Atashpaz-Gargari, E., Hashemzadeh, F., Rajabioun, R., Lucas, C.: Colonial competitive algorithm: a novel approach for PID controller design in MIMO distillation column process. Int. J. Intel. Comput. Cybernet. **1**(3), 337–355 (2008)
4. Kaveh, A., Yousefpoor, H.: Chaos-embedded meta-heuristic algorithms for optimal design of truss structures. Sci. Iranica **29**(6), 2868–2885 (2022). https://doi.org/10.24200/sci.2022.59812.6441
5. Kaveh, A., Yosefpoor, H.: Chaotically enhanced meta-heuristic algorithms for optimal design of truss structures with frequency constraints. Periodica Polytech. Civ. Eng. **66**(3), 900–921 (2022). https://doi.org/10.3311/PPci.20220
6. Kaveh, A., Zarfam, P., Aziminejad, A., Yosefpoor, H.: Comparison of four chaotic meta-heuristic algorithms for optimal design of large-scale truss structures. Iran. J. Sci. Technol. Trans. Civ. Eng. https://doi.org/10.1007/s40996-022-00908-8
7. Ott, E.: Chaos in Dynamical Systems. Cambridge University Press, Cambridge (2002)
8. Bucolo, M., Caponetto, R., Fortuna, L., Frasca, M., Rizzo, A.: Does chaos work better than noise? IEEE Circuit Syst. Mag. (3), 4–19 (2002)
9. Peitgen, H.-O., Jürgens, H., Saupe, D.: "Chaos and Fractals": New Frontiers of Science. Springer, Berlin (2006)
10. Kaveh, A., Sheikholeslami, R., Talatahari, S., Keshvari-Ilkhichi, M.: Chaotic swarming of particles: a new method for size optimization of truss structures. Adv. Eng. Softw. **67**, 136–47 (2014)

Chapter 10
Chaotic Shuffled Frog Leaping Algorithm

10.1 Introduction

To improve the efficiency of meta-heuristic algorithms, a strong and efficient method is to divide the entire population into small complexes and perform parallel optimization in each of them. This idea was first time developed for the meta-heuristic algorithm so-called "Shuffled Complex Evolution" (SCE-UA) from the University of Arizona. In this algorithm, competitive complexes are formed in the first stage. Then by applying geometric maps including reflection and contraction maps, optimization operations are performed on them. In order to improve the complexes as much as possible, their evolution is repeated several times. At the end, by shuffling them, conditions are provided for the next repetitions. Following the three-step proposal of this algorithm, the Shuffled Frog-Leaping Algorithm (SFLA) was invented. In the new algorithm, changes are applied in the complex evolution. In this algorithm, firstly memeplexes are formed by dividing the entire population, then in the complexes evolutionary stage, instead of geometric mappings, the best member of the group and the best member of the entire population are used. If no improvement is obtained from these two scenarios, a random answer is substituted. The movement of the worst answers to the best answer of the Memeplex is considered a local and will play the role of exploitation stage for the algorithm. Also, the movement of the worst answer to the best answer of the whole population will largely play the role of exploration stage. In cases where results are not obtained, random search performs the role of exploration stage. By repeating the complexes evolution, a significant improvement is obtained. At the end, by shuffled them, conditions are provided for the next repetitions. The most important advantage of the algorithm is parallel and simultaneous optimization in the complexes-evolution. But in the movement of the worst member to the best member of the group or the best member of the whole population, random distributions have been used to create diversity in the search space. These random functions in the standard mode of the algorithm have a uniform distribution, and will cause premature convergence and fall into the trap of local optima. In order

A. Kaveh and H. Yousefpoor, *Chaotic Meta-heuristic Algorithms for Optimal Design of Structures*, Studies in Computational Intelligence 1129,
https://doi.org/10.1007/978-3-031-48918-1_10

to improve the results and create interaction between the two stages of exploration and exploitation, chaotic functions replace the random distribution. These functions transfer the local optima to global optima by creating chaotic jumps, and the algorithm gets rid of the trap of local optima. In chaotic Shuffled Frog-Leaping Algorithm (CSFLA), premature convergence is eliminated and we can obtain improved results in a small number of iterations. Also, for problems with high complexity (NP-Hard), non-linear and non-convex, the algorithm should continue processing until reaching the sub optima without stopping at local optima. In this chapter, Logistic, Gaussian, and Piecewise chaos functions have been chosen to form chaotic meta-heuristic algorithms [1]. Also, for the purpose of evaluation, optimization of structures with static loads and well-known structural problems have been considered. For optimization with multi-frequency constraints, logistic and Gaussian chaos functions have resulted in significant improvement and we did not need to check the third chaos function.

10.2 Standard Shuffled Frog-Leaping Algorithm (SFLA)

In the Shuffled Frog-Leaping Algorithm (SFLA), the advantages of two large groups of algorithms, including evolutionary and swarm intelligence algorithms, are combined [2]. The Shuffled Frog-Leaping Algorithm is a memetic or behavioral algorithm and is confronted to the genetic algorithm. The structural units of memetic algorithms are behaviors and in some cases culture. Most of these algorithms are based on Lamarck's theory. In this algorithm, behaviors, thoughts and ideas evolve over time and are transferred to the next generations. In fact, frogs imitate each other, but it is not only pure imitation, and like Lamarck's theory, they also add some things and pass them on to the next generation. This algorithm was proposed by Eusuff et al. [2]. In this algorithm, the behavior of frogs to find food in the wetland has been imitated. The whole population of frogs is divided into smaller groups, each group representing different types of frogs that are spread in different places of the answer space. Then each group of frogs starts a local search in their neighborhood. Each frog is influenced by its group and other groups members. After performing several search steps, in the next step, mating is done and information is shared between all groups. In this algorithm, we denote the original population with F, which includes the number of m complexes and each with n population. Therefore, the total population will be equal to $F = m \times n$. In the next step, a local algorithm called FLA is applied on each of the complexes and improved complexes are obtained. At the end of this stage, the simplexes become shuffle and re-form the overall population. The performance of the local algorithm on each of the complexes plays a major role. In this algorithm, for each complex with n population, we select q number of parents with triangular probability distribution. Now, by determining the best and worst members of the population, (Pb and Pw express these respectively), the improvement of the worst population is done according to Eq. (10.1):

$$StepsizeA = sigma \times \boldsymbol{rand} \otimes (Pb - Pw)$$

$$Pw^{new} = Pw + StepsizeA \tag{10.1}$$

If we do not reach the result in this way, we consider the best member of the whole complexes Pb^x (global best), and the improvement of the worst member of the population is done according to Eq. (10.2):

$$\begin{aligned} StepsizeB &= sigma \times \boldsymbol{rand} \otimes (Pb^x - Pw) \\ Pw^{new} &= Pw + StepsizeB \end{aligned} \tag{10.2}$$

If an improved solution is not obtained in these two strategies, we choose the censorship stage. In this case, we consider a random answer and replace it, and then we return the results on the original complex. The basic steps of this algorithm for standard mode are presented below.

10.3 Basic Steps in Shuffled Frog-Leaping Algorithm

Step 1 Selection of algorithm parameters: Algorithm parameters include the number of decision variables "nvar," the upper and lower bounds of the variables, the maximum number of iterations, the number of complexes "m," the dimensions of each complex "n" and the total population "F."
Step 2 Formation of the initial population: In the search space, frog populations are randomly formed.
Step 3 Sort the answers: By evaluating the objective function, we rank the answers according to merit.
Step 4 Creating complexes by dividing population members: By using the described method, we form m complexes and each with n populations.
Step 5 Improve the members of each complex using the Competitive Complex Evolution algorithm (CCE). We will explain about this algorithm in the following.
Step 6 Combine the members of the improved complexes to create new populations and sort them.
Step 7 Check the termination conditions and repeat step 3 if needed.

Also, the basic steps in the Competitive Complex Evolution (CCE) algorithm, which is used to improve each Complex, include the following:

Step 1 Select the parameters of the evolutionary algorithm of competitive complexes: These parameters include the number of selected parents q from the m population of Complexes, the number of repetitions of the inner ring α, the number of repetitions of the outer ring β and the adjustment coefficient of the change interval σ.
Step 2 Form a subpopulation of parents through weighted random sampling without replacement from the complex population: According to the proposal

of the original version of the algorithm, this probability sampling is done with a triangular distribution.

Step 3 Sort the members of the subpopulation according to the objective function and merit, so that the best member is the first and the worst member is the last member according to Eq. (10.3):

$$sub\_pop = \{u_1, u_2, \ldots, u_q\}$$
$$\begin{cases} first\ member\ is\ the\ best\ member\ of\ population \Rightarrow Pb = U_1 \\ last\ member\ is\ the\ worst\ member\ of\ population \Rightarrow Pw = U_q \end{cases} \quad (10.3)$$

Step 4 Improve the position of the worst frog: Based on relations 36 and 37, a new value is proposed for the position of the worst member of the population, now if the new answer is within the search space, we calculate the value of the objective function for it, and if not, we go to step 5. Also, if the new proposed answer is better than the previous answer, it will be replaced and we will refer to step 7, otherwise we will go to step 5.

Step 5 In this step, to improve the position of the worst frog, we take advantage of the best response of the entire population Pb^x (global best), so we form the new position according to relation 35. If the resulting new answer is within the range of the search space, we evaluate it with the objective function, and otherwise we go to step 6. Also, if the new proposed answer is better than the previous answer, it will be replaced and we will refer to step 7, otherwise we will go to step 6.

Step 6 Censoring incorrigible responses: If after steps 4 and 5 the value of Pw still cannot be improved, we will prevent its release and censor it. In other words, we discard it as an unsuccessful idea and behavior and replace it with a new random idea.

Step 7 Repeating the inner loops: We repeat steps 3–6 for "α" number of times.

Step 8 We replace the members of the complex with new modified members.

Step 9. Repeating the external rings: We repeat steps 2–8 for β times.

10.4 Chaos-Embedded Shuffled Frog-Leaping Algorithm (CSFLA)

In this algorithm, we use the interval strategy between the best and worst answers to determine the corrective steps [3–5]. Therefore, in order to improve the responses of each colony, we will first go to the strategy of the best response available in the same colony, and if no improvement is achieved, in the next step strategy, we will go to the best response related to all colonies. By applying these two strategies, the role of exploration and exploitation is implemented in the algorithm. Now, by replacing the chaos functions in the random selections related to these steps, we will conclude a significant improvement in the performance of the algorithm. The suggested scenarios for this replacement are as follows:

Scenario 1 applying the chaos function in the phase of choosing the best answer from the colony itself: In this proposal, the first CHM1 chaos function is applied in Eqs. (10.1) and (10.4) are concluded.

$$\begin{aligned} StepsizeA &= sigma \times \boldsymbol{CHM1} \otimes (Pb - Pw) \\ Pw^{new} &= Pw + StepsizeA \end{aligned} \tag{10.4}$$

Scenario 2 Applying the chaos function in the stage of choosing the best answer among all colonies: In this proposal, the second CHM2 chaos function is replaced in Eqs. (10.2) and (10.5) are concluded.

$$\begin{aligned} StepsizeB &= sigma \times \boldsymbol{CHM2} \otimes (Pb^{x} - Pw) \\ Pw^{new} &= Pw + StepsizeB \end{aligned} \tag{10.5}$$

Scenario 3 applying chaos functions in both stages simultaneously: In this proposal, two chaos functions are applied simultaneously in Eqs. (10.1) and (10.2).

10.5 Truss Weight Optimization with Static Constraints

To investigate the effectiveness of chaos functions in Shuffled Frog-Leaping Algorithm (SFLA), the Logistic, Gossin and Piecewise chaos functions are selected. By embedding the chaos functions and forming the Chaotic Shuffled Frog-Leaping Algorithm (CSFLA), the competition between the chaos functions is formed to achieve the best optimal weight. Also, in order to reach the best performance of each chaos function in meta-heuristic algorithms, we need three scenarios. In these scenarios, chaos functions replace random distributions corresponding to exploration, exploitation, or both phases simultaneously. In this competition, 9 chaotic models are formed for each structural model and compared with the standard state. Also, in order to form statistical information and increase diversity in the response space, each model is independently evaluated 30 times. To interpret these statistical results, the best weight, the best average and the coefficient of variation are reported. Among a number of results, the use of chaos functions has increased the coefficient of variation. An increase in the coefficient of variation indicates sudden jumps and the growth of the difference between the best and worst answers. Due to the wide scope of the research, there is a strong competition between the chaos functions, and as a result, the possibility of approaching the global optimum increases. In the Logistic chaos map, it most likely converges from local minima to general minima [6]. Therefore, this function is suitable for improving the exploration conditions of algorithms. The Gaussian chaos map is very likely to be within the local minimums and is suitable for improving exploitation conditions [7]. Finally, Piecewise map simultaneously improves both conditions [8]. Therefore, by selecting these chaos maps, the weakness of algorithms of any kind will be repaired and global optimum will not have the opportunity to escape the chaos functions. We can also find out the strengths and weaknesses of the algorithm in the stages of exploration or exploitation.

10.5.1 Formulation of the Structural Optimization Problems

In optimizing the size of truss structures, the main goal is to find the optimal value for the cross-sectional area of the members A_i, in order to optimize the weight of the structure W, and fulfill the limitations related to the response of the structure. Therefore, the optimal design problems are expressed according to the following equations:

$$\begin{cases} Find & X = [x_1, x_2, \ldots, x_n] \\ to\ optimal & Mer(X) = f_{penalty}(X) \times W(X) \\ subject\ to & \begin{cases} \sigma_{\min} \le \sigma_i \le \sigma_{\max} & i = 1, 2, \ldots, nm \\ \delta_{\min} \le \delta_j \le \delta_{\max} & j = 1, 2, \ldots, nn \end{cases} \end{cases} \tag{10.6}$$

where X is the vector embrace the decision variable; the variable x_i is selected between upper and lower bound; n is the number of element, $Mer(X)$ is the merit function; $W(X)$ is the weight of structure; $f_{penalty}(X)$ is the penalty function which results from the violations of the constraints; nm is the number of members forming the topology of structure; nn is the number of nodes except support nodes; σ_i and δ_i are the stress of members and nodal displacement, respectively; min and max mean the lower and upper bound of constraints, respectively. The weight of structure can be expressed as:

$$W(x) = \sum \gamma_i \cdot A_i \cdot L_i \tag{10.7}$$

where γ_i is the material density of member i, L_i is the length of member i, and A_i is the cross-sectional area of member i. The penalty function can be defined as:

$$f_{penalty}(X) = \left[1 + \varepsilon_1 \cdot \sum_{i=1}^{np} (\phi_{\sigma(i)}^k + \phi_{\delta(i)}^k)\right]^{\varepsilon_2} \tag{10.8}$$

where np is the number of multiple loadings; ε_1 is taken as unity; ε_2 is set from 1.5 and increases linearly to 6; φ_σ and φ_δ are the summation of stress and nodal displacement penalize, which are expressed as following equations:

$$\begin{cases} \phi_\sigma = \sum_{i=1}^{nm} \max\left(\left|\frac{\sigma_i}{\bar{\sigma}_i}\right| - 1\right) \\ \phi_\delta = \sum_{i=1}^{nn} \max\left(\left|\frac{\delta_i}{\bar{\delta}_i}\right| - 1\right) \end{cases} \tag{10.9}$$

where σ_i, δ_i are the stress and nodal displacement which are divided by the allowable amount.

10.5.2 Introduction of Selected Chaos Map

To access the chaos dynamic series, sentences of chaos functions are formed sequentially. Then these chaotic series are embedded in the Shuffled Frog-Leaping Algorithm (SFLA). In the Chaotic Shuffled Frog-Leaping Algorithm (CSFLA), the local optima converge to the global optima, and the chaotic series will have a good potential to prevent the algorithm from getting trapped and premature convergence. With chaotic meta-heuristic algorithms, the non-linear and non-convex behavior of the objective function in the optimization of truss structures is easily controlled and adjusted. Among the advantages of chaotic series is non-repetition and non-convergence towards a certain limit. In this algorithm, first the memeplexes are formed by dividing the entire population, then in the evolutionary stage of the complexes, the movement of each frog towards the best member of the same memeplex and the best member of all memeplexes is used. If no improvement is obtained from these two scenarios, a random response is substituted. Performing each of these steps is associated with a random distribution. By replacing the chaos functions in the position related to the random distributions, it jumps the local optima to the global optima and the algorithm is saved from the risk of premature convergence. The improvement of the optimization results of truss structures will be investigated further.

10.5.3 Numerical Examples of Optimal Truss Design

To validate the Chaotic Shuffled Frog-Leaping Algorithm (CSFLA), well-known examples of truss structure optimization are selected as benchmark problems. By comparing the performance of chaotic samples and choosing the best chaos function, a severely challenge is formed between nine chaos responses. Also, in order to know the quality and quantity of answers, the results of the best chaotic model are compared with the standard mode. Due to the variety of reviews, the optimal responses will be largely close to the global optimality. The examples provided are as follows:

- The 52-bar planar truss;
- The 72-bar spatial truss;
- The 120-bar dome shaped truss.

10.5.3.1 A 52-Bar Planar Truss

The topology and nodal numbering of the 52-bar planer truss are shown in Fig. 10.1. The material density is 7860 kg/m^3 and the modulus of elasticity is 207 GPa. The members are subjected to stress limitation 180 MPa in both tension and compression. For loading, the intensity of the loads in the X and Y directions is 100 and 200 kN, respectively, and both loads are applied simultaneously. Also, truss members are classified into 12 groups according to geometric symmetry. The cross-section areas of the members were chosen from the 64 discrete values of the AISC code. In order to ensure the performance of chaos map and algorithms, as well as to increase the accuracy and sensitivity of calculations, each of the modes has been performed independently 30 times and the results related to the best response and the average value of responses are presented in Table 10.1. The graph of these results is also shown in Fig. 10.2. The coefficient of variation of the responses, which indicates the stability and robustness of the responses, has been calculated and another criterion for the efficiency of chaos map and algorithms has been obtained. After reviewing the optimization results for different combinations of algorithms with chaos map and comparing it with the initial and standard mode, a significant improvement in reducing the weight of the 52-bar planar truss results. The enhanced results for Chaotic Shuffled Frog-Leaping Algorithm (CSFLA), by Piecewise chaos map and scenario 1, a weight of 1899.319 kg has the optimal response.

10.5.3.2 A 72-Bar Spatial Truss

The topology and nodal numbering of the 72-bar spatial truss are shown in Fig. 10.3. The material density is 0.1 lb/in^3 and the modulus of elasticity is 10,000 ksi. Permissible members' stress of 25 ksi is assumed. The maximum displacement for each of the upper nodes is limited at 0.25 in in all directions. The two loading conditions are applied according to Table 10.2. Also, truss members are classified into 16 groups according to considering the geometric symmetry. The cross section of the members, which are the same as the decision variables, is considered in the range of 0.1–4 in^2. In order to ensure the performance of chaos map and algorithms, as well as to increase the accuracy and sensitivity of calculations, each of the modes has been performed independently 30 times and the results related to the best response and the average value of responses are presented in Table 10.3. A graph of these results is also shown in Fig. 10.4. The coefficient of variation of the responses, which indicates the stability and robustness of the responses, has been calculated and another criterion for the efficiency of chaos map and algorithms has been obtained.

Examining the optimization results for different combinations of algorithms with chaos map and comparing it with the initial and standard modes, shows a significant improvement in reducing the weight of the 72-bar spatial truss. The enhanced results for Chaotic Shuffled Frog-Leaping Algorithm (CSFLA), by Logistic chaos map and scenario 1, a weight of 379.2456 lb has the optimal response.

Fig. 10.1 Schematic of a 52-bar planar truss

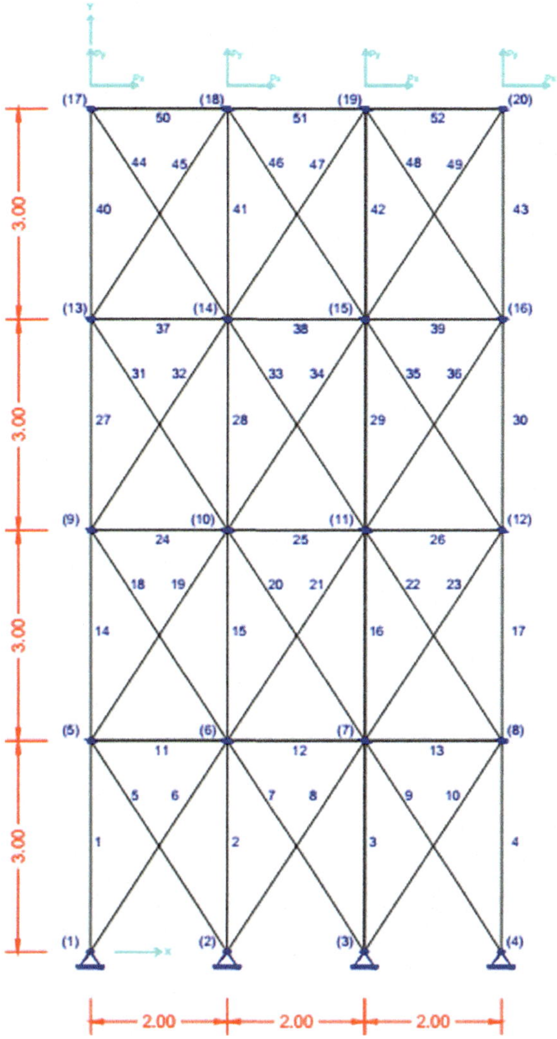

10.5.3.3 A 120-Bar Dome Shaped Truss

The topology and nodal numbering of the 120-bar dome shaped truss shown in Fig. 10.5. The material density is 0.288 lb/in^3 and the modulus of elasticity is 30,450 lb/in^2. Permissible members' tensile and compressive stresses are proposed in accordance with AISC ASD code.

The maximum deformation for each node in all directions is set at 0.1969 in. All non-abutment nodes are affected by vertical loading, with intensities of − 13.49 kips in node one and − 6.744 kips in spheres 2–14 and 2.248 kips in the other nodes, respectively. Also, truss members are classified into seven groups according to their

Table 10.1 Optimal design comparison for the 52-bar planar truss

Number group	Element group	Standard	Logistic map			Gauss map			Piecewise map		
		SFLA	CSFLA 21	CSFLA 22	CSFLA 23	CSFLA 31	CSFLA 32	CSFLA 33	CSFLA 41	CSFLA 42	CSFLA 43
1	A_1–A_4	4658.055	4658.055	4658.055	4658.055	4658.055	4658.055	4658.055	4658.055	4658.055	4658.055
2	A_5–A_{10}	1161.288	1161.288	1161.288	1161.288	1161.288	1161.288	1161.288	1161.288	1161.288	1161.288
3	A_{11}–A_{13}	645.16	494.193	363.225	494.193	494.193	363.225	494.193	494.193	506.451	494.193
4	A_{14}–A_{17}	3303.219	3303.219	3303.219	3303.219	3303.219	3303.219	3303.219	3303.219	3303.219	3303.219
5	A_{18}–A_{23}	1008.385	1008.385	939.998	1008.385	1008.385	1008.385	1008.385	939.998	939.998	1008.385
6	A_{24}–A_{26}	388.386	388.386	494.193	388.386	388.386	494.193	645.16	506.451	494.193	494.193
7	A_{27}–A_{30}	2290.318	2290.318	2238.705	2290.318	2290.318	2238.705	2238.705	2238.705	2238.705	2238.705
8	A_{31}–A_{36}	1008.385	1008.385	1008.385	1008.385	1008.385	1008.385	1008.385	1008.385	1008.385	1008.385
9	A_{37}–A_{39}	388.386	388.386	388.386	494.193	388.386	494.193	388.386	388.386	494.193	388.386
10	A_{40}–A_{43}	1283.868	1283.868	1283.868	1283.868	1283.868	1283.868	1283.868	1283.868	1283.868	1283.868
11	A_{44}–A_{49}	1161.288	1161.288	1161.288	1161.288	1161.288	1161.288	1161.288	1161.288	1161.288	1161.288
12	A_{50}–A_{52}	498.193	494.193	729.256	494.193	506.451	494.193	506.451	506.451	506.451	645.16
Best weight		1917.526	1910.217	1904.806	1914.659	1910.795	1909.791	1917.489	**1899.319**	1903.761	1916.911
Mean weight		1924.774	1916.243	1909.416	1921.284	1915.04	1913.171	1922.045	**1901.681**	1909.008	1921.836
Coefficient variation		0.13564	0.11537	0.10162	0.13178	0.10224	0.10153	0.10841	**0.097682**	0.11194	0.12491

The bold items in the table indicate the optimal values for the best weight, the best mean and the best coefficient of variation

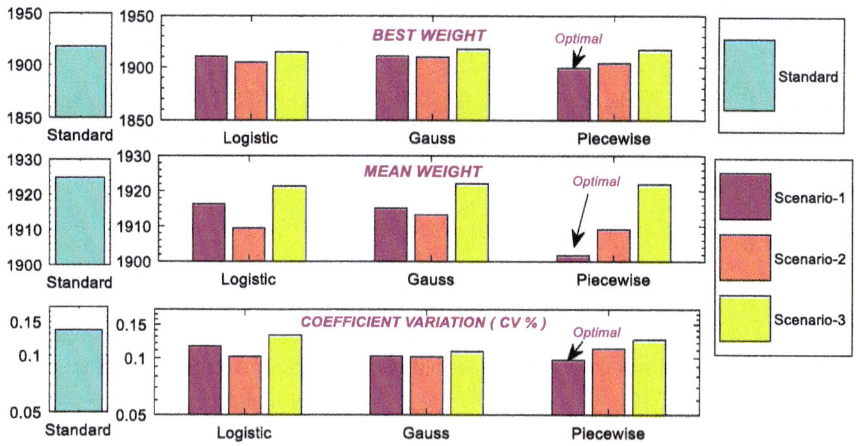

Fig. 10.2 Optimization results in standard and chaos map for the 52-bar planar truss

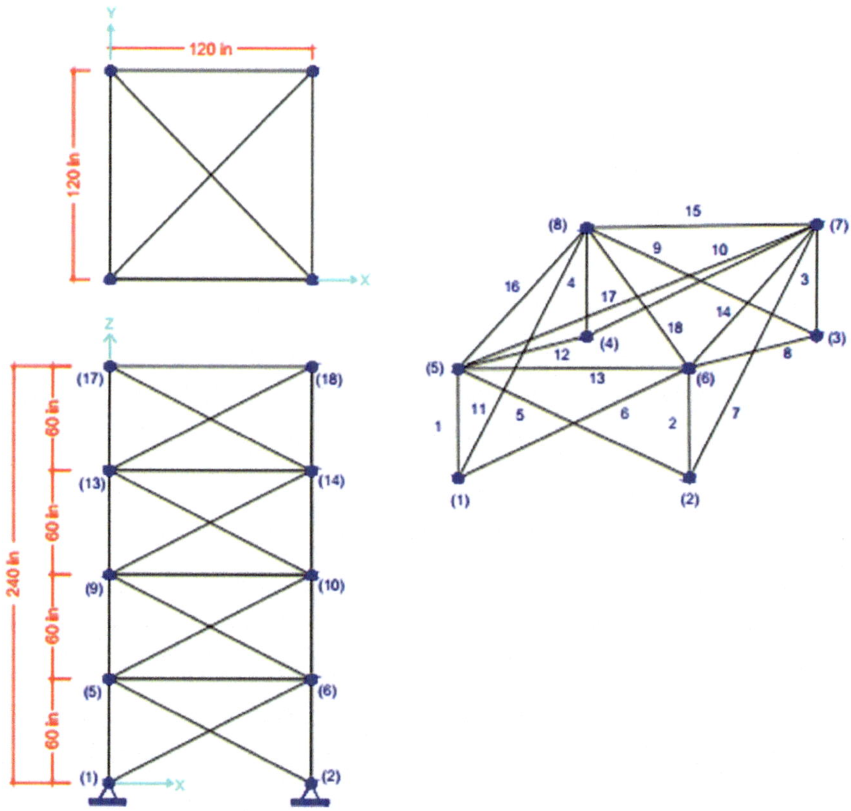

Fig. 10.3 Schematic of a 72-bar spatial truss and typical story

Table 10.2 Loading conditions for the 72-bar spatial truss

Node	Load case 1			Load case 2		
	F_x (kips)	F_y (kips)	F_z (kips)	F_x (kips)	F_y (kips)	F_z (kips)
17	5	5	– 5	0	0	– 5
18	0	0	0	0	0	– 5
19	0	0	0	0	0	– 5
20	0	0	0	0	0	– 5

shape and geometric symmetry. The minimum cross-sectional area of the members, which is the lower limit of the decision variables, is 0.775 in^2, and the maximum cross-sectional area is 20 in^2. The allowable stress in tension and pressure is determined based on the following equations.

$$\sigma_i = \begin{cases} 0.6F_y \; for \; \sigma_i \geq 0 \\ \sigma_i^- \quad for \; \sigma_i < 0 \end{cases} \tag{10.10}$$

For compressive stresses, we have the following equations:

$$\sigma_i^- = \begin{cases} (1 - \frac{\lambda^2}{2C^2})F_y/FS \; for \; \lambda < C \\ \frac{12\pi^2 E}{23\lambda^2} \qquad for \; \lambda \geq C \end{cases} \tag{10.11}$$

In this equation we have:

$$FS = (\frac{5}{3} + \frac{3\lambda}{8C} - \frac{\lambda^3}{8C^3}); \; C = \sqrt{(\frac{2\pi^2 E}{F_y})}; \; \lambda = \frac{kl}{r};$$
$$r = aA^b; \; a = 0.4993; \; b = 0.6777 \tag{10.12}$$

The components used include E modulus of elasticity, F_y the yield stress of steel, C the ratio of the limit slenderness which separates the elastic and inelastic buckling region in comparison with the existing slandering λ. Also, k is the effective length factor of the l and r is the radius of rotation of the limb.

In order to ensure the performance of chaos map and algorithms, as well as to increase the accuracy and sensitivity of calculations, each of the modes has been performed independently 30 times and the results related to the best response and the average value of responses are presented in Table 10.4. A graph of these results is shown in Fig. 10.6. Also, the coefficient of variation of the responses, which indicates the stability and robustness of the responses, has been calculated and another criterion for the efficiency of chaos map and algorithms has been obtained. The minimum and maximum cross-sectional area of all members is 0.775 in^2 and 20 in^2 respectively. By examining the optimization results for different combinations of algorithms chaos map and comparing it with standard mode, a significant improvement in reducing the weight of the 120-bar dome shaped truss is achieved. The enhanced results for

Table 10.3 Optimal design comparison for the 72-bar spatial truss

Number group	Element group	Standard	Logistic map			Gauss map			Piecewise map		
		SFLA	CSFLA 21	CSFLA 22	CSFLA 23	CSFLA 31	CSFLA 32	CSFLA 33	CSFLA 41	CSFLA 42	CSFLA 43
1	A_1–A_4	1.704	2.0333	1.8838	1.7342	1.8668	2.0909	1.974	1.7968	1.8785	1.8679
2	A_5–A_{12}	0.43536	0.49321	0.51771	0.49818	0.49805	0.48419	0.49875	0.49277	0.51506	0.5141
3	A_{13}–A_{16}	0.13211	0.10141	0.1017	0.11057	0.10072	0.10044	0.10066	0.10008	0.10195	0.10032
4	A_{17}–A_{18}	0.14147	0.11672	0.13031	0.10013	0.10274	0.10408	0.10072	0.10022	0.10138	0.10016
5	A_{19}–A_{22}	1.3687	1.4262	1.3291	1.1856	1.2966	1.1923	1.3135	1.3005	1.3363	1.2879
6	A_{23}–A_{30}	0.57001	0.47578	0.45837	0.51815	0.52677	0.50052	0.50708	0.51276	0.50295	0.52743
7	A_{31}–A_{34}	0.23509	0.1041	0.102	0.10104	0.10732	0.10001	0.10136	0.10008	0.10061	0.10027
8	A_{35}–A_{36}	0.12921	0.11034	0.10634	0.10035	0.10151	0.10858	0.10117	0.10253	0.10043	0.10021
9	A_{37}–A_{40}	0.50087	0.58248	0.55946	0.55573	0.57457	0.50173	0.48487	0.55385	0.50169	0.55526
10	A_{41}–A_{48}	0.44493	0.54439	0.5457	0.53052	0.48579	0.513	0.52355	0.53111	0.51263	0.51865
11	A_{49}–A_{52}	0.20845	0.11593	0.11272	0.1031	0.10241	0.10091	0.10013	0.10034	0.10058	0.10007
12	A_{53}–A_{54}	0.33062	0.11859	0.14729	0.14581	0.12025	0.1769	0.11155	0.10025	0.10409	0.10024
13	A_{55}–A_{58}	0.15318	0.15807	0.15718	0.15277	0.15665	0.15334	0.15754	0.15752	0.15668	0.15745
14	A_{59}–A_{66}	0.48464	0.48147	0.51867	0.55421	0.56828	0.53564	0.54013	0.53638	0.5463	0.5195
15	A_{67}–A_{70}	0.36945	0.34271	0.38799	0.40547	0.39855	0.44173	0.40895	0.40576	0.41533	0.40762
16	A_{71}–A_{72}	0.4882	0.61059	0.57221	0.61025	0.53955	0.53744	0.54731	0.62594	0.5567	0.57944
Best weight		379.9653	**379.2456**	379.5775	379.6237	379.9598	379.892	379.7902	379.3374	379.643	379.9275
Mean weight		407.6924	381.717	381.6647	384.5879	382.4042	381.1685	381.2064	380.1245	**380.0933**	380.4563
CV		0.3476	0.63125	0.51677	1.2676	0.85377	0.35557	0.32102	0.12756	**0.090428**	0.15624

The bold items in the table indicate the optimal values for the best weight, the best mean and the best coefficient of variation

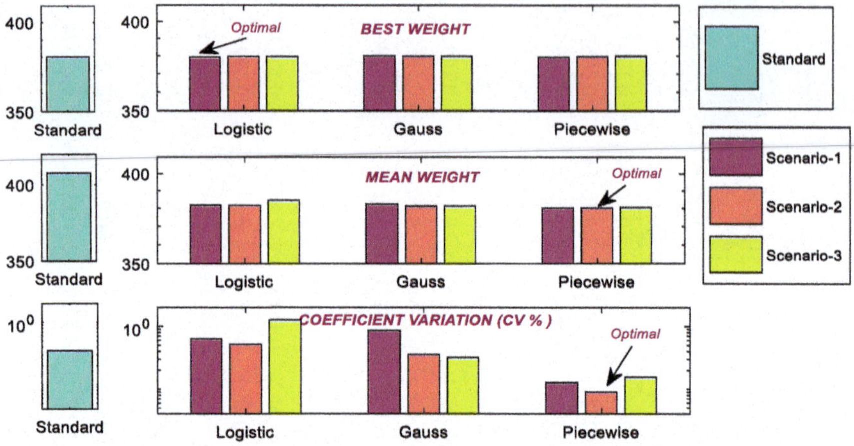

Fig. 10.4 Optimization results in standard mode and chaos map for the 72-bar spatial truss

Chaotic Shuffled Frog-Leaping Algorithm (CSFLA), by Logistic chaos map and scenario 2, a weight of 33,150.88 lb has the optimal response.

10.6 Truss Weight Optimization with Multiple Frequency Constraints

Natural frequencies of structures play a significant role in their dynamic behavior. When the frequency of incoming loads such as earthquake and wind are equal to them, the destructive phenomenon of resonance will happen. For this reason, one of the limitations in the optimization of structures will be determining a specific range for the natural frequencies of the structure. The components of the objective function in the frequency Constraints are associated with non-linear and non-convex modes, so the optimization with meta-heuristic algorithms in the standard mode, in most cases, stops at the local optimum. In the Chaotic Shuffled Frog-Leaping Algorithm (CSFLA), by embedding chaotic dynamic series, including Logistic and Gaussian chaos map, a significant improvement in the optimization results is achieved. These chaotic dynamic series can replace random distributions in colonial competition strategies. In this algorithm strategies, first, memeplexes are formed by dividing the whole population, then in the evolutionary stage of the complexes from the movement of each frog to the best member of the same memeplex and the best member of all memeplexes is used. If no improvement is obtained from these two strategies, a random response is substituted. By replacing the dynamic functions of chaos in the random distribution of these strategies, a balance is achieved between them, and the algorithm is saved from the trap of local optima, and local optima converge towards global optima. For each investigation, 6 chaotic modes are considered and compared

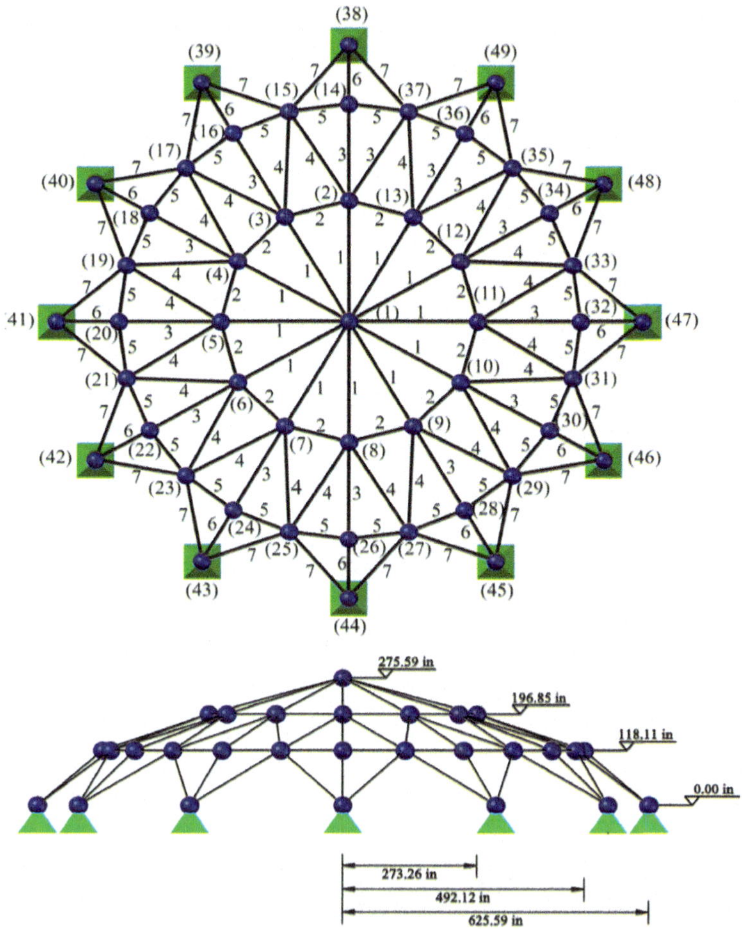

Fig. 10.5 Schematic of a 120-bar dome shaped truss

with the standard mode. In order to form statistical information and increase diversity in research, each model is independently evaluated 30 times. Finally, to analyze the statistical results, the best weight, the best average and the coefficient of variation are reported. Also, useful information is obtained about the weakness and strength of the algorithm and how to improve the results.

Table 10.4 Optimal design comparison for the 120-bar dome shaped truss

Number group	Standard	Logistic map			Gauss map			Piecewise map		
	SFLA	CSFLA 21	CSFLA 22	CSFLA 23	CSFLA 31	CSFLA 32	CSFLA 33	CSFLA 41	CSFLA 42	CSFLA 43
1	3.02444	3.02286	3.02503	3.02379	3.02455	3.02352	3.02334	3.0247	3.0239	3.02524
2	14.8198	14.7985	14.5471	15.1841	14.8652	14.6679	14.4783	14.8519	14.8214	14.3854
3	5.08948	4.99662	5.32093	5.00308	5.12642	4.99544	5.06597	5.12575	4.95975	5.26959
4	3.13522	3.03682	3.01691	3.08057	3.05768	3.04696	3.06751	3.05345	3.07105	3.05142
5	8.44909	8.63559	8.51234	8.39628	8.44675	8.64561	8.70894	8.44553	8.55944	8.56996
6	3.28865	3.28854	3.28677	3.28241	3.31821	3.36909	3.30234	3.33854	3.39524	3.34889
7	2.49618	2.49891	2.49527	2.49767	2.49618	2.49772	2.49784	2.49592	2.49723	2.49473
Best weight	33,249.66	33,161.87	**33,150.88**	33,214.3	33,182.98	33,172.01	33,186.86	33,179.81	33,199.39	33,179.31
Mean weight	33,253.02	33,170.67	**33,169.31**	33,235.68	33,196.08	33,205.57	33,224.42	33,235.05	33,220.57	33,292.53
Coefficient variation	**0.010729**	0.032714	0.05095	0.041964	0.033713	0.069327	0.1026	0.12327	0.077249	0.2839

The bold items in the table indicate the optimal values for the best weight, the best mean and the best coefficient of variation

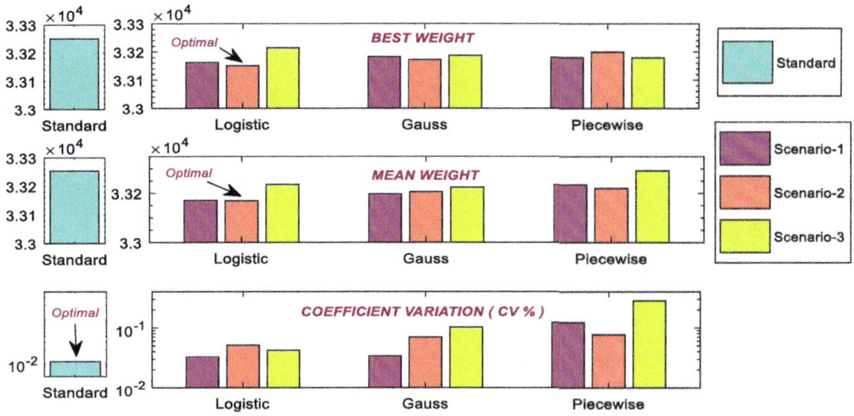

Fig. 10.6 Optimization results in standard mode and chaos map for the 120-bar dome shaped truss

10.6.1 Formulation of the Structural Optimization with Frequency Constraints

In structural optimization problems the main objective is to minimize the weight of the structures under some constraints. In this class of optimization problems, the design variables can be considered as: the cross-sectional areas of elements and/or the coordinates of nodes [9]. The optimization problem for a truss structure can be stated as Eq. (10.13).

$$\begin{array}{ll} Find & \{X\} = [x_1, x_2, \ldots, x_{ng}] \\ to\ minimize & W(\{X\}) = \sum_{i=1}^{nm} \rho_i \cdot A_i \cdot L_i \\ subjected\ to & \begin{cases} g_j(\{X\}) \leq 0, & j = 1, 2, \ldots, nc \\ x_{i\,\min} \leq x_i \leq x_{i\,\max} \end{cases} \end{array} \tag{10.13}$$

The most common and convenient method in metaheuristic algorithms to handle constraints is to use penalty functions. In this method, a constrained optimization problem is transformed into an unconstrained one by multiplying a certain value to objective function based on the amount of constraint violation appears in a problem. Therefore, the merit (or pseudo-objective) function which should be minimized is defined as:

$$Mer(X) = W(X) \times f_{penalty}(X) \tag{10.14}$$

where Mer(X) is the merit function; W(X) is the objective function (weight of truss structure); $f_{penalty}(X)$ is the penalty function which results from the violations of the constraints corresponding to the response of the structure:

$$f_{penalty}(X) = 1 + \gamma_p \cdot \sum_{j=1}^{m} \max(o, g_j(X)) = 1 + \gamma_p \times G(X) \quad (10.15)$$

where γ_p is penalty parameter, and G(X) is the constraint function. In dynamic constraint problem of a structure, the behavioral constraints are the restricted natural frequencies that are defined as follows:

$$if \begin{cases} \omega_n(X) \leq \omega_{upper,n} \\ \Rightarrow g_n(X) = \dfrac{\omega_n(X)}{\omega_{upper,n}} - 1 \leq 0 \quad for\ some\ natural\ frequencies\ n \\ \omega_m(X) \geq \omega_{lower,m} \\ \Rightarrow g_m(X) = 1 - \dfrac{\omega_m(X)}{\omega_{lower,m}} \leq 0 \quad for\ some\ natural\ frequencies\ m \end{cases} \quad (10.16)$$

where ω_n and ω_m are the nth and mth natural frequency, respectively; $\omega_{lower,n}$ and $\omega_{upper,m}$ are lower bound and upper bound of the nth and mth natural frequency, respectively. As can be seen from Eq. (10.16), the penalty function method consists of a penalty parameter where the feasible solutions are dependent on right tuning of these parameters. By setting small values of the penalty parameter, the pseudo-objective function becomes less sensitive to optimization constraints and the optimization process may converge to a violated solution. Conversely, setting the penalty parameter to a large value led to amplify the effect of optimization constraints and the optimization process may get stuck in a local optimum.

10.6.2 Numerical Examples of Optimal Truss Design

To determine the efficiency of the models in the Chaotic Shuffled Frog-Leaping Algorithm (CSFLA), the competition between 6 chaotic modes takes place. The best chaotic state is compared with the standard value to provide the final improvement. The used examples are selected from the well-known examples of truss optimization with layout and size optimization as criteria problems. Due to the diversity in the surveys, the optimal responses will be very close to the global optimum. The examples provided are as follows:

- The planar 10-bar truss;
- The simply supported 37-bar planar truss;
- 200-bar planar truss structure.

10.6.2.1 A Planar 10-Bar Truss

The planar 10-bar truss as depicted in Fig. 10.7 is a well-known benchmark problem in the field of weight optimization of the structures with frequency constraints. This truss only considers "optimizing the size of the sections" and the geometric shape of the structure is constant during the optimization process. The non-structural concentrated mass of 454 kg affected the free nodes. The mechanical properties of the structure are: The density of materials is 2770 kg/m^3, the modulus of elasticity is 69,800 MPa, the frequency limits of the structure in the first, second and third modes are greater than 7, 15 and 20 Hz, respectively. The cross-sectional area of each member is selected as an independent decision variable. Therefore, ten design variables with a lower limit of 0.645 cm^2 and an upper limit of 50 cm^2 are considered. In order to ensure the performance of chaotic functions and algorithms, as well as to increase the accuracy and sensitivity of calculations, each of the modes has been performed independently 20 times and the results related to the best response and the average value of responses are presented in Statistical Table 10.5. Also, the coefficient of variation of responses, which is a measure of the robustness of responses, has been calculated and used to compare the efficiency of chaos maps and standard algorithms. For quick access to optimization information, a bar chart of each component is shown in Fig. 10.8.

Examining the optimization results for different combinations of algorithms with chaos maps and comparing it with the standard mode, shows a significant improvement in reducing the weight of the planar 10-bar truss. The enhanced results for Chaotic Shuffled Frog-Leaping Algorithm (CSFLA), by Logistic chaos map and scenario 3, a weight of 521.044 kg has the optimal response.

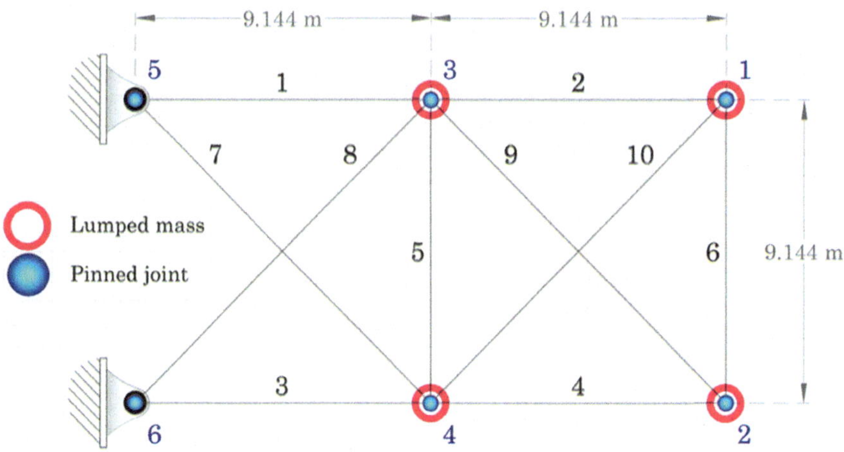

Fig. 10.7 Schematic of a planar 10-bar truss

Table 10.5 Optimal design comparison for the planar 10-bar truss

Number group	Standard	Logistic map			Gauss map		
	SFLA	CSFLA 21	CSFLA 22	CSFLA 23	CSFLA 31	CSFLA 32	CSFLA 33
A_1	33.9838	34.6288	34.882	37.9216	35.9871	36.2851	37.1624
A_2	14.037	14.839	14.9213	18.8548	10.1738	9.74954	17.6613
A_3	36.1756	35.0803	35.187	36.7634	38.5916	37.7406	35.7864
A_4	15.6942	14.2121	13.8511	9.7625	18.3453	18.4088	9.9331
A_5	0.646737	0.645574	0.645006	4.70946	5.05578	5.06712	5.10718
A_6	4.56014	4.58704	4.59821	4.07707	3.95804	3.96216	3.93953
A_7	21.7299	24.5837	25.1969	21.8032	21.9955	21.2607	21.8204
A_8	25.2785	23.1239	22.8199	19.9173	19.9587	20.9932	22.2706
A_9	12.6807	12.1321	12.0229	11.2156	12.9954	13.5713	11.4373
A_{10}	12.6331	13.0818	12.8407	13.2649	11.8807	11.2119	13.2403
Best weight (kg)	525.261	524.61	524.695	**521.044**	523.356	521.821	523.912
Mean weight (kg)	529.904	529.5	526.836	**524.274**	525.505	524.448	527.882
Coefficient variation	0.7736	0.5164	0.5102	0.5306	**0.4282**	0.4322	0.5577
ω_1 (HZ)	7.0001	7	7.0054	7.0001	7	7.00051	7
ω_2 (HZ)	16.1398	16.1552	16.1583	17.5327	17.6763	17.5364	17.6592
ω_3 (HZ)	20.0001	20	20	20	20	20	20

The bold items in the table indicate the optimal values for the best weight, the best mean and the best coefficient of variation

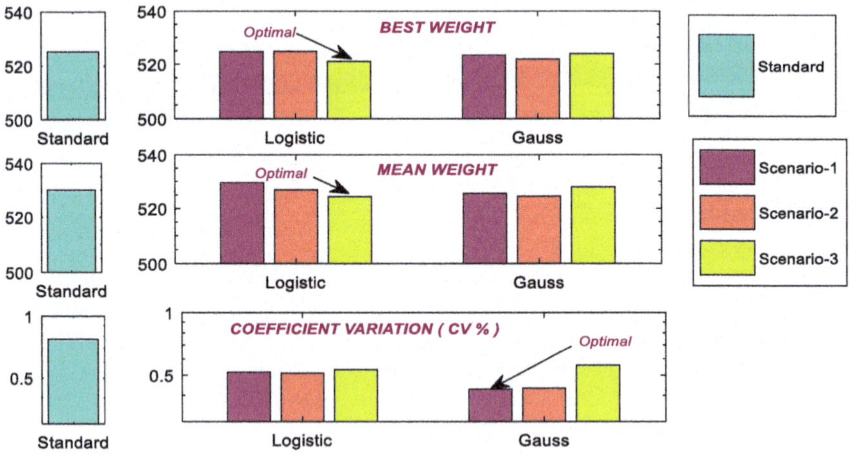

Fig. 10.8 Optimization results in standard mode and chaos map for the planar 10-bar truss

10.6.2.2 A Simply Supported 37-Bar Planar Truss

A 37-member planar truss with a simple abutment as shown in Fig. 10.9 is a well-known benchmark problem for optimizing the weight and shape of a truss with a frequency limit. This truss considers both optimizations of the section size and geometric coordinates of the nodes and the geometric shape of the structure is determined during the optimization process. The lower edge member is considered to be a rod member with a rectangular cross-section and an area of 40 cm^2. The other members are modeled as rod members.

These decision variables are grouped according to symmetry in the geometric shape. The y-axis coordinates form the shape variables in the upper year, taking into account the symmetry. In the lower axis, a non-structural concentrated mass of 10 kg affects all free nodes. The mechanical properties of the structure are: The density of the materials is 7800 kg/m^3, the modulus of elasticity is 21,000 MPa, the frequency limits of the structure in the first, second and third modes are greater than 20, 40 and 60 Hz, respectively. For the cross section of the members, the lower limit is 1 cm^2 and the upper limit is 10 cm^2. The total size and shape decision variables are 19 variables, of which 14 variables are related to the size and cross-sectional area of the members and 5 variables are related to the shape of the structure and the coordinates of the nodes. There are also three frequency limits. In order to ensure the performance of turbulent functions and algorithms, as well as to increase the accuracy and sensitivity of calculations, each of the independent modes has been performed 20 times and the results related to the best response and the average value of responses. Is presented in statistical Table 10.6. Also, the coefficient of variation of responses, which is a measure of the robustness and robustness of responses, has been calculated and used to compare the efficiency of turbulence functions and algorithms. For quick access to optimization information, a bar chart of each component is shown in Fig. 10.10. Examining the optimization results for different combinations of algorithms with turbulence functions and comparing it with the standard mode, shows a significant improvement in reducing the weight of A simply supported 37-bar planar truss. The enhanced results for Chaotic Shuffled Frog-Leaping Algorithm (CSFLA), by Logistic chaos map and scenario 3, a weight of 355.683 kg has the optimal response.

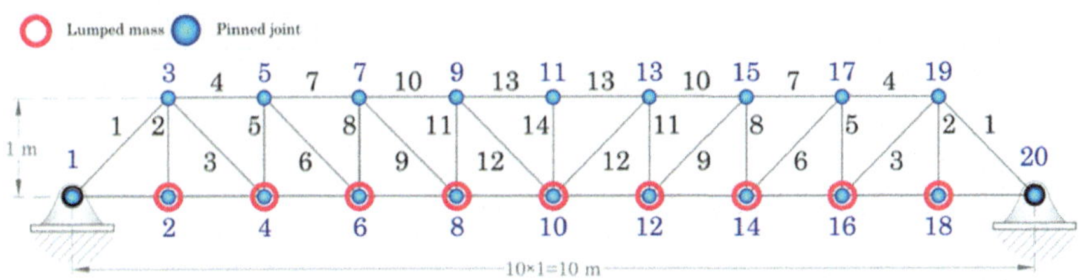

Fig. 10.9 Schematic of a simply supported 37-bar planar truss

Table 10.6 Optimal design comparison for the simply supported 37-bar planar truss

Decision variable	Standard	Logistic map			Gauss map		
	SFLA	CSFLA 21	CSFLA 22	CSFLA 23	CSFLA 31	CSFLA 32	CSFLA 33
Y_3, Y_{19} (m)	0.87225	0.91615	0.88734	0.89502	0.90021	0.92732	0.88713
Y_5, Y_{17} (m)	1.2239	1.3049	1.2439	1.2234	1.2589	1.2602	1.2238
Y_7, Y_{15} (m)	1.4126	1.481	1.3898	1.346	1.4028	1.3856	1.3453
Y_9, Y_{13} (m)	1.532	1.6268	1.4778	1.4546	1.5197	1.4939	1.4371
Y_{11} (m)	1.6129	1.6963	1.5393	1.5337	1.6043	1.5763	1.5153
A_1, A_{27}	3.2002	2.8568	2.9047	2.7566	2.5954	2.558	2.7658
A_2, A_{26}	1.0202	1.0063	1.0057	1.0017	1.0196	1.0017	1.0003
A_3, A_{24}	1.0034	1.0242	1.0013	1.0031	1.0026	1.0001	1.0002
A_4, A_{25}	2.991	2.6342	2.6468	2.4206	2.3871	2.4218	2.5612
A_5, A_{23}	1.0734	1.1743	1.1702	1.0792	1.0887	1.1287	1.0911
A_6, A_{21}	1.197	1.2268	1.2192	1.1864	1.2379	1.1949	1.2052
A_7, A_{22}	2.8867	2.4945	2.6144	2.4445	2.5737	2.3856	2.4695
A_8, A_{20}	1.4175	1.3818	1.297	1.2636	1.3563	1.2753	1.3179
A_9, A_{18}	1.3973	1.5031	1.4686	1.4895	1.5599	1.5134	1.4451
A_{10}, A_{19}	3.2172	2.6977	2.6775	2.4279	2.4353	2.367	2.4163
A_{11}, A_{17}	1.2906	1.262	1.122	1.1677	1.2203	1.1789	1.1117
A_{12}, A_{15}	1.2064	1.3403	1.2761	1.3596	1.3497	1.3608	1.3367
A_{13}, A_{16}	2.5953	2.7032	2.6343	2.34	2.3904	2.2686	2.3951
A_{14}	1.0235	1.0125	1.0008	1.0011	1.0021	1.0001	1.0001
Best weight (kg)	360.447	359.944	357.736	**355.683**	356.944	355.818	355.792
Mean weight (kg)	361.102	360.513	359.206	356.752	357.978	**356.735**	357.9754
Coefficient variation	0.2068	**0.1513**	0.2447	0.2004	0.1947	0.1829	0.4083
ω_1 (HZ)	20	20.001	20.0042	20.07	20	20.0042	20.0012
ω_2 (HZ)	40	40.002	40.001	40	40.0012	40.0052	40.0012
ω_3 (HZ)	60	60.001	60.0052	60.02	60	60	60.0007

The bold items in the table indicate the optimal values for the best weight, the best mean and the best coefficient of variation

10.6.2.3 A 200-Bar Planar Truss Structure

The 200-bar planar truss structure as shown in Fig. 10.11 is a well-known benchmark problem for weight optimization with frequency limits. This truss only considers "optimizing the size of the sections" and the geometric shape of the structure is constant during the optimization process. The decision variables related to the

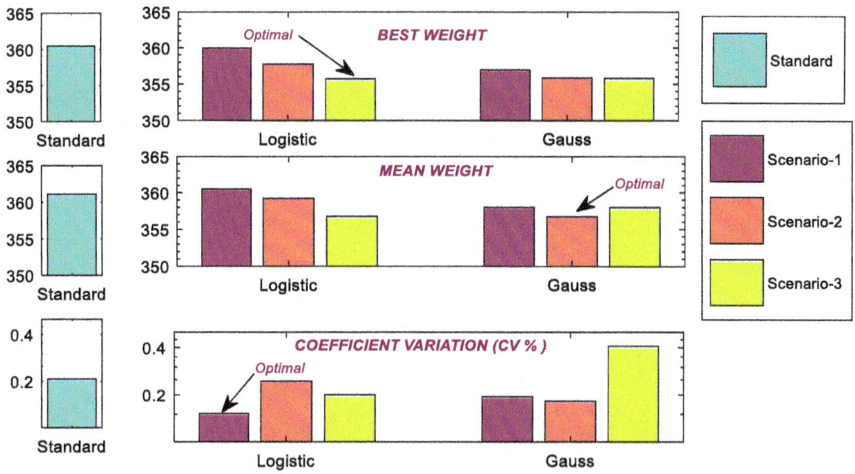

Fig. 10.10 Optimization results in standard mode and chaos map for the simply supported 37-bar planar truss

sections size of the members are classified into 29 groups according Table 10.7. Non-structural concentrated mass in nodes 1–5 and in the amount of 100 kg affect the structure. The mechanical characteristics of the structure are: material density 7860 kg/m^3, modulus of elasticity 210,000 MPa, frequency limits of the structure in the first, second and third modes are greater than 5, 10 and 15 Hz, respectively. For the cross section of the members, the lower limit range is 0.1 cm^2 and the upper limit is 25 cm^2. In order to ensure the performance of turbulent functions and algorithms, as well as to increase the accuracy and sensitivity of calculations, each of the modes has been performed independently 30 times and the results related to the best response and the average value of responses are presented in Statistical Table 10.8. Also, the coefficient of change of responses, which is a measure of the robustness and robustness of responses, has been calculated and used to compare the efficiency of turbulence functions and algorithms. For quick access to optimization information, a bar chart of each component is shown in Fig. 10.12.

Examining the optimization results for different combinations of algorithms with turbulence functions and comparing it with the standard mode, shows a significant and significant improvement in reducing the weight of the 200-bar planar truss structure. The enhanced results for Chaotic Shuffled Frog-Leaping Algorithm (CSFLA), by Logistic chaos map and scenario 3, a weight of 2024.084 kg has the optimal response.

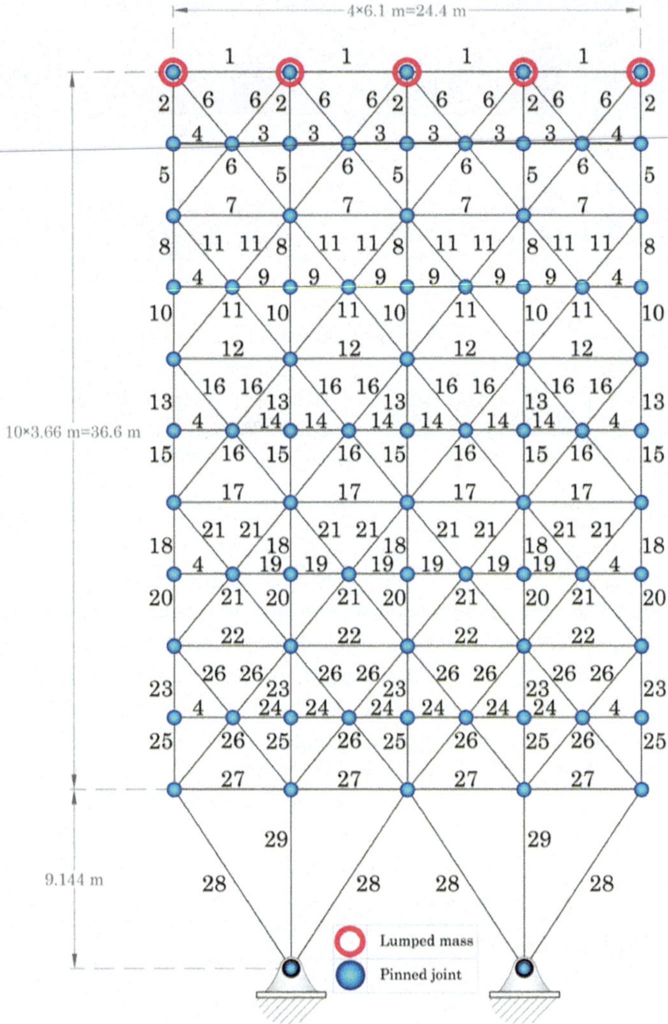

Fig. 10.11 Schematic of a 200-bar planar truss structure

10.7 Concluding Remarks

The basis of the Shuffled Frog-Leaping Algorithm (SFLA), is simulating the behavior of frogs to find food in the wetland. This algorithm has the advantages of both large groups of meta-heuristic algorithms, i.e., evolutionary groups and collective intelligence. The whole population of frogs is divided into memeplex groups, each group represents different types of frogs that are spread in different places of the answer space. Each group of these frogs then searches based on the best member of their memeplex. If the results do not improve, the search is based on the best member of

Table 10.7 Classification of members in the 200-bar planar truss structure

Group	Element group	Group	Element group
A_1	1, 2, 3, 4	A_{16}	82, 83, 85, 86, 88, 89, 91, 92, 103, 104, 106, 107, 109, 110, 112, 113
A_2	5, 8, 11, 14, 17	A_{17}	115, 116, 117, 118
A_3	19, 20, 21, 22, 23, 24	A_{18}	119, 122, 125, 128, 131
A_4	18, 25, 56, 63, 94, 101, 132, 139, 170, 177	A_{19}	133, 134, 135, 136, 137, 138
A_5	26, 29, 32, 35, 38	A_{20}	140, 143, 146, 149, 152
A_6	6, 7, 9, 10, 12, 13, 15, 16, 27, 28, 30, 31, 33, 34, 36, 37	A_{21}	120, 121, 123, 124, 126, 127, 129, 130, 141, 142, 144, 145, 147, 148, 150, 151
A_7	39, 40, 41, 42	A_{22}	153, 154, 155, 156
A_8	43, 46, 49, 52, 55	A_{23}	157, 160, 163, 166, 169
A_9	57, 58, 59, 60, 61, 62	A_{24}	171, 172, 173, 174, 175, 176
A_{10}	64, 67, 70, 73, 76	A_{25}	178, 181, 184, 187, 190
A_{11}	44, 45, 47, 48, 50, 51, 53, 54, 65, 66, 68, 69, 71, 72, 74, 75	A_{26}	158, 159, 161, 162, 164, 165, 167, 168, 179, 180, 182, 183, 185, 186, 188, 189
A_{12}	77, 78, 79, 80	A_{27}	191, 192, 193, 194
A_{13}	81, 84, 87, 90, 93	A_{28}	195, 197, 198, 200
A_{14}	95, 96, 97, 98, 99, 100	A_{29}	196, 199
A_{15}	102, 105, 108, 111, 114		

the whole set. In this step, the algorithm uses the strategy of the distance between the best and the worst answer to determine the corrective steps. Therefore, in order to improve the responses of each memeplex, we first go to the best response of the same memeplex, and if no improvement is achieved, we go to the best response of all memeplexes in the next strategy. With these two strategies, the exploration and exploitation steps are implemented in the algorithm. In these two stages of the strategy, random distributions have been used to create diversity. In this case, the uniform random distribution, it causes premature convergence in the algorithm. Therefore, the algorithm is stuck in the trap of local optima and stagnation is achieved. Chaotic functions converge the results towards the global optimum by creating sudden jumps and create a balance between search and extraction. In the frog jump algorithm, by embedding the Logistic, Gaussian and Piecewise chaos functions, a significant improvement in the optimization of problems. has been achieved in complex problems with multi-frequency constraints, the standard mode of the algorithm is associated with stagnation and stoppage, and by embedding chaos functions, a significant improvement in the optimization results has been achieved. In the competition between 9 chaotic models, the best answer is selected and compared with the standard state.

Table 10.8 Optimal design comparison for the 200-bar planar truss structure

Number group	Standard	Logistic map			Gauss map		
	SFLA	CSFLA 21	CSFLA 22	CSFLA 23	CSFLA 31	CSFLA 32	CSFLA 33
A_1	0.438196	0.320822	0.296752	0.445951	0.316746	0.408928	0.288308
A_2	0.652842	0.441576	0.46114	1.33645	0.436047	0.451832	0.553494
A_3	0.328543	0.100098	0.1	0.750166	0.100005	0.264877	0.114668
A_4	0.375015	0.100323	0.1	0.804303	0.1	0.1613	0.121633
A_5	0.680928	0.505316	0.49406	2.01661	0.504105	0.466654	0.615656
A_6	0.875196	0.808912	0.817324	0.486177	0.812114	0.804145	0.746749
A_7	0.170041	0.1	0.1	0.635807	0.100076	0.324079	0.268593
A_8	1.08424	1.43848	1.44704	1.72778	1.36841	1.30769	1.59452
A_9	0.184511	0.100034	0.10003	0.328264	0.100008	0.220682	0.385528
A_{10}	1.53887	1.70613	1.55525	2.30047	1.56218	1.67023	1.33189
A_{11}	0.967819	1.14744	1.15461	1.13084	1.15023	1.30099	1.27598
A_{12}	0.349921	0.121297	0.136157	0.484084	0.1	0.156065	0.281423
A_{13}	2.84816	2.9623	3.01704	1.76343	3.05204	3.16771	3.19282
A_{14}	0.247625	0.100243	0.100019	2.14056	0.100807	0.115638	0.119007
A_{15}	3.01965	3.25243	3.31087	1.92556	3.2694	2.97506	3.27524
A_{16}	1.40608	1.59962	1.56567	1.6027	1.56817	1.59148	1.42854
A_{17}	0.67016	0.291268	0.28491	0.479393	0.100024	0.22095	0.652634
A_{18}	4.20899	5.07353	5.06151	3.50493	5.21271	5.09572	4.29139
A_{19}	0.510358	0.1	0.107078	1.12626	0.101927	0.459611	0.215942
A_{20}	6.12385	5.32592	5.36521	3.66082	5.46103	5.51439	4.67435

(continued)

Table 10.8 (continued)

Number group	Standard	Logistic map			Gauss map		
	SFLA	CSFLA 21	CSFLA 22	CSFLA 23	CSFLA 31	CSFLA 32	CSFLA 33
A_{21}	2.2326	2.12231	2.13125	2.57621	2.06489	2.19036	2.09528
A_{22}	0.647795	0.804711	0.675661	1.69479	0.7197	0.537327	1.51148
A_{23}	8.28751	7.51474	7.71088	6.96652	7.62278	7.02392	7.92729
A_{24}	0.802519	0.100181	0.100659	2.85498	0.100281	0.231476	1.26288
A_{25}	8.77402	7.82922	7.80412	8.07202	7.94721	7.60048	7.72996
A_{26}	3.18971	2.8901	2.76555	1.54013	2.86366	3.02955	2.89139
A_{27}	9.17026	10.321	10.473	8.79468	10.2803	11.2682	9.80325
A_{28}	22.8365	20.976	21.2716	16.7021	21.3152	20.2884	18.6853
A_{29}	10.2741	11.33	10.9333	11.5787	11.057	11.8917	13.229
Best Weigh	2249.103	2157.654	2156.931	**2024.084**	2157.925	2182.835	2125.286
Mean Weight	2382.391	2356.243	2158.728	**2157.794**	2204.748	2159.232	2215.902
Coefficient Variation	5.212	**0.0519**	0.1102	7.916	0.0605	1.504	1.619
ω_1 (HZ)	5.00593	5.0025	5.0001	5	5.00003	5.0046	5.0014
ω_2 (HZ)	14.0948	12.3519	12.1493	14.8264	12.2811	13.4042	12.5054
ω_3 (HZ)	15.8139	15.0827	15.013	15.5885	15.015	15.2363	15.232

The bold items in the table indicate the optimal values for the best weight, the best mean and the best coefficient of variation

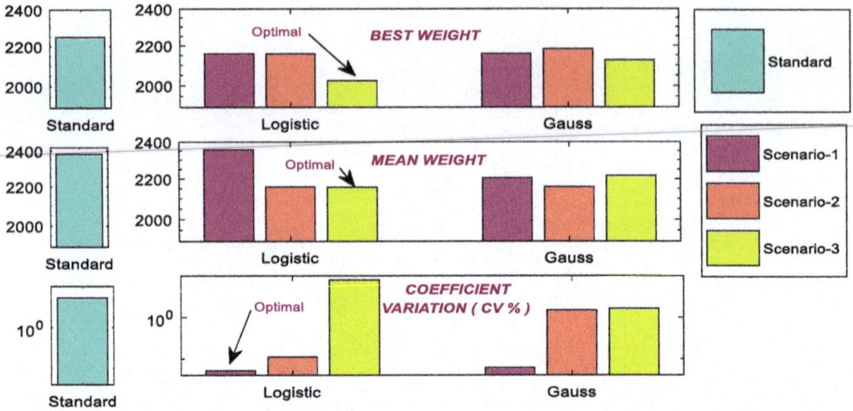

Fig. 10.12 Optimization results in standard mode and chaos map for the 200-bar planar truss structure

References

1. Kaveh, A.: Advances in Metaheuristic Algorithms for Optimal Design of Structures, 2nd edn. Springer, Cham, Switzerland (2021). https://doi.org/10.1007/978-3-319-05549-7
2. Eusuff, M., Lansey, K., Pasha, F.: Shuffled frog-leaping algorithm: a memetic meta-heuristic for discrete optimization. Eng. Optim. **38**(2), 129–154 (2006). https://doi.org/10.1080/03052150500384759
3. Kaveh, A., Yousefpoor, H.: Chaos-embedded metaheuristic algorithms for optimal design of truss structures. Sci. Iranica **29**(6), 2868–2885 (2022). https://doi.org/10.24200/sci.2022.59812.6441
4. Kaveh, A., Yosefpoor, H.: Chaotically enhanced metaheuristic algorithms for optimal design of truss structures with frequency constraints. Periodica Polytech. Civ. Eng. **66**(3), 900–921 (2022). https://doi.org/10.3311/PPci.20220
5. Kaveh, A., Zarfam, P., Aziminejad, A., Yosefpoor, H.: Comparison of four chaotic metaheuristic algorithms for optimal design of large-scale truss structures. Iran. J. Sci. Technol. Trans. Civ. Eng. https://doi.org/10.1007/s40996-022-00908-8
6. Ott, E.: Chaos in Dynamical Systems. Cambridge University Press, Cambridge (2002)
7. Bucolo, M., Caponetto, R., Fortuna, L., Frasca, M., Rizzo, A.: Does chaos work better than noise? IEEE Circuit Syst. Mag. **2**(3), 4–19 (2002)
8. Peitgen, H.-O., Jürgens, H., Saupe, D.: "Chaos and Fractals": New Frontiers of Science. Springer, Berlin (2006)
9. Kaveh, A., Sheikholeslami, R., Talatahari, S., Keshvari-Ilkhichi, M.: Chaotic swarming of particles: a new method for size optimization of truss structures. Adv. Eng. Softw. **67**, 136–47 (2014)

Chapter 11
Chaotic Particle Swarm Optimization

11.1 Introduction

Particle Swarm Optimization (PSO) is one of the most important algorithms in the field of swarm intelligence. This algorithm is inspired by social behavior, group of fishes and birds. In the Particle Swarm optimization, all members of the population communicate with each other and solve the problem through information exchange [1]. In this algorithm, each member of the population is called a particle, these particles are distributed in the search space of the function that aims to optimize it. The position of each particle is checked by calculating the objective function. Then, using the combination of information about its current location, the best location it has been in before, and also the best location for all particles, a direction to move is selected. After all the particles have updated their position, one step of the algorithm is finished. These steps are repeated until the conditions of the problem are met. In fact, the set of particles that are searching for the optimum of a function can be considered as a group of birds that are looking for food. The basis of the Particle Swarm Optimization (PSO) can be summarized as follows: "At each moment each particle adjusts its location in the search space according to the best location it has ever been in and the best location of all particles."

In the implementation of the algorithm, first a random initial population is formed within the range of decision variables [2]. These primary particles start to move in the search space of the problem and each time by estimating the position based on the fitness function, they seek to find better positions. In order to achieve the best position, they need two memories. One memory to store the best position of each particle in the past and the other memory to store the best position for all particles until now. Using the information stored in the memories, the particles decide for the next path of their movement. The decision components to determine the new position of each particle are presented in Fig. 11.1. In this decision, we combine a coefficient of the previous path, a coefficient of the best local position that the particle has experienced so far, and a coefficient of the best global position of all the particles, and we get access to the

A. Kaveh and H. Yousefpoor, *Chaotic Meta-heuristic Algorithms for Optimal Design of Structures*, Studies in Computational Intelligence 1129,
https://doi.org/10.1007/978-3-031-48918-1_11

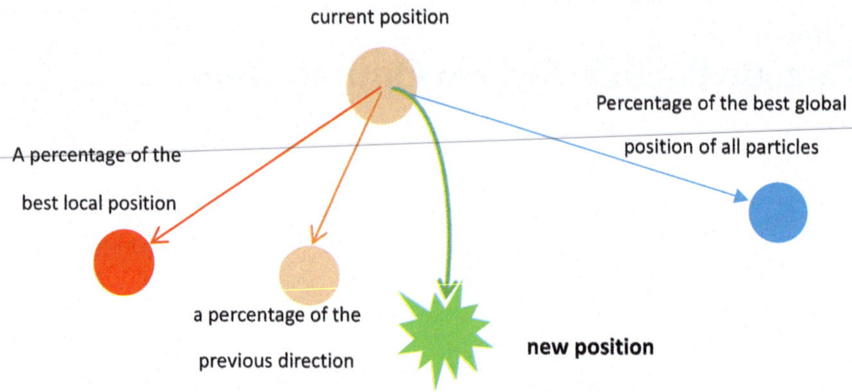

Fig. 11.1 Decision components in determining the new position of each particle in PSO

new position of the particle. Also, two random functions with uniform distribution have been proposed to create variety in local and national positioning. These functions play a decisive role in the interaction between the exploration and exploitation stages. For non-linear, non-convex and NP-hard problems, the proposed functions of the algorithm in the standard mode face problems such as early convergence, getting stuck in the trap of local optima and imbalance between the exploration and exploitation stages. Chaotic functions have the potential to save the algorithm from local optima by creating chaotic jumps. By embedded these functions in the algorithm and forming a Chaotic Particle Swarm optimization, premature convergence is resolved and we can obtain improved results with a low number of iterations. In optimization problems with multi-frequency constraints, Gaussian, Liebovitch and Piecewise chaos functions have brought significant improvement. In the rest of this chapter, after introducing the Particle Swarm Optimization (PSO), the basic steps of the algorithm will be presented in the standard mode, and then the Chaotic Particle Swarm Optimization (CPSO) will be formed with three scenarios. At the end, with famous examples of optimization of structures with multi-frequency constraints, the effectiveness of chaos functions will be challenged.

11.2 Standard Particle Swarm Optimization (PSO)

This algorithm is the most famous and most widely used meta-heuristic algorithm after the genetic algorithm. The domain is influenced by the algorithm with continuous variables and its idea was introduced by Kennedy and Eberhart [3]. In this algorithm, a number of particles are scattered in the search space, which is evaluated by applying each particle in the objective function, its criteria and value. By combining the previous velocity, the best position for each period of populations, and the best position for all population periods, the new position of each particle can

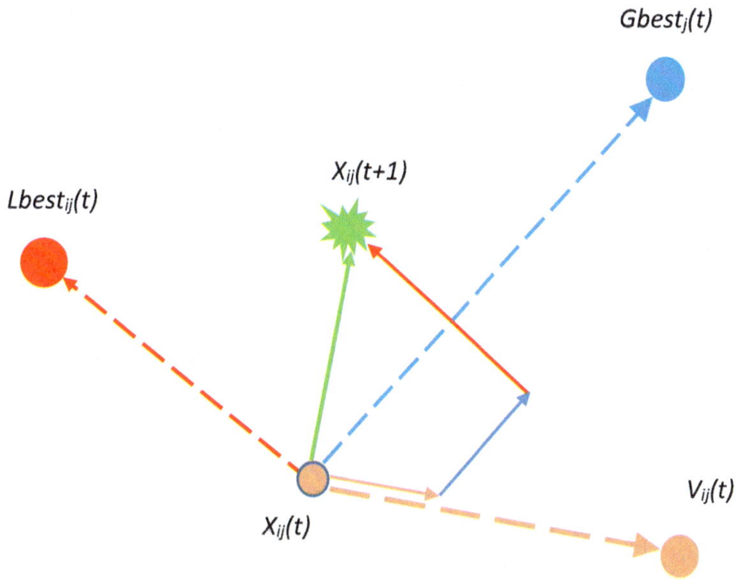

Fig. 11.2 Method to determine the new position in the particle for PSO

be determined. In this case, to move from the existing position to the new position, the velocity vector is formed and added to the existing position. The components that participate in the velocity vector are: the coefficient of the previous velocity, the coefficient of the best local position of the course, and the coefficient of the best position of the whole best global position of the course. The vector composition of these components is shown in Fig. 11.2.

11.3 Basic Steps in Particle Swarm Optimization

Step 1. Random formation of the initial population of particles and evaluation of each of them.
Step 2. Determine the best particle for each population period and the best particle for the global population period.
Step 3. Update the velocity vector and position vector based on the Eqs. (11.1) and (11.2):

$$\begin{aligned} v_{ij}(t+1) = {} & W \times v_{ij}(t) \\ & + C_1 \times \boldsymbol{rand_{1j}(t)} \times \{Lbest_{ij}(t) - X_{ij}(t)\} \\ & + C_2 \times \boldsymbol{rand_{2j}(t)} \times \{Gbest_j(t) - X_{ij}(t)\} \end{aligned} \tag{11.1}$$

$$X_{ij}(t+1) = X_{ij}(t) + v_{ij}(t+1) \tag{11.2}$$

In these relations W, the coefficient of inertia is usually chosen in the range of 0.4–0.9. C_1 and C_2 are also personal and collective learning coefficients, respectively. In the original version, the algorithm is proposed for both coefficients of zero to two.

Step 4. Check the termination conditions and repeat the operation from step 2 if necessary.

11.4 Chaos-Embedded Particle Swarm Optimization (CPSO)

This algorithm consists of two important positions, including the best position of each period and the best position of all periods, and to increase the variety in the search space, each is accompanied by a random coefficient [4–6]. By replacing the chaos functions in the random selection of each of the periods, we will see a significant improvement in the performance of the algorithm. The proposed scenarios for this replacement are as follows:

Scenario 1. Insert the chaos function as the coefficient of the best position of each period: In this case, the first chaos function **CHM1** in Eq. (11.1) replaces the random distribution that the result will be Eq. (11.3):

$$\begin{aligned} v_{ij}(t+1) = {} & W \times v_{ij}(t) \\ & + C_1 \times \boldsymbol{CHM_{1j}(t)} \times \{Lbest_{ij}(t) - X_{ij}(t)\} \\ & + C_2 \times \boldsymbol{rand_{2j}(t)} \times \{Gbest_j(t) - X_{ij}(t)\} \end{aligned} \tag{11.3}$$

Scenario 2. Positioning the chaos function as the coefficient of the best position of all periods: In this case, the second chaos function **CHM2** in Eq. (11.1). replaces the random distribution and the result will be Eq. (11.4):

$$\begin{aligned} v_{ij}(t+1) = {} & W \times v_{ij}(t) \\ & + C_1 \times \boldsymbol{rand_{1j}(t)} \times \{Lbest_{ij}(t) - X_{ij}(t)\} \\ & + C_2 \times \boldsymbol{CHM_{2j}(t)} \times \{Gbest_j(t) - X_{ij}(t)\} \end{aligned} \tag{11.4}$$

Scenario 3. Placing both turbulence functions as coefficients of the best position of each period and all periods: In this case, both chaos maps simultaneously replace the random distribution in Eq. (11.1).

11.5 Truss Weight Optimization with Multiple Frequency Constraints

In order to protect structures from the effects of destructive phenomena such as earthquakes and wind, their natural frequency should be constrained within a certain range. The shape of the objective function in the frequency constraint is strongly nonlinear and non-convex, therefore, the optimization using meta-heuristic algorithms in the standard mode stops at the local optimum in most cases. In the Chaotic Particle Swarm Optimization (CPSO), by embedding chaotic dynamic series, including Gaussian, Liebovitch and piecewise chaos series, a significant improvement in the optimization results is achieved. These chaotic dynamic series can replace random distributions in Particle Swarm Optimization (PSO), strategies. In the strategies of this algorithm to move to a new position, the best position of each stage and the best position of all stages are considered. This movement takes place with a uniform random distribution. By replacing the chaotic dynamic functions in the random distribution of these strategies, a balance is achieved between exploration and exploitation and the algorithm is saved from the local optimal trap and the local optimal converges to the global optimal. In each investigation, 9 disturbed states are considered and compared with the standard state. In order to form statistical information and increase diversity in research, each model is independently evaluated 30 times. Finally, to analyze the statistical results, the best weight, the best average and the coefficient of variation are reported. Also, useful information is obtained about the weakness and strength of the algorithm and how to improve the results. In the chaos maps, Liebovitch converges with a very high probability from a local minimum to a general minimum. Therefore, this function is suitable for improving the exploration conditions of algorithms. The chaos Gaussian map is most likely in the local minimum range and is suitable for improving exploitation conditions. Finally, the chaotic map of Piecewise simultaneously improves both exploration and exploitation conditions. Therefore, by selecting these chaotic functions, the weakness of algorithms of any kind is improved. The numerical distribution of these chaos maps for 100 iterations is presented in Fig. 11.3.

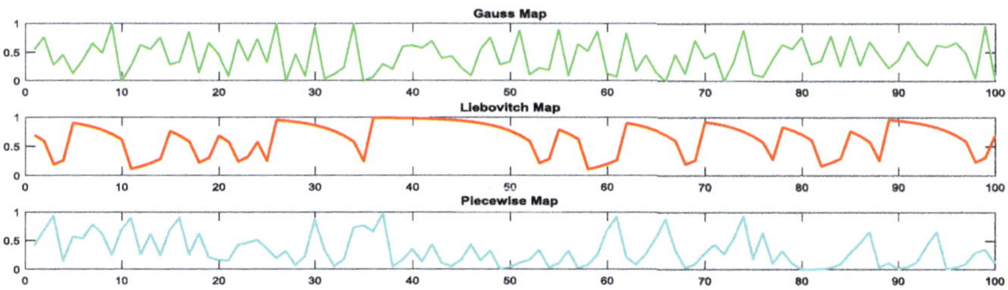

Fig. 11.3 Distribution of numerical values of chaos functions in 100 repetitions

11.5.1 Gaussian Map

Using this function in nonlinear dynamic behaviors has good results [7]. The statements of chaotic sequences in the Gaussian function are obtained according to the Eq. (11.5):

$$X_{n+1} = \begin{cases} 0 & X_n = 0 \\ \frac{1}{X_n} - \left[\frac{1}{X_n}\right] & X_n \neq 0 \end{cases} \tag{11.5}$$

11.5.2 Liebovitch Map

This function consists of three separate linear rules and there will be no common or repetitive sentences in these intervals [8]. The chaotic sequence sentences in this function are expressed by the Eq. (11.6):

$$X_{n+1} = \begin{cases} \alpha_1 X_n & 0 < X_n \leq d_1 \\ \frac{d_2 - X_n}{d_2 - d_1}: & d_1 < X_n \leq d_2 \\ 1 - \alpha_2 \times (1 - X_n) & d_2 < X_n \leq 1 \end{cases} \tag{11.6}$$

Here the values $d_1 = 1/3$ and $d_2 = 2/3$ are selected. A_1 and α_2 are also calculated based on the Eqs. (11.7) and (11.8):

$$\alpha_1 = \frac{d_2}{d_1} \times (1 - (d_2 - d_1)) \tag{11.7}$$

$$\alpha_2 = \frac{1}{d_2 - 1}((d_2 - 1) - d_1(d_2 - d_1)) \tag{11.8}$$

11.5.3 Piecewise Map

The piecewise map is the same as the Liebovitch map of three criteria, and the value of P is considered as the control parameter [9]. The range of changes of P is in the range of 0–0.5, which in this study we have considered 1/3, and the following category of equations indicates the relationship between the sentences of chaotic sequences in this function that expressed by the Eq. (11.9):

$$X_{n+1} = \begin{cases} \frac{X_n}{p} & 0 \le X_n < p \\ \frac{X_n - p}{0.5 - p} & p \le X_n < \frac{1}{2} \\ \frac{1 - p - X_n}{0.5 - p} & \frac{1}{2} \le X_n < 1 - p \\ \frac{1 - X_n}{p} & 1 - p \le X_n < 1 \end{cases} \tag{11.9}$$

11.6 Formulation of Coincidently Size and Layout Optimization of Truss Structures

In coincidently size and topology optimization problem two goals should be achieved: (1) optimal shape or topology of a ground structure should be found, (2) search for the optimal cross sections of the optimized shape for topology must be performed. Therefore, the problem starts with the ground structure, which is composed of all possible nodes and members [10]. Then, the node layout and the cross-sectional areas are found such that the cost of the structure is minimized. This optimization problem can formally be stated as follows:

$$Minimize \quad f(X) = \sum_{i=1}^{nm} \rho_i \cdot A_i \cdot L_i + \sum_{j=1}^{nn} b_j \tag{11.10}$$

$$subject\ to:$$

$$\begin{cases} \sigma_{i\min} \le \sigma_{il} \le \sigma_{i\max} \\ -\sigma_i \le -\sigma_i^E & i = 1, 2, \ldots, nm \quad l = 1, 2, \ldots, k \\ A_i \ge 0 \end{cases}$$

$$\begin{cases} \delta_{k\min} \le \delta_{kl} \le \delta_{k\max} \\ b_j \ge 0 & j = 1, 2, \ldots, nn \quad k = 1, 2, \ldots, kc \end{cases}$$

$$\begin{cases} \omega_m \le \omega_m^* \ for\ some\ frequencies\ m \\ \omega_n \ge \omega_n^* \ for\ some\ frequencies\ n \end{cases} \tag{11.11}$$

where f(x) is the structural cost; nm and nn denote the number of members and nodes of the ground structure, respectively; ρ_i, A_i and Li are density, cross sectional area and length of the ith member, respectively; b_j is the cost of the jth node; kc and lc are the number of displacement constraints and loading conditions, respectively; σ_{il} is the stress of the ith member under lth loading condition and σ_{imin} and σ_{imax} are its lower and upper bounds, respectively; δ_{kl} is the displacement of the kth degree of freedom under the lth loading condition, δ_{kmin} and δ_{kmax} are the corresponding lower and upper limits respectively; σ_i^E is the stress at which the ith member buckles, i.e. Euler buckling stress; ω_m is the mth natural frequency of the structure and ω_m^* is its upper bound. ω_n is the nth natural frequency of the structure and ω_n^* is its lower bound.

11.7 Numerical Examples of Optimal Truss Design

To determine the efficiency of Chaotic Particle Swarm Optimization (CPSO), the well-known examples of truss structures with layout and size optimization have been used as benchmark problems. By creating a challenge between chaotic modes and triple scenarios, a significant improvement in results is achieved. In this study, 10 independent answers compete with each other. Due to the variety of checks, the optimal responses will be very close to the global optimal. The examples provided are as follows:

- The planar 10-bar truss;
- The simply supported 37-bar planar truss;
- The 120-bar spatial dome
- 200-bar planar truss structure.

11.7.1 A Planar 10-Bar Truss

The planar 10-bar truss as depicted in Fig. 11.4 is a well-known benchmark problem in the field of weight optimization of the structures with frequency constraints. This truss only considers "optimizing the size of the sections" and the geometric shape of the structure is constant during the optimization process. The non-structural concentrated mass of 454 kg affected the free nodes.

The mechanical properties of the structure are: The density of materials is 2770 kg/m^3, the modulus of elasticity is 69,800 MPa, the frequency limits of the structure in the first, second and third modes are greater than 7, 15 and 20 Hz, respectively. The cross-sectional area of each member is selected as an independent decision variable.

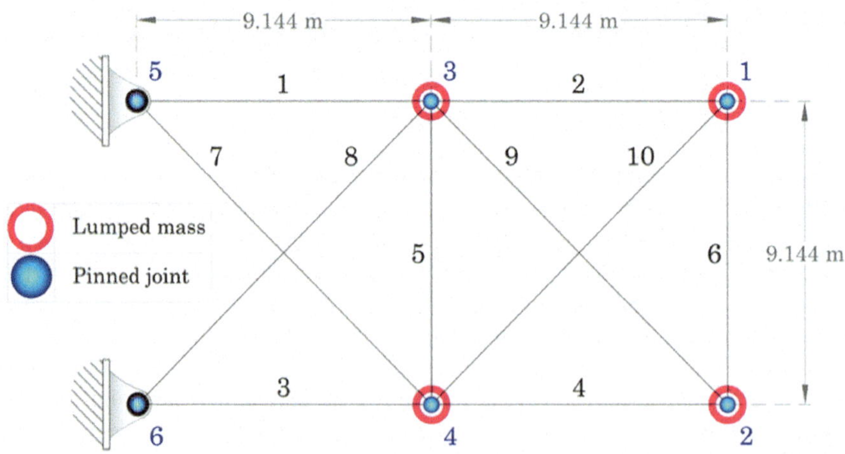

Fig. 11.4 Schematic of a planar 10-bar truss

Therefore, ten design variables with a lower limit of 0.645 cm^2 and an upper limit of 50 cm^2 are considered. In order to ensure the performance of chaotic functions and algorithms, as well as to increase the accuracy and sensitivity of calculations, each of the modes has been performed independently 30 times and the results related to the best response and the average value of responses are presented in Statistical Table 11.1. Also, the coefficient of variation of responses, which is a measure of the robustness of responses, has been calculated and used to compare the efficiency of chaos maps and standard algorithms. For quick access to optimization information, a bar chart of each component is shown in Fig. 11.5. Examining the optimization results for different combinations of algorithms with chaos maps and comparing it with the standard mode, shows a significant improvement in reducing the weight of the planar 10-bar truss. The enhanced results for Chaotic Particle Swarm Optimization (CPSO), by Liebovitch chaos map and scenario 3, a weight of 523.8687 kg has the optimal response.

11.7.2 A Simply Supported 37-Bar Planar Truss

A 37-member planar truss with a simple abutment as shown in Fig. 11.6 is a well-known benchmark problem for optimizing the weight and shape of a truss with a frequency limit. This truss considers both optimizations of the section size and geometric coordinates of the nodes and the geometric shape of the structure is determined during the optimization process. The lower edge member is considered to be a rod member with a rectangular cross-section and an area of 40 cm^2. The other members are modeled as rod members. These decision variables are grouped according to symmetry in the geometric shape. The y-axis coordinates form the shape variables in the upper year, taking into account the symmetry. In the lower axis, a non-structural concentrated mass of 10 kg affects all free nodes. The mechanical properties of the structure are: The density of the materials is 7800 kg/m^3, the modulus of elasticity is 21,000 MPa, the frequency limits of the structure in the first, second and third modes are greater than 20, 40 and 60 Hz, respectively. For the cross section of the members, the lower limit is 1 cm^2 and the upper limit is 10 cm^2. The total size and shape decision variables are 19 variables, of which 14 variables are related to the size and cross-sectional area of the members and 5 variables are related to the shape of the structure and the coordinates of the nodes.

There are also three frequency limits. In order to ensure the performance of turbulent functions and algorithms, as well as to increase the accuracy and sensitivity of calculations, each of the independent modes has been performed 20 times and the results related to the best response and the average value of responses. Is presented in statistical Table 11.2.

Also, the coefficient of variation of responses, which is a measure of the robustness and robustness of responses, has been calculated and used to compare the efficiency of turbulence functions and algorithms. For quick access to optimization information, a bar chart of each component is shown in Fig. 11.7. Examining the optimization results

Table 11.1 Optimal design comparison for the planar 10-bar truss

Number group	Standard	Gauss map			Liebovitch map			Piecewise map		
	PSO	CPSO 21	CPSO 22	CPSO 23	CPSO 31	CPSO 32	CPSO 33	CPSO 41	CPSO 42	CPSO 43
A_1	35.259	35.0948	35.0437	35.3284	35.0324	35.2096	34.9832	35.2941	35.2092	35.0093
A_2	14.5824	14.2811	15.5476	15.0577	13.9615	14.6239	14.4111	14.3202	14.4762	13.9671
A_3	35.0544	34.9958	35.4554	35.1815	34.9914	35.0478	35.4391	35.1627	34.9692	35.0325
A_4	14.7349	14.4285	14.4274	14.8277	14.7889	14.6929	15.4087	15.0217	13.9171	14.5278
A_5	0.645	0.645	0.645	0.645	0.645573	0.645771	0.650398	0.645	0.645	0.645
A_6	4.56394	4.59729	4.53114	4.53192	4.59819	4.56507	4.62847	4.56218	4.61202	4.60913
A_7	23.5806	23.8936	24.7277	23.7861	23.0705	23.7476	22.8	22.9775	24.8677	23.4941
A_8	23.9312	24.1665	22.2437	23.2404	24.959	23.7701	24.3772	24.5353	23.5268	24.7795
A_9	12.4843	12.3232	11.984	12.2462	12.8658	12.1191	12.6025	12.5758	11.9209	12.6917
A_{10}	12.2845	12.4855	12.8188	12.5013	12.0141	12.602	11.8543	12.0955	12.7533	12.1115
Best weight (kg)	524.4591	524.5474	524.6986	524.5	524.6305	524.1717	**523.8687**	524.5347	524.7221	524.6553
Mean weight (kg)	525.4801	**524.9422**	527.1482	525.7127	525.8038	525.0732	526.1518	526.3564	528.6349	527.6562
Coefficient variation	0.21273	**0.086986**	0.36712	0.29837	0.27701	0.20874	0.49741	0.45951	0.43922	0.41012
ω_1 (HZ)	7	7	7.0012	7.0037	7.0001	7	7.0001	7.0028	7.0087	7.0101
ω_2 (HZ)	16.1923	16.1125	16.1011	16.2105	16.1023	16.3152	16.2177	16.015	16.001	16.0270
ω_3 (HZ)	20	20	20	20.0045	20.0011	20.0029	20.0001	20	20.0012	20

The bold items in the table indicate the optimal values for the best weight, the best mean and the best coefficient of variation

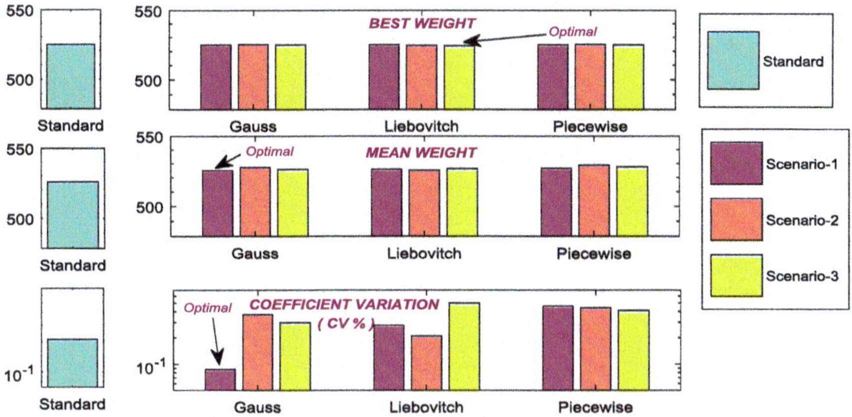

Fig. 11.5 Optimization results in standard mode and selection of chaos map for the planar 10-bar truss

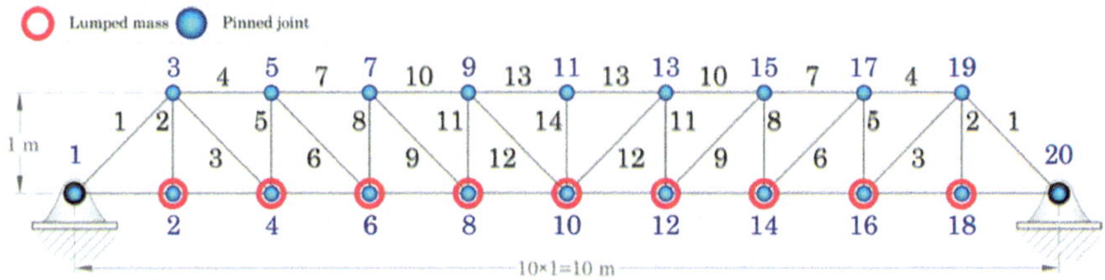

Fig. 11.6 Schematic of a simply supported 37-bar planar truss

for different combinations of algorithms with turbulence functions and comparing it with the standard mode, shows a significant improvement in reducing the weight of A simply supported 37-bar planar truss. The enhanced results for Chaotic Particle Swarm Optimization (CPSO), by Liebovitch chaos map and scenario 2, a weight of 359.7637 kg has the optimal response.

11.7.3 A 120-Bar Spatial Dome

The 120-bar spatial dome as shown in Fig. 11.8 is a well-known benchmark problem with weight-limit optimization. This truss only considers "optimizing the size of the sections" and the geometric shape of the structure is constant during the optimization process.

The decision variables related to the size of the members' sections and according to the symmetry in the geometric shape of the dome along the X and Y axes, are classified into 7 groups. Non-structural concentrated mass in all free nodes affects the structure. Their values are 3 kg in node 1500 kg in nodes 2–13 and 100 kg in

Table 11.2 Optimal design comparison for the simply supported 37-bar truss

Decision variable	Standard	Gauss map			Liebovitch map			Piecewise map		
	PSO	CPSO 21	CPSO 22	CPSO 23	CPSO 31	CPSO 32	CPSO 33	CPSO 41	CPSO 42	CPSO 43
Y_3, Y_{19} (m)	0.91606	0.93755	0.96678	0.92902	0.96626	0.943	0.99393	0.90309	0.96787	0.92307
Y_5, Y_{17} (m)	1.2886	1.3196	1.3473	1.31	1.3525	1.3247	1.3699	1.2719	1.3493	1.3102
Y_7, Y_{15} (m)	1.4806	1.5151	1.5282	1.514	1.5338	1.5119	1.5382	1.4554	1.5279	1.5135
Y_9, Y_{13} (m)	1.6318	1.6464	1.6576	1.6538	1.6458	1.6277	1.6723	1.5869	1.656	1.6476
Y_{11} (m)	1.7031	1.7206	1.7291	1.7339	1.7165	1.7003	1.7472	1.6574	1.7303	1.7177
A_1, A_{27}	3.0557	2.9362	2.786	3.0367	2.7545	2.8346	2.8506	3.0409	2.861	2.8574
A_2, A_{26}	1.0121	1.0045	1.0005	1.0034	1.0009	1.0008	1.038	1.0002	1.022	1.0009
A_3, A_{24}	1	1	1.0001	1	1.0005	1	1.0046	1	1	1
A_4, A_{25}	2.7751	2.545	2.4788	2.7044	2.5636	2.7427	2.4969	2.7599	2.6324	2.6358
A_5, A_{23}	1.1295	1.207	1.1958	1.1686	1.2109	1.1396	1.1329	1.0956	1.1978	1.1879
A_6, A_{21}	1.1923	1.1884	1.2703	1.1695	1.257	1.3037	1.1874	1.1348	1.2783	1.241
A_7, A_{22}	2.531	2.6178	2.5429	2.6098	2.5389	2.6397	2.5353	2.8209	2.4579	2.6388
A_8, A_{20}	1.392	1.4359	1.4253	1.4402	1.3893	1.3658	1.3486	1.3522	1.3649	1.4342
A_9, A_{18}	1.4774	1.4999	1.4608	1.5137	1.5318	1.5023	1.5757	1.6267	1.4438	1.5169
A_{10}, A_{19}	2.5635	2.4754	2.6174	2.3593	2.5882	2.4246	2.477	2.7575	2.488	2.464
A_{11}, A_{17}	1.1428	1.2279	1.2348	1.1699	1.2845	1.1184	1.2322	1.2305	1.2534	1.2408
A_{12}, A_{15}	1.4469	1.3292	1.3315	1.3863	1.2386	1.3815	1.2792	1.2486	1.3015	1.3522
A_{13}, A_{16}	2.5783	2.5167	2.4663	2.4321	2.5099	2.5625	2.5125	2.6518	2.5212	2.4493
A_{14}	1	1	1.0002	1	1	1.0007	1	1	1	1

(continued)

Table 11.2 (continued)

Decision variable	Standard	Gauss map			Liebovitch map			Piecewise map		
	PSO	CPSO 21	CPSO 22	CPSO 23	CPSO 31	CPSO 32	CPSO 33	CPSO 41	CPSO 42	CPSO 43
Best weight (kg)	359.9386	359.8147	359.8283	359.8734	359.8657	**359.7637**	359.8648	360.0338	359.8243	359.8418
Mean weight (kg)	360.1691	360.0518	360.1642	359.9823	360.0039	359.9635	359.961	360.1809	**359.9044**	359.9807
Coefficient variation	0.057778	0.059567	0.058111	0.053771	0.035223	0.068019	0.022057	0.036882	**0.021497**	0.0381
ω_1 (HZ)	20	20.0012	20.0054	20.0479	20	20	20.0058	20.0001	20.1011	20
ω_2 (HZ)	40.002	40.058	40.0001	40	40	40	40.0146	40	40	40.0001
ω_3 (HZ)	60	60.0025	60.0087	60	60	60	60.0054	60.0005	60.0180	60.0051

The bold items in the table indicate the optimal values for the best weight, the best mean and the best coefficient of variation

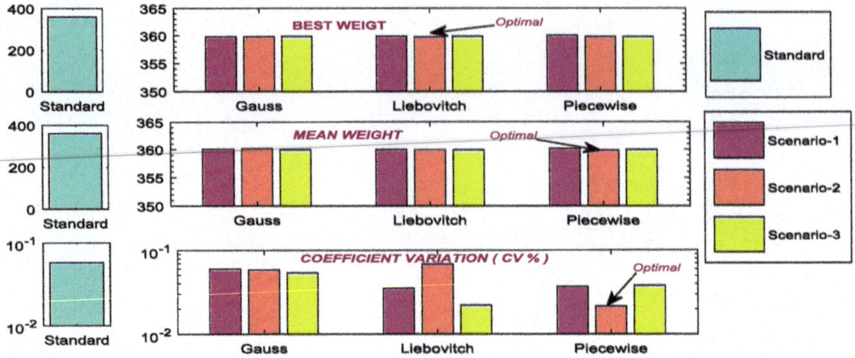

Fig. 11.7 Optimization results in standard mode and selection of chaos map for the simply supported 37-bar planer truss

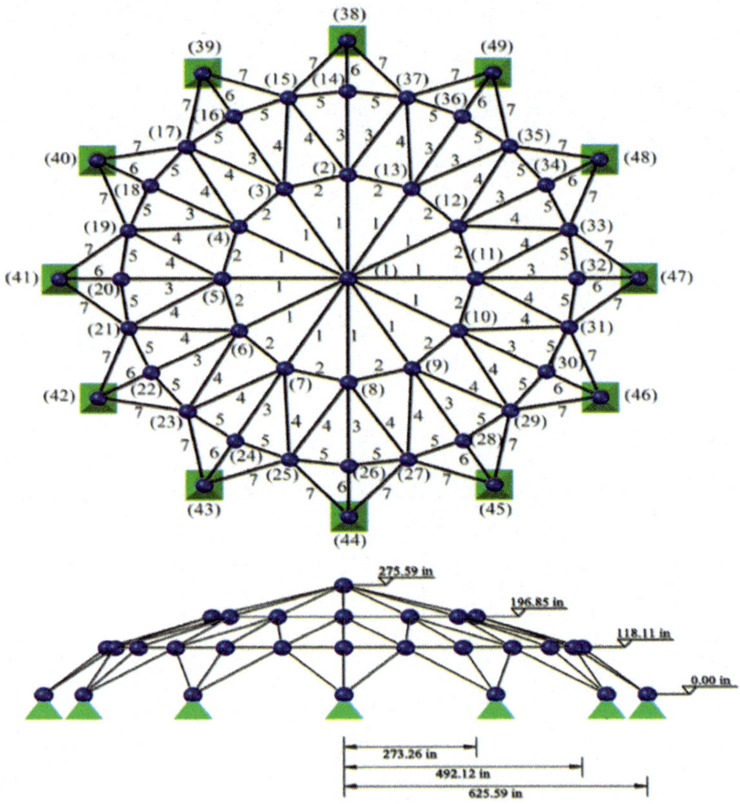

Fig. 11.8 Schematic of a 120-bar spatial dome

other nodes. The mechanical characteristics of the structure are: material density 7971.81 kg/m^3, modulus of elasticity 210,000 MPa, frequency limits of the structure in the first and second modes are greater than 9 and 11 Hz, respectively. For the cross section of the members, the range of the lower limit is 1 cm^2 and the upper limit is 129.3 cm^2. In order to ensure the performance of chaos map and algorithms, as well as to increase the accuracy and sensitivity of calculations, each of the modes has been performed independently 20 times and the results related to the best response and the average value of responses are presented in Statistical Table 11.3. Also, the coefficient of change of responses, which is a measure of the robustness and robustness of responses, has been calculated and used to compare the efficiency of turbulence functions and algorithms. For quick access to optimization information, the bar chart of each component is shown in Fig. 11.9. Examining the optimization results for different combinations of algorithms with turbulence functions and comparing it with the standard mode, shows a significant and significant improvement in reducing the weight of the 120-bar spatial dome. The enhanced results for Chaotic Particle Swarm Optimization (CPSO), by Liebovitch chaos map and scenario 1, a weight of 8709.136 kg has the optimal response.

11.7.4 A 200-Bar Planar Truss Structure

The 200-bar planar truss structure as shown in Fig. 11.10 is a well-known benchmark problem for weight optimization with frequency limits. This truss only considers "optimizing the size of the sections" and the geometric shape of the structure is constant during the optimization process. The decision variables related to the sections size of the members are classified into 29 groups according Table 11.4. Non-structural concentrated mass in nodes 1–5 and in the amount of 100 kg affect the structure. The mechanical characteristics of the structure are: material density 7860 kg/m^3, modulus of elasticity 210,000 MPa, frequency limits of the structure in the first, second and third modes are greater than 5, 10 and 15 Hz, respectively. For the cross section of the members, the lower limit range is 0.1 cm^2 and the upper limit is 25 cm^2. In order to ensure the performance of turbulent functions and algorithms, as well as to increase the accuracy and sensitivity of calculations, each of the modes has been performed independently 20 times and the results related to the best response and the average value of responses are presented in Statistical Table 11.5.

Also, the coefficient of change of responses, which is a measure of the robustness and robustness of responses, has been calculated and used to compare the efficiency of turbulence functions and algorithms. For quick access to optimization information, a bar chart of each component is shown in Fig. 11.11. Examining the optimization results for different combinations of algorithms with turbulence functions and comparing it with the standard mode, shows a significant and significant improvement in reducing the weight of the 200-bar planar truss structure. The enhanced results for Chaotic Particle Swarm Optimization (CPSO), by Piecewise chaos map and scenario 2, a weight of 2156.858 kg has the optimal response.

Table 11.3 Optimal design comparison for the 120-bar spatial dome

Decision	Standard	Gauss map			Liebovitch map			Piecewise map		
Variable	PSO	CPSO 21	CPSO 22	CPSO 23	CPSO 31	CPSO 32	CPSO 33	CPSO 41	CPSO 42	CPSO 43
A_1	19.3795	19.6444	19.5778	19.7077	19.6213	19.7024	19.5092	19.6382	19.5402	19.6519
A_2	40.7548	39.8476	40.1836	39.8299	39.9903	39.4796	40.4854	40.2618	40.5187	39.9484
A_3	10.2669	10.6842	10.405	10.6634	10.7143	10.7795	10.5667	10.5456	10.6457	10.4102
A_4	20.8168	21.182	21.2354	21.0454	21.2248	21.2914	21.5774	21.4902	20.9151	21.0051
A_5	10.0788	9.85836	9.86971	9.76361	9.79745	9.9324	9.72978	9.61367	9.71524	9.90123
A_6	12.2283	11.6953	12.1361	11.6275	11.6808	11.6087	12.0571	11.9929	11.6501	11.9817
A_7	15.0136	14.8425	14.6855	15.0443	14.7918	14.7694	14.3433	14.5097	15.097	14.9929
Best weight (kg)	8713.132	8709.171	8710.103	8709.749	**8709.136**	8710.882	8714.338	8712.972	8709.7	8709.895
Mean weight (kg)	8721.294	8715.666	8714.166	8716.759	8713.901	**8713.795**	8721.388	8717.515	8719.514	8718.659
Coefficient variation	0.10558	0.052448	**0.031257**	0.071588	0.046184	0.03573	0.086145	0.042645	0.10432	0.09619
ω_1 (HZ)	9	9	9.0011	9.0021	9	9	9.0101	9.0270	9.0035	9.0010
ω_2 (HZ)	11	11	11.0019	11.0028	11	11	11.0254	11.0047	11.0101	11.0001

The bold items in the table indicate the optimal values for the best weight, the best mean and the best coefficient of variation

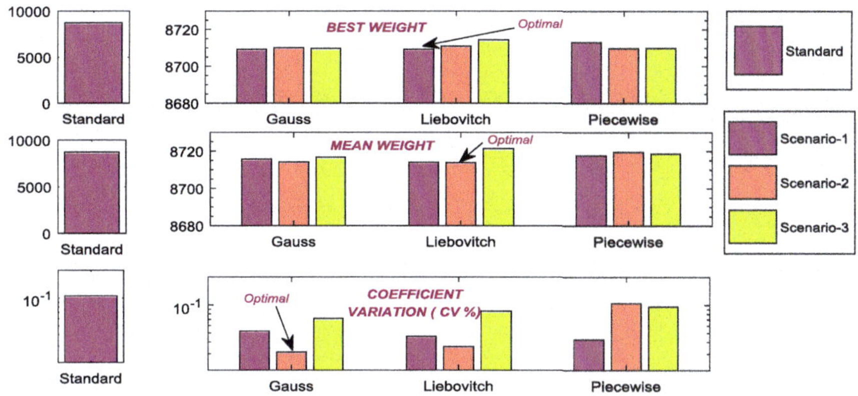

Fig. 11.9 Optimization results in standard mode and selection of chaos map for the 120-bar spatial dome

11.8 Concluding Remarks

In this algorithm, a number of particles are spread in the search space, and by applying each particle to the objective function, its criteria and value are determined. The new position of each particle can be determined by combining the previous velocity vector, the best position for each population period and the best position of all population periods. In this regard, to move from the existing position to the new position, the velocity vector is formed and added to the existing position. The components that contribute to the speed vector are: a coefficient of the previous speed, a coefficient of the best local position of the period and a coefficient of the best position of all periods. This algorithm consists of two important positions, including the best position of each course and the best position of all courses, and in order to increase the diversity in the search space, each is accompanied by a random coefficient. By replacing the chaos functions in the random selection of each period, we will witness a significant improvement in the performance of the algorithm. This algorithm consists of two important strategies, including the best position of each particle and the best position of all particles, and each is accompanied by a random coefficient in order to increase diversity in the search space. These strategies are responsible for the search and extraction stages. Their random selection, although it increases the diversity in the search space, it causes early convergence and decreases the optimization speed. Getting stuck in the trap of local optima is natural for the standard mode of the algorithm. By embedding chaos functions in the algorithm, chaotic jumps are formed and the solutions converge towards the overall optimality. Also, there is a balance between the search and extraction stages. By embedding Gossin, Leibovich and Pace-Weiss chaos functions, a significant improvement in optimization has been achieved. With extensive competition between 9 chaotic models, the best response is compared to the standard case. By embedding chaos functions, we can learn about the strengths and weaknesses of algorithms.

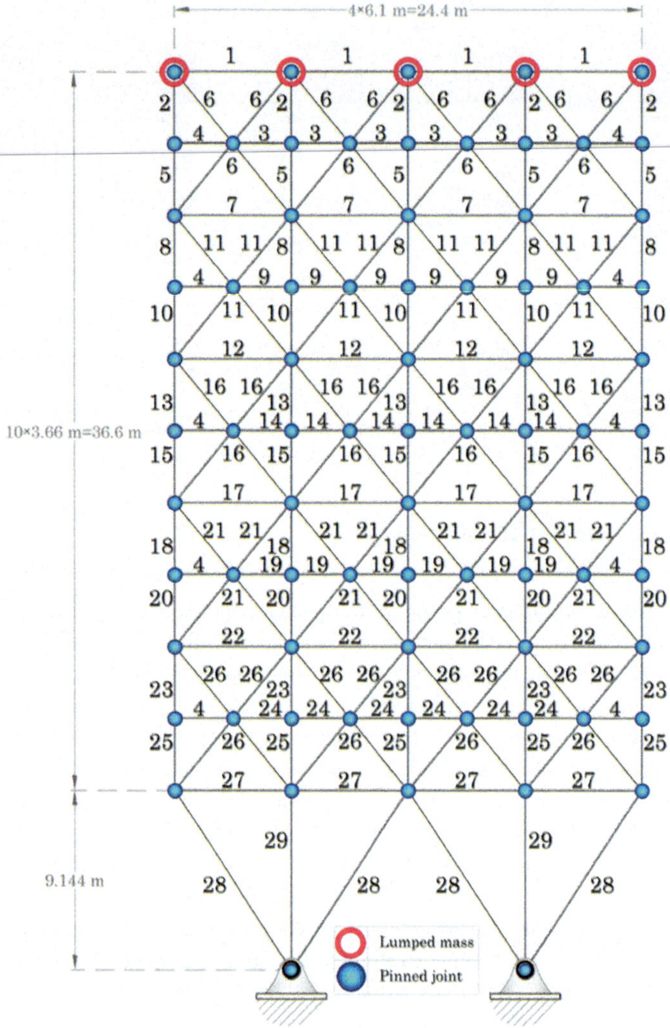

Fig. 11.10 Schematic of a 200-bar planar truss structure

Table 11.4 Classification of members in the 200-bar planar truss structure

Group	Element group	Group	Element group
A_1	1, 2, 3, 4	A_{16}	82, 83, 85, 86, 88, 89, 91, 92, 103, 104, 106, 107, 109, 110, 112, 113
A_2	5, 8, 11, 14, 17	A_{17}	115, 116, 117, 118
A_3	19, 20, 21, 22, 23, 24	A_{18}	119, 122, 125, 128, 131
A_4	18, 25, 56, 63, 94, 101, 132, 139, 170, 177	A_{19}	133, 134, 135, 136, 137, 138
A_5	26, 29, 32, 35, 38	A_{20}	140, 143, 146, 149, 152
A_6	6, 7, 9, 10, 12, 13, 15, 16, 27, 28, 30, 31, 33, 34, 36, 37	A_{21}	120, 121, 123, 124, 126, 127, 129, 130, 141, 142, 144, 145, 147, 148, 150, 151
A_7	39, 40, 41, 42	A_{22}	153, 154, 155, 156
A_8	43, 46, 49, 52, 55	A_{23}	157, 160, 163, 166, 169
A_9	57, 58, 59, 60, 61, 62	A_{24}	171, 172, 173, 174, 175, 176
A_{10}	64, 67, 70, 73, 76	A_{25}	178, 181, 184, 187, 190
A_{11}	44, 45, 47, 48, 50, 51, 53, 54, 65, 66, 68, 69, 71, 72, 74, 75	A_{26}	158, 159, 161, 162, 164, 165, 167, 168, 179, 180, 182, 183, 185, 186, 188, 189
A_{12}	77, 78, 79, 80	A_{27}	191, 192, 193, 194
A_{13}	81, 84, 87, 90, 93	A_{28}	195, 197, 198, 200
A_{14}	95, 96, 97, 98, 99, 100	A_{29}	196, 199
A_{15}	102, 105, 108, 111, 114		

Table 11.5 Optimal design comparison for the 200-bar planar truss structure

Number group	Standard	Gauss map			Liebovitch map			Piecewise map		
	PSO	CPSO 21	CPSO 22	CPSO 23	CPSO 31	CPSO 32	CPSO 33	CPSO 41	CPSO 42	CPSO 43
A_1	0.399749	0.300714	0.285416	0.302919	0.329312	0.252109	0.275649	0.306822	0.302063	0.331738
A_2	0.444018	0.487028	0.45969	0.455289	0.437424	0.6174	0.53805	0.45384	0.454298	0.446853
A_3	0.100015	0.100382	0.100009	0.100019	0.102373	0.100053	0.109993	0.1	0.100002	0.100061
A_4	0.102799	0.100013	0.1	0.100032	0.10001	0.100058	0.100019	0.100054	0.100009	0.104032
A_5	0.626832	0.520951	0.533541	0.52157	0.552121	0.60206	0.758105	0.519895	0.50404	0.500564
A_6	0.766369	0.81995	0.812665	0.817878	0.840882	0.776802	0.815062	0.826241	0.82176	0.814764
A_7	0.153687	0.100001	0.100473	0.103767	0.1	0.100005	0.100028	0.100017	0.100297	0.100078
A_8	1.35883	1.42365	1.43379	1.41952	1.40794	1.36105	1.70409	1.4243	1.42481	1.40633
A_9	0.101596	0.100261	0.100631	0.100039	0.102547	0.101161	0.110581	0.1	0.100167	0.1
A_{10}	1.60654	1.58252	1.61149	1.58241	1.43223	1.6202	1.65231	1.59397	1.58657	1.5741
A_{11}	1.18293	1.16679	1.1562	1.17507	1.12753	1.06618	1.2099	1.15281	1.14708	1.09042
A_{12}	0.164803	0.120587	0.114341	0.129209	0.101129	0.216426	0.112688	0.100308	0.125689	0.135314
A_{13}	3.37842	2.80757	3.08398	2.99326	3.06805	2.80859	2.6649	2.99326	3.01281	3.08775
A_{14}	0.100046	0.105857	0.100519	0.100171	0.100842	0.122789	0.102539	0.103248	0.102418	0.100026
A_{15}	3.78244	3.31833	3.19943	3.25082	3.19697	2.92814	3.58056	3.3063	3.28965	3.29915
A_{16}	1.55488	1.60853	1.55814	1.59914	1.55596	1.67065	1.71048	1.56763	1.58039	1.6002
A_{17}	0.254805	0.288016	0.263865	0.242181	0.186575	0.313125	0.100003	0.269978	0.214955	0.237313
A_{18}	4.93185	5.09667	5.17235	5.07989	5.30433	4.68168	5.05229	5.16548	5.15085	4.90231
A_{19}	0.101278	0.106401	0.102483	0.101136	0.100003	0.227643	0.276579	0.100638	0.100002	0.100763
A_{20}	5.38894	5.35331	5.39318	5.4956	5.38465	5.1408	6.70669	5.5009	5.61334	5.34998

(continued)

Table 11.5 (continued)

Number group	Standard	Gauss map			Liebovitch map			Piecewise map		
	PSO	CPSO 21	CPSO 22	CPSO 23	CPSO 31	CPSO 32	CPSO 33	CPSO 41	CPSO 42	CPSO 43
A_{21}	2.11414	2.1852	2.10591	2.08478	2.04959	2.04461	2.13393	2.11919	2.07797	2.13783
A_{22}	0.748555	0.747612	0.760202	0.732358	0.976692	0.812115	0.12335	0.68439	0.6979	0.643091
A_{23}	6.8166	7.90071	7.58912	7.68163	7.25963	8.09801	8.45921	7.77777	7.74148	7.65735
A_{24}	0.10133	0.151302	0.110966	0.155374	0.121364	0.38136	0.178433	0.188833	0.122323	0.333256
A_{25}	8.20108	7.84085	8.02421	7.8719	7.70375	8.08195	7.96771	8.08467	8.09707	7.87871
A_{26}	3.08458	2.90167	2.81619	2.88237	2.81883	2.95628	2.91668	2.86819	2.8186	3.06711
A_{27}	10.6055	10.2685	10.2189	10.208	9.98539	11.0421	8.68784	10.0997	10.5128	10.0275
A_{28}	20.5876	20.8016	21.2143	21.0229	21.4013	20.4845	20.9298	20.9686	21.0633	20.6351
A_{29}	11.6354	11.1379	11.0415	11.1761	11.9791	12.1661	12.8963	11.1265	10.7336	12.1313
Best Weigh	2170.827	2158.078	2157.138	2157.047	2161.826	2168.983	2190.149	2157.19	**2156.858**	2160.828
Mean Weight	2173.746	2158.985	2160.605	2158.22	2170.161	2183.684	2227.386	2157.856	**2157.656**	2240.553
Coefficient Variation	0.14449	0.055937	0.12816	0.045259	0.32028	0.54788	2.0575	**0.025187**	0.032237	4.8003
ω_1 (HZ)	5.0009	5.0104	5.0024	5.0162	5.0001	5.0297	5	5	5	5.0437
ω_2 (HZ)	12.2005	12.0250	12.0001	12.2732	12.1924	12.0120	12.9140	12.5249	12.2061	12.7351
ω_3 (HZ)	15.3945	15.0041	15.0264	15.0017	15.2436	15.1941	15.8721	15.3816	15.0124	15.0054

The bold items in the table indicate the optimal values for the best weight, the best mean and the best coefficient of variation

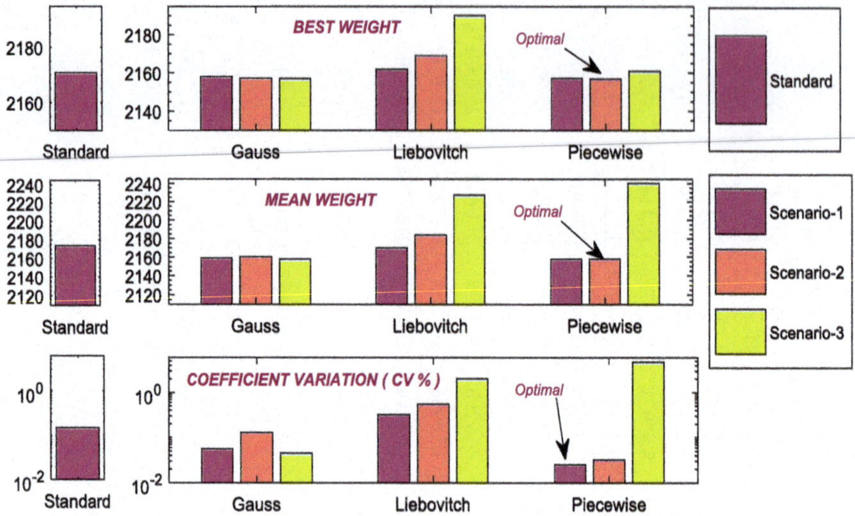

Fig. 11.11 Optimization results in standard mode and selection of chaos map for the 200-bar planar truss structure

References

1. Kaveh, A.: Advances in Metaheuristic Algorithms for Optimal Design of Structures, 2nd edn. Springer, Cham, Switzerland (2021). https://doi.org/10.1007/978-3-319-05549-7
2. Kaveh, A., Zolghadr, A.: Democratic PSO for truss layout and size optimization with frequency constraints. Comput. Struct. **130**, 10–21 (2014). https://doi.org/10.1016/j.compstruc.2013.09.002
3. Kennedy, J., Eberhart, R.: Particle swarm optimization. In: Proceedings of ICNN'95—International Conference on Neural Networks, Perth, WA, Australia, pp. 1942–1948 (1995)
4. Kaveh, A., Yousefpoor, H.: Chaos-embedded metaheuristic algorithms for optimal design of truss structures. Sci. Iranica **29**(6), 2868–2885 (2022). https://doi.org/10.24200/sci.2022.59812.6441
5. Kaveh, A., Yosefpoor, H.: Chaotically enhanced metaheuristic algorithms for optimal design of truss structures with frequency constraints. Periodica Polytech. Civ. Eng. **66**(3), 900–921 (2022). https://doi.org/10.3311/PPci.20220
6. Kaveh, A., Zarfam, P., Aziminejad, A., Yosefpoor, H.: Comparison of four chaotic metaheuristic algorithms for optimal design of large-scale truss structures. Iran. J. Sci. Technol. Trans. Civ. Eng. https://doi.org/10.1007/s40996-022-00908-8
7. Bucolo, M., Caponetto, R., Fortuna, L., Frasca, M., Rizzo, A.: Does chaos work better than noise? IEEE Circuit Syst. Mag. **2**(3), 4–19 (2002)
8. Erramilli, A., Singh, R., Pruthi, P.: Modeling Packet Traffic with Chaotic Maps. KTH, Stockholm (1994)
9. Peitgen, H.-O., Jürgens, H., Saupe, D.: "Chaos and Fractals": New Frontiers of Science. Springer, Berlin (2006)
10. Kaveh, A., Sheikholeslami, R., Talatahari, S., Keshvari-Ilkhichi, M.: Chaotic swarming of particles: a new method for size optimization of truss structures. Adv. Eng. Softw. **67**, 136–147 (2014)

Chapter 12
Chaotic Tug-of-War Optimization

12.1 Introduction

In fact, no specific time or place can be identified as the origin of tug-of-war. However, it can be said that the tug-of-war competition originates from ancient celebrations and ceremonies that existed in many parts of the world [1]. The leaders of the teams had a lot of power and connected the teams with their hands. The rules and regulations of the game of tug-of-war come with a wide variety. But the basic part of the game has been kept unchanged. In this competition, each member of the team holds the rope tightly and pulls it towards them. Each solution is considered as a team and all population forms a league. The merit of each team is proportional to the pull weight, and is determined by the objective function. Accordingly, the best team has the most weight and the worst team has the least weight. During one period of the game, the lightest team loses the competition and is drawn to the heaviest team. According to Newton's second law, the lightest team accelerates towards the heaviest team Fig. 12.1. The displacement position in each step can be accessed based on the equation of motion with constant acceleration Eq. (12.1) describes Newton's laws of motion in mechanics. This relationship is suggested to determine the position of the rope.

$$x_i^{new} = \frac{1}{2}at^2 + x_i^{old} \tag{12.1}$$

If we consider this relationship as the moving steps of team i towards team j, we can rewrite it as form Eq. (12.2).

$$stepsize_{ij} = \frac{1}{2}a_{ij}\Delta T^2 + \alpha\boldsymbol{\beta}(Lb - Ub) \odot randn \tag{12.2}$$

A. Kaveh and H. Yousefpoor, *Chaotic Meta-heuristic Algorithms for Optimal Design of Structures*, Studies in Computational Intelligence 1129,
https://doi.org/10.1007/978-3-031-48918-1_12

Fig. 12.1 A competing team in a tug-of-war sport competition

The coefficient β is known as the scaling factor and it controls the steps of the proposed response. Also, the random function vector randn (1, n) is applied with normal distribution and sentence by sentence multiplication. These coefficients play a decisive role in creating diversity, determining the location of local optima and global optima. The proposed functions of the algorithm in the standard mode, for non-linear, non-convex and NP-Hard problems, will be associated with premature convergence problems. In such cases, the optimization operation is caught in the trap of local optima and the imbalance between exploration and exploitation occurs. Chaotic functions, by creating chaotic jumps, have the necessary potential to free the algorithm from the trap of local optima. By embedded these functions in the algorithm and forming a Chaotic Tug-of-War Optimization (CTWO), premature convergence is eliminated and we can obtain improved results with a low number of iterations. In optimization problems with multi-frequency constraints, Logistic and Gaussian chaos functions have brought significant improvement. In the rest of this chapter, after introducing the Tug-of-War Optimization (TWO), the basic steps of the algorithm will be presented in the standard mode, and then the Chaotic Tug-of-War Optimization (CTWO) will be formed with three scenarios. Finally, with well-known optimization examples of multi-frequency constraints, the efficiency of chaos functions will be challenged.

12.2 Standard Tug of War Optimization (TWO)

This algorithm was presented by Kaveh and Zolghadr [2]. In this algorithm, it is inspired by the game of tug-of-war and is based on the population. Like other meta-heuristic algorithms, a set of initial answers is selected from the decision space and each solution is considered as a team, and all of them form a league. In each iteration of the algorithm, the evaluation of the teams is determined and sorted based on merit. In this competition, the best team has the most weight and the worst team has the least weight. Both competing teams are pulling the rope, the light team loses the competition and moves to the heavy team.

12.3 Basic Steps in Tug of War Optimization

Step 1 Determining the parameters of the algorithm: The parameters include the following.

- The number of teams participating in the competition, which is represented by the symbol nT
- the number of members of each team or T, which is known as the league matrix.

In this way, in each step, it consists of evaluating the values of the vectors corresponding to the objective function and the penalized objective function.

Step 2 Determining the weights in the tug-of-war competition: Each solution is known as a team from the league, and the numerical value of its weight in these competitions is determined according to the Eq. (12.3):

$$W_i = \frac{PFit_i - \min(PFit)}{\min(PFit) - \max(PFit)} + 1 \tag{12.3}$$

In this regard, the penalized objective function along with its maximum and minimum value is considered to determine the weight of each team. Based on this relationship, the weight of each team is placed between 1 and 2. Numerical value 2 is the best and heaviest team.

Step 3 Competing between teams: Each team in the league competes with all other teams. To move to its new position in each period of repetition, the tensile force applied by each team is proportional to the frictional force at rest. In the modeling, the value of the coefficient of friction is assumed to be one, and the pulling force between the two teams i and j can be the maximum of the following two values:

$$F_{p,ij} = \max\{W_i \mu_s, W_j \mu_s\} \tag{12.4}$$

Finally, the resulting force on team i in the face of heavier team j is obtained as follows:

$$F_{r,ij} = F_{p,ij} - W_i \mu_k \tag{12.5}$$

The amount of acceleration in the movement of team i towards team j results from the following equation:

$$a_{ij} = \frac{F_{r,ij}}{W_i \mu_k} g_{ij}, \; g_{ij} = T_j - T_i \tag{12.6}$$

In this relationship, g is the acceleration constant of gravity, which is obtained from the difference of the position vector of the two teams. Also, the displacement amount of team i after competing with team j is determined according to the following relationship:

$$stepsize_{ij} = \frac{1}{2} a_{ij} \Delta T^2 + \alpha \boldsymbol{\beta} (Lb - Ub) \odot randn \tag{12.7}$$

The second term of this relationship considers random cases in the amount of team displacement. This term can consider that part of the search space that team i travels before being stopped by team j. The α coefficient gradually reduces the possible effects and is selected in the range of [0.99, 0.9]. Larger values for α slow down the convergence rate and give enough time for the proposed answers to fully explore the search space. Also, β is the scaling factor that can be selected in the range of [1, 0]. This parameter controls the steps of the proposed answers, and in cases where we need more accuracy for searching, the β parameter is selected with smaller steps. Also, Lb and Ub are the lower limit and upper limit of the search space and their differences express the allowed range for design variables. The multiplication used is of member-by-member type and random values with standard normal distribution are selected for randn vector. In this relationship, the multiplication used is member-by-member type and a random vector with normal distribution is selected. We can define ΔT time periods as 1. It should be noted that in cases where j is lighter than i, the values related to the displacement of team i are considered zero. Finally, the total relocation of team i is summarized as follows:

$$stepsize_{ij} = \sum_{j=1}^{nT} stepsize_{ij}, \quad i \neq j \tag{12.8}$$

The new position of team i is as follows:

$$T_i^{new} = T_i + stepsize_i \tag{12.9}$$

Step 4 Updating the position of league teams: After the competition of the league teams, the results should be updated. For this, the new results are replaced if they are better than the current values.
Step 5 The possibility of the results leaving the range of the control decision space and if in some cases the positions of the teams are out of this search range, it is reconstructed based on the following relationship.

$$T_{ij} = bestT_j + (\frac{\boldsymbol{randn}}{NIT_s})(bestT_j - T_{ij}) \tag{12.10}$$

In this regard, bestT is the best team so far and NITs is a repeat number.
Step 6 The termination conditions are checked and if necessary, the competition between the league teams is repeated.

12.4 Chaos-Embedded Tug of War Optimization (CTWO)

In order to determine the new position of the league teams, two exploration and exploitation strategies have been considered in the scale factor β and the search space limitation [3–5]. By replacing the chaos maps in the random selections related to these steps, we will see a significant improvement in the performance of the algorithm. The suggested scenarios for this replacement are as follows:

Scenario 1 Replacing the chaos function in choosing the scale factor β: In this case, the first CHM1 chaos map is replaced in Eq. (12.7). The result will be as below:

$$stepsize_{ij} = \frac{1}{2}a_{ij}\Delta T^2 + \alpha \cdot \boldsymbol{CHM1} \otimes (Lb - Ub) \otimes randn(1, n) \tag{12.11}$$

Scenario 2 Replacing the chaos map in applying the search space limitation: In this case, the second CHM2 chaos map is replaced in Eq. (12.10). The result will be as below:

$$T_{ij} = bestT_j + (\frac{\boldsymbol{CHM2}}{NIT_s})(bestT_j - T_{ij}) \tag{12.12}$$

Scenario 3 placing chaos functions in both stages: In this case, two chaos maps are simultaneously replaced in Eqs. (12.7) and (12.10).

12.5 Truss Weight Optimization with Multiple Frequency Constraints

For dynamic control of structures, their natural frequency must be limited to a certain range so that the destructive phenomenon of resonance does not occur. The mathematical model for frequency limitation consists of implicit and complex variables, and optimization using meta-heuristic algorithms in the standard mode is faced with premature convergence in most cases. In Chaotic Tug of War Optimization (CTWO), by embedding chaotic dynamic series, including Logistic and Gaussian series, a significant improvement in optimization results is achieved. These series can be replaced in state of random distributions in Tug of War Optimization (TWO) strategies. In the strategies of this algorithm, the movement of the lighter team towards the heavier team and the acceleration of this movement in each stage and the updating of the results at the end of each league are done with a uniform random distribution. Therefore, by replacing chaotic dynamic functions in the random distribution of these strategies, the balance between exploration and exploitation is achieved and the algorithm is saved from the local optimality trap and finally the local optimality converges to the global optimality. In this study, for each model, 6 chaotic modes are considered and compared with the standard mode. In order to form statistical information and increase diversity in research, each model is independently evaluated 30 times. Finally, to analyze the statistical results, the best weight, the best average and the coefficient of variation are reported. Also, useful information is obtained about the weakness and strength of the algorithm and how to improve the results. In the Logistic chaos map, it most likely converges from local minima to general minima. Therefore, this function is suitable for improving the exploration conditions of algorithms. The Gaussian chaos map is very likely to be within the local minimums and is suitable for improving exploitation conditions. Therefore, by selecting these chaos maps, the weakness of algorithms of any kind will be repaired and global optimum will not have the opportunity to escape the chaos functions. The numerical distribution of these chaos map for a period of 100 iterations is shown in Fig. 12.2. In the following, we will examine the equations of chaos maps. How to apply then to metaheuristic algorithms is shown in the flowchart of Fig. 12.3.

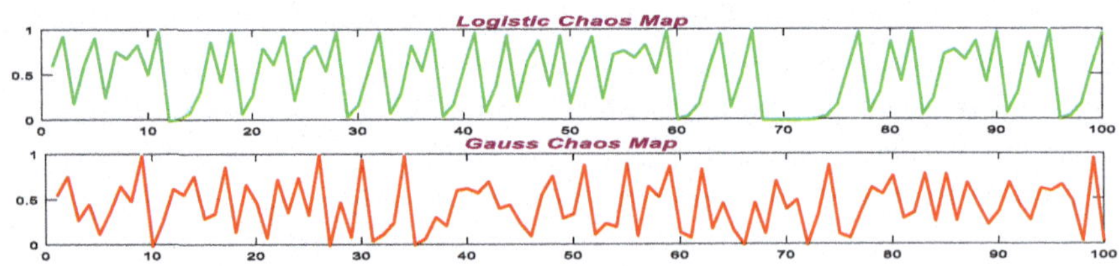

Fig. 12.2 Chaotic value distribution during 100 iterations

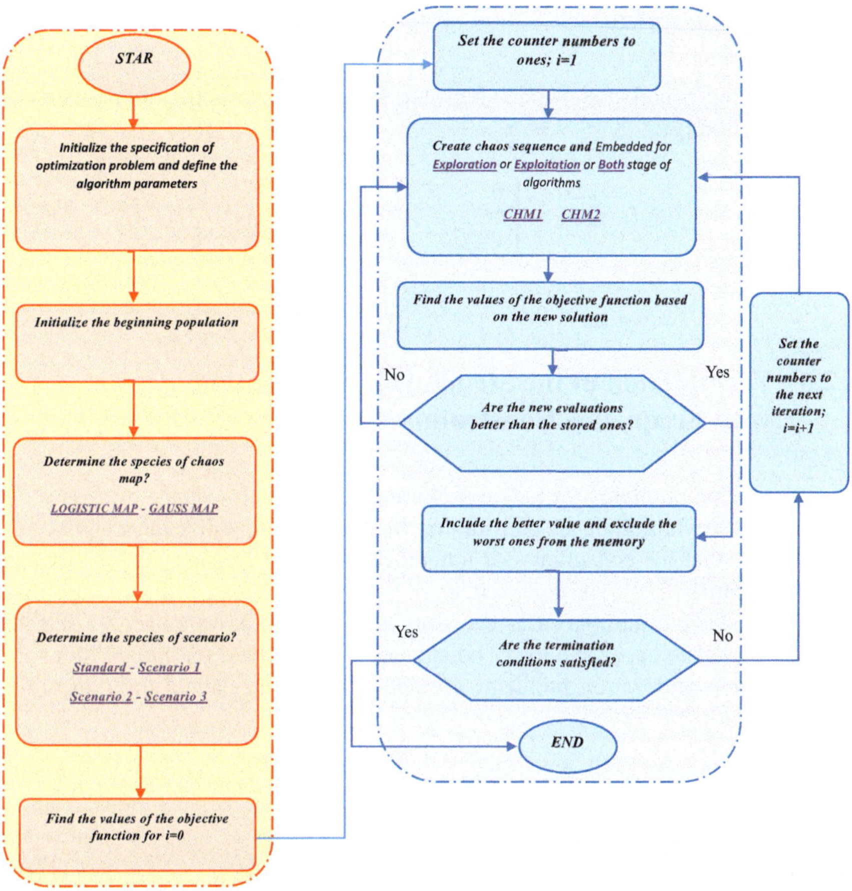

Fig. 12.3 Flowchart for the chaos algorithm

12.5.1 Logistics Map

This map appears in nonlinear dynamic behaviors related to biological populations [6]. The sentences of chaotic sequences in the logistic map are obtained according to the following equations:

$$CHM_{k+1} = \mathrm{a} \times CHM_k(1 - CHM_k) \tag{12.13}$$

$$CHM_{k+1} \in (0, 1); \quad CHM_0 \in (0, 1); \quad CHM_0 \notin (0, 0.25, 0.5, 0.75, 1) \tag{12.14}$$

In the performed studies, a = 4 has been considered.

12.5.2 Gauss Map

The following equation shows the between sentences of chaotic sequences in the Gauss function [7]:

$$CHM_{k+1} = \begin{cases} 0 & CHM_k = 0 \\ \frac{1}{CHM_k} - \left[\frac{1}{CHM_k}\right] & CHM_k \neq 0 \end{cases} \quad (12.15)$$

12.6 Formulation of the Structural Optimization with Frequency Constraints

In optimization problems for the cross section and geometric shape of trusses that are associated with frequency constraints, the goal is to minimize the weight of the structure so that the constraints for a number of natural frequencies for vibration modes are satisfied [8]. The cross section of the members along with the coordinates of some nodes are introduced as decision variables. These variables are selected continuously. The upper and lower bounds are also specified for variables in some cases. These optimization problems are defined in mathematical form according Eq. (12.16):

$$\begin{array}{ll} Find & X = \{A, S\}, \quad A = \{A_1, A_2, \ldots, A_{na}\}, \\ & S = \{S_1, S_2, \ldots, S_{ns}\} \\ to\ Minimize: & W(A, S) = \sum_{i=1}^{nm} \rho_i \times L_i(S) \times A_i \\ subjected\ to: & \begin{cases} \tilde{\omega}_i^L \leq \omega_i \leq \tilde{\omega}_i^U & i = 1, 2, \ldots, n\omega \\ A_j^L \leq A_j \leq A_j^U & j = 1, 2, \ldots, na \\ S_k^L \leq S_k \leq S_k^U & k = 1, 2, \ldots, ns \end{cases} \end{array} \quad (12.16)$$

In this relation, X is the vector of decision variables, A is the variable of the cross-section of the members, na is the symbol to represent the number of variables of the cross-section, A_i is the value of the cross-section of the ith variable, S is the variable of shape and arrangement, ns is the number of shape variables with the same coordinates Nodes are, S_i is the numerical value of the ith variable, W expresses the weight of the truss, nm indicates the total number of members, ρ_i the specific gravity of the material belonging to the ith member of the truss, L_i the length corresponding to the ith member which can be expressed through the variables Determine, ω_i expression of the ith natural frequency of the truss, ω_i^L and ω_i^u, respectively, represent the lower limit and the upper limit of the fixed frequency of the base, $n\omega$ indicates the total number of frequency restrictions, Aj^L and Aj^u, respectively, express the lower limit and the upper limit of the A_j cross section, and S_k^L and S_k^U to Arrange the lower limit and limit of kth variable S_k. Given that metaheuristic algorithms are used for unbounded problems; we use the penalty function in modeling to convert the bound to unbound state. In this method, if no violation has been committed, the amount of the fine will be zero, otherwise, if there is a violation, the amount of the penalty function will be obtained from Eqs. (12.17)–(12.20):

$$V_i = \begin{cases} 0 & if\ \tilde{\omega}_i^L \le \omega_i \le \tilde{\omega}_i^U \\ 1 - \frac{\omega_i}{\tilde{\omega}_i^L}\ if\ \tilde{\omega}_i^L > \omega_i \\ 1 - \frac{\tilde{\omega}_i^U}{\omega_i}\ if\ \omega_i > \tilde{\omega}_i^U \end{cases} \tag{12.17}$$

$$\gamma = \sum_{i=1}^{n\omega} V_i \tag{12.18}$$

$$F_{penalty} = (1 + \varepsilon_1 \times \gamma)^{\varepsilon_2} \tag{12.19}$$

$$to\ Minimize\quad Mer(A, S) = W(A, S) \times F_{penalty} \tag{12.20}$$

In these relations γ represents the set of violations and ε_1 and ε_2 are selected based on the search and extraction ratio. In this study, ε_1 units and ε_2 are selected with incremental linear changes in the range of 1.3–3 at the end of the iteration. Finally, "*Mer*" is the merit function or the objective function after the penalty is imposed.

12.6.1 Numerical Examples of Optimal Truss Design

In Chaotic Tug of War Optimization (CTWO), to determine the efficiency, known examples of truss structures have been used with layout and size optimization as benchmark problems. A significant improvement in the results is obtained by creating a competition between the perturbed states in the triple scenarios. In this challenge, 7 independent answers compete with each other. Due to the variety of checks, the optimal responses will be very close to the global optimal. The examples provided are as follows:

- The 52-bar dome-like truss
- The 72-bar spatial truss
- The 120-bar spatial dome
- 200-bar planar truss structure.

12.6.1.1 A 52-Bar Dome-Like Truss

The 52-bar dome-like truss as shown in Fig. 12.4 is a well-known benchmark problem for optimizing the weight and shape of trusses with frequency constraints. This truss

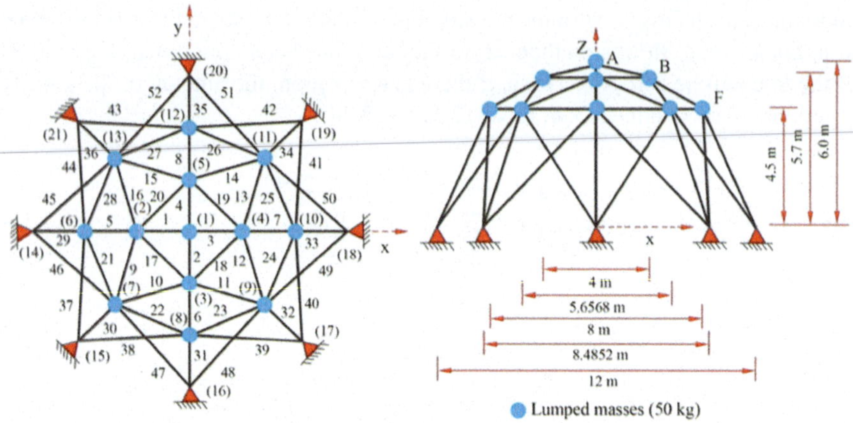

(a) (b)

Fig. 12.4 Schematic of a 52-bar dome-like truss

considers both optimizations of the section size and geometric coordinates of the nodes and the geometric shape of the structure is determined during the optimization process. The decision variables related to the section size are classified into 8 groups according to the symmetry in the geometric shape. Geometric coordinates of all symmetric free nodes can be changed by 2 m from the initial position along the coordinate axes.

In this case, the number of decision variables related to the shape of the structure and the geometric coordinates of the nodes is limited to 5 variables, and the sum of the variables, including shape and size, will be 13 variables. In all free nodes, a non-structural concentrated mass of 50 kg has affected all free nodes. The mechanical characteristics of the structure are: density of materials 7800 kg/m^3, modulus of elasticity 210,000 MPa, frequency limitations of the structure in the first mode are less than 15.916 Hz and in the second mode are greater than 28.648 Hz. For the cross section of the members, the lower limit is 1 cm^2 and the upper limit is 10 cm^2. In order to ensure the performance of turbulent functions and algorithms, as well as to increase the accuracy and sensitivity of calculations, each of the modes has been performed independently 20 times and the results related to the best response and the average value of responses are presented in Statistical Table 12.1. Also, the coefficient of variation of responses, which is a measure of the robustness and robustness of responses, has been calculated and used to compare the efficiency of turbulence functions and algorithms. For quick access to optimization information, a bar chart of each component is shown in Fig. 12.5. Examining the optimization results for different combinations of algorithms with turbulence functions and comparing it with the standard mode, shows a significant and significant improvement in reducing the weight of the 52-bar dome-like truss. The enhanced results for Chaotic Tug of War Optimization (CTWO), by Logistic chaos map and scenario 1, a weight of 193.165 kg has the optimal response.

Table 12.1 Optimal design comparison for the 52-bar dome-like truss

Decision variable	Standard	Logistic map			Gauss map		
	TWO	CTWO 21	CTWO 22	CTWO 23	CTWO 31	CTWO 32	CTWO 33
Z_A (m)	6.2249	5.9453	5.9186	5.645	5.871	5.9686	5.5115
X_B (m)	2.3133	2.2653	2.2671	2.1674	2.2692	2.1909	1.9854
Z_B (m)	3.8057	3.7218	3.7412	3.7192	3.7163	3.7923	3.7447
X_F (m)	4.0495	3.9684	3.9888	3.9159	3.9666	3.9862	3.8447
Z_F (m)	2.5002	2.5	2.5	2.5	2.5	2.5	2.5
A_1 (cm^2)	1.0012	1	1	1	1	1	1
A_2 (cm^2)	1.1222	1.1187	1.1109	1.1787	1.1113	1.1527	1.3252
A_3 (cm^2)	1.2466	1.2221	1.2257	1.2529	1.2217	1.2273	1.3057
A_4 (cm^2)	1.3687	1.4376	1.4513	1.5014	1.4665	1.4188	1.5152
A_5 (cm^2)	1.4937	1.4036	1.4218	1.403	1.4094	1.4153	1.4134
A_6 (cm^2)	1.0004	1	1	1	1	1	1
A_7 (cm^2)	1.4778	1.5917	1.5487	1.5883	1.5783	1.5571	1.5956
A_8 (cm^2)	1.4	1.3707	1.3944	1.3911	1.3818	1.3978	1.3954
Best weight (kg)	194.066	**193.165**	193.29	193.66	193.256	193.431	194.658
Mean weight (kg)	196.172	196.761	**193.531**	194.002	193.817	194.929	201.675
Coefficient variation	1.0556	2.2447	**0.1278**	0.1285	0.2736	1.2944	4.2939
ω_1 (HZ)	11.6148	11.5242	11.6053	11.7218	11.3487	11.2834	11.8516
ω_2 (HZ)	28.6506	28.9879	28.7232	28.8523	28.8624	28.9711	28.8627

The bold items in the table indicate the optimal values for the best weight, the best mean and the best coefficient of variation

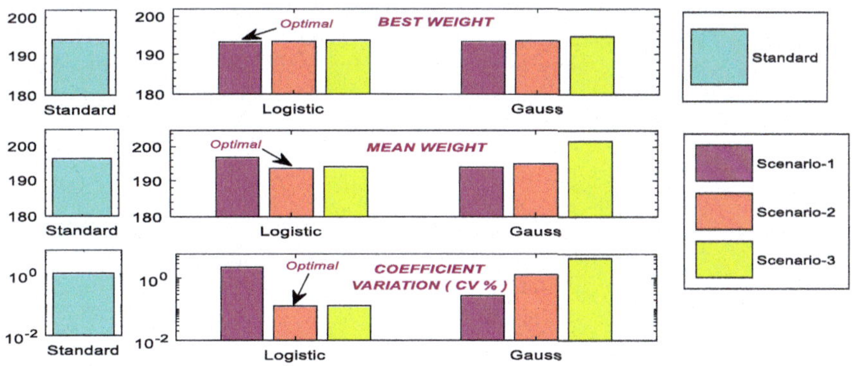

Fig. 12.5 Optimization results in standard mode and selection of chaos map for the 52-bar dome-like truss

12.6.1.2 A 72-Bar Spatial Truss

The 72-bar spatial truss as shown in Fig. 12.6 is a well-known benchmark problem with weight-limit optimization. This truss only considers "optimizing the size of the sections" and the geometric shape of the structure is constant during the optimization process (Table 12.2).

For quick access to optimization information, the bar chart of each component is shown in Fig. 12.7.

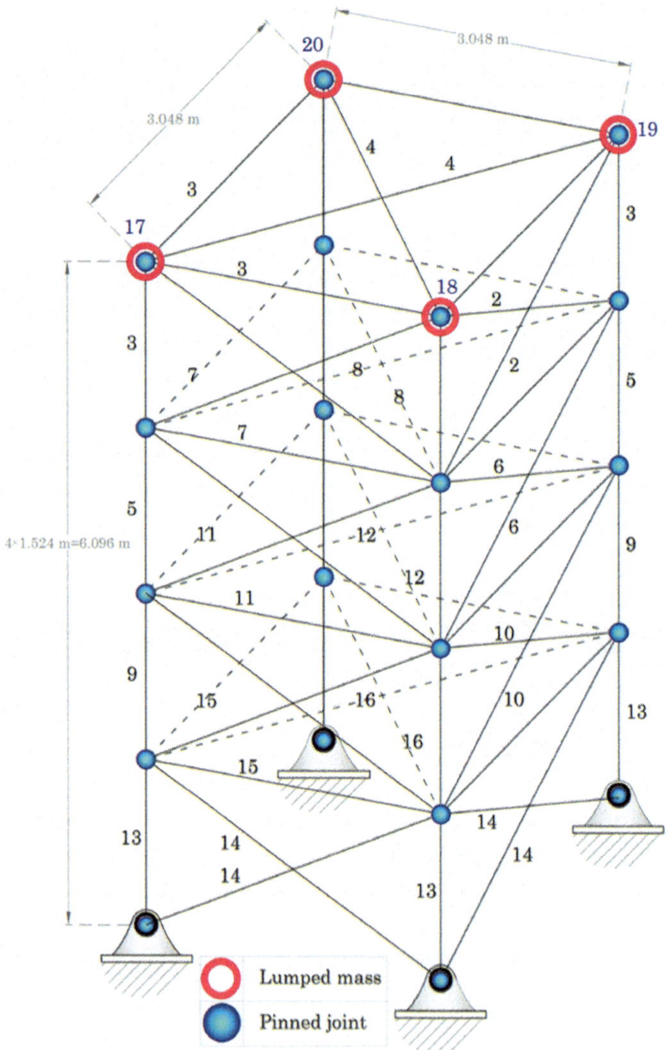

Fig. 12.6 Schematic of a 72-bar spatial truss

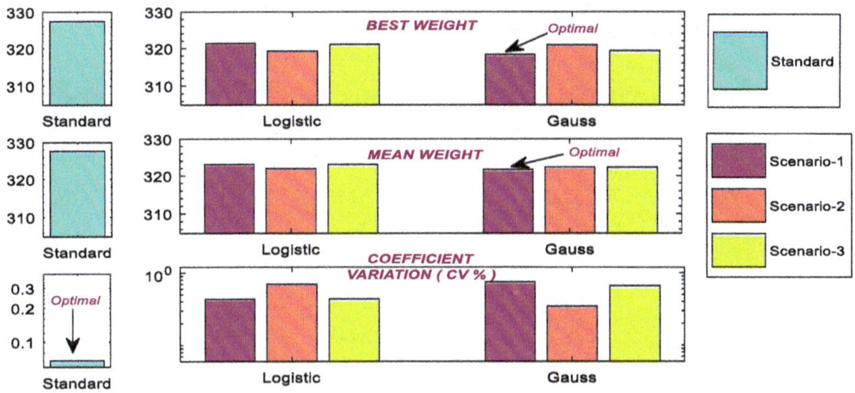

Fig. 12.7 Optimization results in standard mode and selection of chaos map for the 72-bar spatial truss

The decision variables related to the section size are classified into 16 groups according to the symmetry in the geometric shape. In nodes 1–4, the concentrated non-structural mass of 2270 kg has affected the truss. The mechanical characteristics of the structure are: material density 2770 kg/m^3, modulus of elasticity 69,800 MPa, frequency limits of the structure in the first mode are equal to 4 Hz and in the second mode are greater than 6 Hz. For the cross section of the members, the lower limit is 0.645 cm^2 and the upper limit is 20 cm^2. Examining the optimization results for different combinations of algorithms with turbulence functions and comparing it with the standard mode, shows a significant and significant improvement in reducing the weight of the 72-bar spatial truss. The enhanced results for Chaotic Tug of War Optimization (CTWO), by Gauss chaos map and scenario 1, a weight of 318.319 kg has the optimal response.

12.6.1.3 A 120-Bar Spatial Dome

The 120-bar spatial dome as shown in Fig. 12.8 is a well-known benchmark problem with weight-limit optimization. This truss only considers "optimizing the size of the sections" and the geometric shape of the structure is constant during the optimization process. The decision variables related to the size of the members' sections and according to the symmetry in the geometric shape of the dome along the X and Y axes, are classified into 7 groups. Non-structural concentrated mass in all free nodes affects the structure. Their values are 3 kg in node 1500 kg in nodes 2–13 and 100 kg in other nodes. The mechanical characteristics of the structure are: material density 7971.81 kg/m^3, modulus of elasticity 210,000 MPa, frequency limits of the structure in the first and second modes are greater than 9 and 11 Hz, respectively. For the cross section of the members, the range of the lower limit is 1 cm^2 and the upper limit is 129.3 cm^2.

Table 12.2 Optimal design comparison for the 72-bar spatial truss

Element group	Standard	Logistic map			Gauss map		
	TWO	CTWO 21	CTWO 22	CTWO 23	CTWO 31	CTWO 32	CTWO 33
A_1–A_4	3.77059	3.33102	3.89578	3.60998	3.80538	3.87535	4.39291
A_5–A_{12}	7.90361	7.75122	7.68004	7.81463	7.59803	7.79542	7.62284
A_{13}–A_{16}	0.517435	0.5	0.5	0.5	0.5	0.5	0.5
A_{17}–A_{18}	0.597535	0.559113	0.530277	0.544924	0.51942	0.543731	0.522797
A_{19}–A_{22}	8.66822	7.78833	9.00549	9.14049	9.71854	8.53073	8.88645
A_{23}–A_{30}	8.07275	7.88883	7.69843	7.8211	7.71128	7.80473	7.7193
A_{31}–A_{34}	0.535659	0.5	0.5	0.5	0.5	0.5	0.5
A_{35}–A_{36}	0.527344	0.5	0.5	0.5	0.5	0.5	0.5
A_{37}–A_{40}	13.1673	14.2373	12.9573	12.8807	13.0683	13.314	13.3116
A_{41}–A_{48}	8.19612	7.82316	7.80131	7.93901	7.76494	7.82546	7.94737
A_{49}–A_{52}	0.5	0.500001	0.5	0.5	0.5	0.5	0.5
A_{53}–A_{54}	0.5	0.5	0.5	0.5	0.5	0.5	0.5
A_{55}–A_{58}	16.5831	17.8106	17.5853	17.5152	17.2218	17.3758	17.0249
A_{59}–A_{66}	8.15471	7.88627	7.8481	7.7506	7.74807	7.89521	7.71295
A_{67}–A_{70}	0.5	0.5	0.5	0.5	0.5	0.5	0.5
A_{71}–A_{72}	0.503967	0.5	0.5	0.5	0.5	0.5	0.5
Best weight (kg)	327.38	321.303	319.272	321.051	**318.319**	320.929	319.353
Mean weight (kg)	327.592	323.073	321.894	323.014	**321.653**	322.313	322.243
Coefficient variation	**0.0684**	0.4326	0.6951	0.4388	0.7464	0.3448	0.6612
ω_1 (HZ)	4	4.001	4.0240	4.0021	4	4	4.0247
ω_3 (HZ)	6.001	6.0231	6.0021	6.0428	6.0387	6.0271	6.005

The bold items in the table indicate the optimal values for the best weight, the best mean and the best coefficient of variation

In order to ensure the performance of chaos map and algorithms, as well as to increase the accuracy and sensitivity of calculations, each of the modes has been performed independently 30 times and the results related to the best response and the average value of responses are presented in Statistical Table 12.3. Also, the coefficient of change of responses, which is a measure of the robustness and robustness of responses, has been calculated and used to compare the efficiency of turbulence functions and algorithms. For quick access to optimization information, the bar chart of each component is shown in Fig. 12.9. Examining the optimization results for different combinations of algorithms with turbulence functions and comparing it with the standard mode, shows a significant and significant improvement in reducing

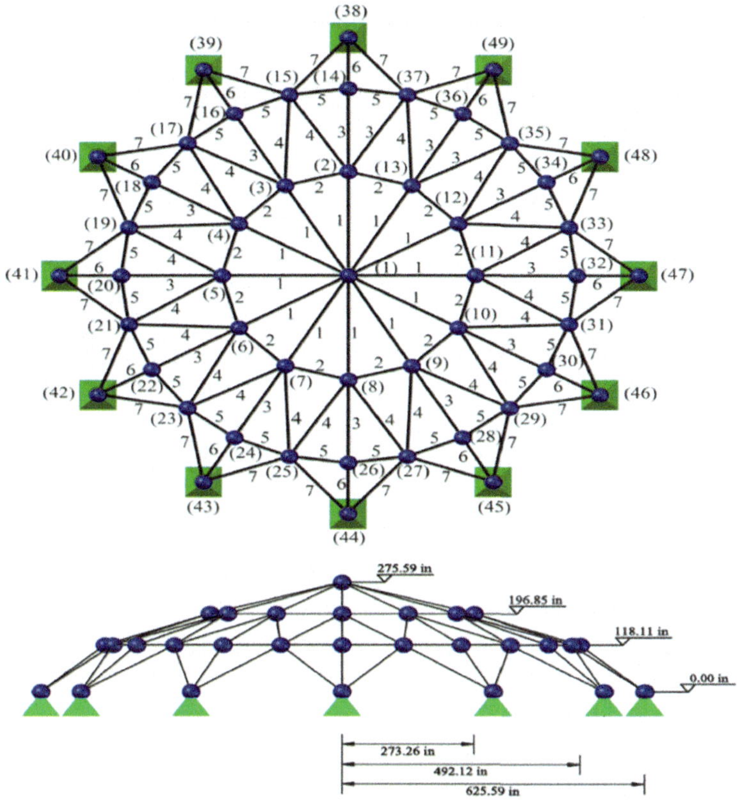

Fig. 12.8 Schematic of a 120-bar spatial dome

the weight of the 120-bar spatial dome. The enhanced results for Chaotic Tug of War Optimization (CTWO), by Logistic chaos map and scenario 1, a weight of 8708.525 kg has the optimal response.

12.6.1.4 A 200-Bar Planar Truss Structure

The 200-bar planar truss structure as shown in Fig. 12.10 is a well-known benchmark problem for weight optimization with frequency limits. This truss only considers "optimizing the size of the sections" and the geometric shape of the structure is constant during the optimization process. The decision variables related to the sections size of the members are classified into 29 groups according Table 12.4. Non-structural concentrated mass in nodes 1–5 and in the amount of 100 kg affect the structure. The mechanical characteristics of the structure are: material density 7860 kg/m^3, modulus of elasticity 210,000 MPa, frequency limits of the structure in the first, second and third modes are greater than 5, 10 and 15 Hz, respectively.

Table 12.3 Optimal design comparison for the 120-bar spatial dome

Decision Variable	Standard TWO	Logistic map CTWO 21	CTWO 22	CTWO 23	Gauss map CTWO 31	CTWO 32	CTWO 33
A_1	19.3559	19.5219	19.2691	19.3281	19.4821	19.5555	19.2832
A_2	41.0144	40.2956	41.9686	41.0106	40.3506	40.2037	41.0679
A_3	10.569	10.6032	10.6493	10.4351	10.6179	10.623	10.796
A_4	20.9229	21.1208	20.7767	21.1707	20.946	21.1157	20.9873
A_5	9.94109	9.87263	9.70749	9.92585	9.95306	9.85177	9.89651
A_6	11.8617	11.8217	11.7082	12.1643	11.7726	11.7782	11.6061
A_7	14.9312	14.8197	15.0762	14.6048	14.9981	14.8518	14.8574
Best Weight(kg)	8715.187	**8708.525**	8724.61	8710.925	8710.711	8708.553	8709.415
Mean Weight(kg)	8715.516	**8709.419**	8729.013	8714.213	8714.145	8709.62	8712.938
Coefficient Variation	**0.0039**	0.0089	0.0316	0.0329	0.0245	0.0165	0.0229
ω_1 (HZ)	9	9.0015	9.0001	9.0231	9	9	9.0249
ω_2 (HZ)	11	11	11.0008	11.0247	11.0734	11.0075	11.0068

The bold items in the table indicate the optimal values for the best weight, the best mean and the best coefficient of variation

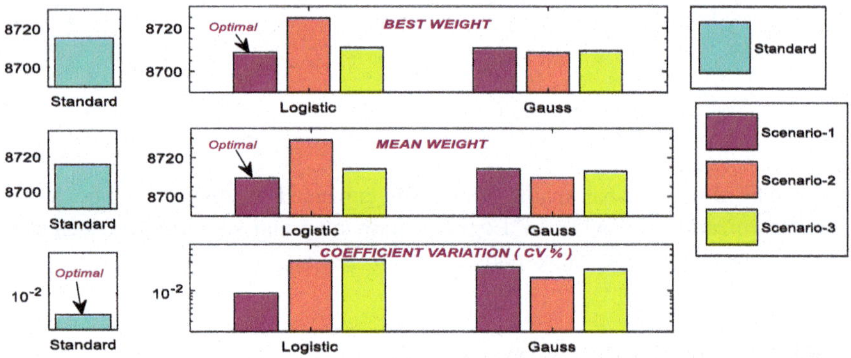

Fig. 12.9 Optimization results in standard mode and selection of chaos map for the 120-bar spatial dome

For the cross section of the members, the lower limit range is 0.1 cm^2 and the upper limit is 25 cm^2. In order to ensure the performance of turbulent functions and algorithms, as well as to increase the accuracy and sensitivity of calculations, each of the modes has been performed independently 30 times and the results related to the best response and the average value of responses are presented in Statistical Table 12.5.

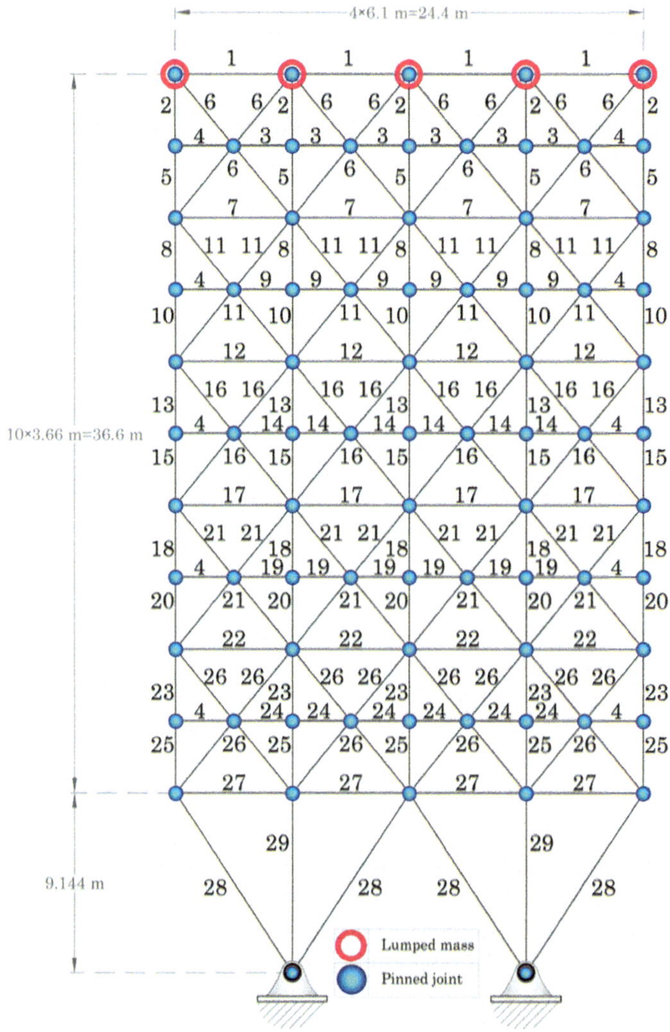

Fig. 12.10 Schematic of a 200-bar planar truss structure

Also, the coefficient of change of responses, which is a measure of the robustness and robustness of responses, has been calculated and used to compare the efficiency of turbulence functions and algorithms. For quick access to optimization information, a bar chart of each component is shown in Fig. 12.11. Examining the optimization results for different combinations of algorithms with turbulence functions and comparing it with the standard mode, shows a significant and significant improvement in reducing the weight of the 200-bar planar truss structure. The enhanced results for Chaotic Tug of War Optimization (CTWO), by Logistic chaos map and scenario 1, a weight of 2072.411 kg has the optimal response.

Table 12.4 Classification of members in the 200-bar planar truss structure

Group	Element group	Group	Element group
A_1	1, 2, 3, 4	A_{16}	82, 83, 85, 86, 88, 89, 91, 92, 103, 104, 106, 107, 109, 110, 112, 113
A_2	5, 8, 11, 14, 17	A_{17}	115, 116, 117, 118
A_3	19, 20, 21, 22, 23, 24	A_{18}	119, 122, 125, 128, 131
A_4	18, 25, 56, 63, 94, 101, 132, 139, 170, 177	A_{19}	133, 134, 135, 136, 137, 138
A_5	26, 29, 32, 35, 38	A_{20}	140, 143, 146, 149, 152
A_6	6, 7, 9, 10, 12, 13, 15, 16, 27, 28, 30, 31, 33, 34, 36, 37	A_{21}	120, 121, 123, 124, 126, 127, 129, 130, 141, 142, 144, 145, 147, 148, 150, 151
A_7	39, 40, 41, 42	A_{22}	153, 154, 155, 156
A_8	43, 46, 49, 52, 55	A_{23}	157, 160, 163, 166, 169
A_9	57, 58, 59, 60, 61, 62	A_{24}	171, 172, 173, 174, 175, 176
A_{10}	64, 67, 70, 73, 76	A_{25}	178, 181, 184, 187, 190
A_{11}	44, 45, 47, 48, 50, 51, 53, 54, 65, 66, 68, 69, 71, 72, 74, 75	A_{26}	158, 159, 161, 162, 164, 165, 167, 168, 179, 180, 182, 183, 185, 186, 188, 189
A_{12}	77, 78, 79, 80	A_{27}	191, 192, 193, 194
A_{13}	81, 84, 87, 90, 93	A_{28}	195, 197, 198, 200
A_{14}	95, 96, 97, 98, 99, 100	A_{29}	196, 199
A_{15}	102, 105, 108, 111, 114		

12.7 Concluding Remarks

This algorithm is inspired by the game of tug-of-war and is based on population. Like other meta-heuristic algorithms, a set of initial answers is selected from the decision space and each solution is considered as a team. And all of them form a league. In each iteration of the algorithm, the evaluation of the teams is determined and sorted based on merit. In this competition, the best team has the most weight and the worst team has the least weight. Both competing teams are pulling the rope, the light team loses the competition and moves to the heavy team. In the algorithm, both exploration and exploitation strategies are associated with a random distribution. The existence of random distributions in these strategies is, on the one hand, to create diversity in the search space, but it causes the optimization operation to stagnate and stop at the local optimal position. Therefore, chaotic dynamic functions have a high potential to save artificial bee colony algorithm. By embedding the logistic and Gaussian chaos functions, the local optimum converges to the global optimum and early convergence is eliminated. By forming chaotic models, 6 different chaotic modes compete with each other and the best answer is compared with the standard mode. For problems with multi-frequency, size, and layout optimization constraints, the standard tug-of-war game mode faces a broad stagnation that is resolved by

Table 12.5 Optimal design comparison for the 200-bar planar truss structure

Number group	Standard	Logistic map			Gauss map		
	TWO	CTWO 21	CTWO 22	CTWO 23	CTWO 31	CTWO 32	CTWO 33
A_1	0.319865	0.699854	0.70801	0.313875	0.320327	0.319705	0.70457
A_2	0.554731	0.562731	0.558837	0.494198	0.527374	0.521829	0.535734
A_3	0.1	0.1	0.1	0.1	0.1	0.1	0.1
A_4	0.1	0.1	0.1	0.1	0.1	0.1	0.1
A_5	0.594559	0.609864	0.660405	0.564889	0.572964	0.560164	0.570597
A_6	0.76968	0.725795	0.71226	0.76276	0.744691	0.75087	0.717978
A_7	0.1	0.120645	0.113793	0.1	0.1	0.1	0.133673
A_8	1.39991	1.37876	1.2496	1.36574	2.03683	1.29357	1.22818
A_9	0.1	0.1	0.1	0.1	0.1	0.1	0.1
A_{10}	1.53847	1.44833	1.39848	1.5163	1.49205	1.93858	1.35199
A_{11}	1.03448	0.978094	0.928986	1.01455	0.994051	1.00471	0.958867
A_{12}	0.103886	0.100289	0.1	0.100884	0.1	0.1	0.1
A_{13}	2.94455	2.8191	3.22221	3.51618	4.77313	2.67087	8.83899
A_{14}	0.100012	0.100304	4.67262	0.1	0.138395	0.1	0.100744
A_{15}	3.71269	6.77599	2.46258	2.8917	4.97975	4.19844	6.86766
A_{16}	1.35716	1.28198	1.25936	1.33138	1.32706	1.31621	1.27851
A_{17}	0.19751	0.100551	0.1	0.100866	0.1	0.103038	0.1
A_{18}	5.40289	4.29142	4.85459	5.73314	6.3621	6.33482	4.2832
A_{19}	0.240169	1.54112	3.62108	0.22122	0.170513	0.1	2.11635
A_{20}	4.37479	5.99401	9.13332	6.87857	5.79268	5.98238	6.56809

(continued)

Table 12.5 (continued)

Number group	Standard	Logistic map			Gauss map		
	TWO	CTWO 21	CTWO 22	CTWO 23	CTWO 31	CTWO 32	CTWO 33
A_{21}	1.81173	1.789	1.75284	1.76451	1.73799	1.65921	1.80952
A_{22}	0.434362	0.955487	0.730935	0.336823	0.291861	5.07609	0.910087
A_{23}	10.5743	11.646	10.5408	8.09885	7.23178	6.12241	10.5382
A_{24}	4.29855	3.38336	0.706386	0.961018	4.22432	0.911396	4.85394
A_{25}	7.40296	10.3571	10.5464	8.48018	8.21322	6.5315	7.19623
A_{26}	2.78712	3.1443	2.65172	2.58009	3.13655	3.0721	2.97621
A_{27}	10.5087	5.24572	5.91573	9.65421	7.65129	8.01126	8.08613
A_{28}	16.2625	13.5857	15.2808	18.0819	14.7851	17.8945	14.0453
A_{29}	21.4908	17.6261	16.0235	15.546	17.867	15.55	14.5652
Best Weigh	2184.379	**2072.411**	2118.812	2089.843	2081.136	2112.734	2147.054
Mean Weight	2412.92	2306.277	**2273.443**	2306.992	2354.293	2373.989	2382.391
Coefficient Variation	8.324	9.535	**5.331**	6.2002	7.541	7.675	11.079
ω_1 (HZ)	5	5.0017	5.0023	5	5	5.0037	5.0104
ω_2 (HZ)	12.8130	12.0015	12.6580	12.5176	13.1870	12.2870	12.8240
ω_3 (HZ)	15.268	15.009	15.0253	15.8124	15.0001	15.3281	15.6743

The bold items in the table indicate the optimal values for the best weight, the best mean and the best coefficient of variation

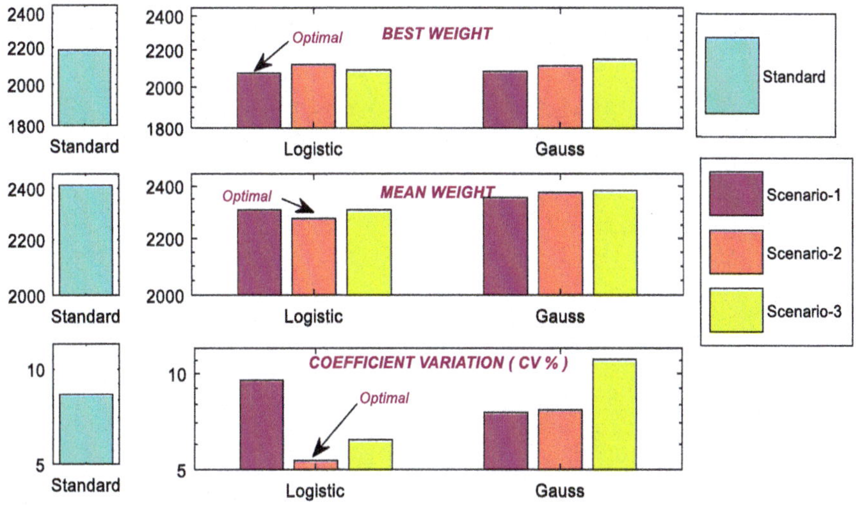

Fig. 12.11 Optimization results in standard mode and selection of chaos map for the 200-bar planar truss structure

chaotic jumps. Also, in chaotic methods, the weaknesses and strengths of algorithms in dealing with optimization problems are determined.

References

1. Kaveh, A.: Advances in Metaheuristic Algorithms for Optimal Design of Structures, 2nd edn. Springer, Cham, Switzerland (2021). https://doi.org/10.1007/978-3-319-05549-7
2. Kaveh, A., Zolghadr, A.: Tug of war optimization: a new metaheuristic algorithm. Int J Optim Civil Eng **6**(4), 469–493 (2016)
3. Kaveh, A., Yousefpoor, H.: Chaos-embedded metaheuristic algorithms for optimal design of truss structures. Sci. Iranica **29**(6), 2868–2885 (2022). https://doi.org/10.24200/sci.2022.59812.6441
4. Kaveh, A., Yosefpoor, H.: Chaotically enhanced metaheuristic algorithms for optimal design of truss structures with frequency constraints. Periodica Polytech. Civ. Eng. **66**(3), 900–921 (2022). https://doi.org/10.3311/PPci.20220
5. Kaveh, A., Zarfam, P., Aziminejad, A., Yosefpoor, H.: Comparison of four chaotic metaheuristic algorithms for optimal design of large-scale truss structures. Iran. J. Sci. Technol. Trans. Civ. Eng. https://doi.org/10.1007/s40996-022-00908-8
6. Ott, E.: Chaos in Dynamical Systems. Cambridge University Press, Cambridge (2002)
7. Bucolo, M., Caponetto, R., Fortuna, L., Frasca, M., Rizzo, A.: Does chaos work better than noise? IEEE Circuit Syst. Mag. **2**(3), 4–19 (2002)
8. Kaveh, A., Sheikholeslami, R., Talatahari, S., Keshvari-Ilkhichi, M.: Chaotic swarming of particles: a new method for size optimization of truss structures. Adv. Eng. Softw. **67**, 136–147 (2014)

Chapter 13
Chaotic Thermal Exchange Optimization

13.1 Introduction

The Thermal Exchange Optimization (TEO), is one of the physical algorithms and is inspired by Newton's law of cooling. This algorithm was introduced in 2017 by Kaveh et al. According to Newton's law of cooling, the rate of heat loss of an object is proportional to the temperature difference between the object and the surrounding environment. In this algorithm, each particle is considered as a cooling or heating object, and by assuming the other agent as the environment, heat exchange is done between the two. After heat exchange, the new temperature of the object is its position in the search space. Initial responses within the range of decision variables are formed randomly. These proposed candidates are evaluated and sorted by applying the objective function. Then the heat exchanger population is divided into two parts. All particles are affected by ambient temperature. If the object is located in the first half, then the second part is considered as the external environment, and under its influence, they exchange heat, and vice versa. The best responses, which are located in the first half and have a high temperature, are cooled by moving slightly towards the particles with a lower temperature. The particles with worse conditions that make up the second half of the body and have a low temperature are heated by moving towards the high temperature particles. To maintain elitism, a special memory is considered to store the best-known solutions. Up to this stage, the calculations are based on specific and definite relationships. In order to escape from the trap of local optima and premature maturity of the two phases of exploration and exploitation, the heat exchange of particles should be considered [1]. In each of the phases, random distribution has been used to create variety of answers in the search space. The basic steps of the algorithm are presented below.

A. Kaveh and H. Yousefpoor, *Chaotic Meta-heuristic Algorithms for Optimal Design of Structures*, Studies in Computational Intelligence 1129,
https://doi.org/10.1007/978-3-031-48918-1_13

13.2 Standard Thermal Exchange Optimization (TEO)

This algorithm was presented by Kaveh et al. [2], inspired by Newton's law of cooling. In this algorithm, the physical relations related to the thermal exchange between the object and the surrounding environment are used and it is based on the population. The use of this algorithm for structural problems has brought significant improvement. Newton's law of cooling states that the rate of heat loss between an object and the surrounding environment is proportional to their temperature difference. Therefore, the population of heat exchanging particles is separated into two parts and the first half is placed in heat exchange with the second half.

13.3 Basic Steps in Thermal Exchange Optimization

Step 1 Determining the initial values of the algorithm: The thermal exchange algorithm starts with the introduction of the initial proposed solutions. These solutions are selected within the scope of the search space. The number of elements selected for thermal exchange is equal to *nTO*. Then the objective functions and penalized objective functions corresponding to each of the solutions are determined.

Step 2 Creating groups: In order to carry out the thermal transfer process, first the introduced components are sorted in ascending form based on the penalized objective function and then they are classified into two groups with the same number of components. In this step, the components of the first group (i = 1, 2, …, nTO/2) exchange thermal with the components of the second half (i = nTO/ 2 + 1, …, n).

Step 3 thermal transfer and updating the position of the components: Based on the thermal exchange relationship of the elements with the surrounding environment, the final position of the elements is determined by the following important heat exchange relationship.

$$newTO(i) = envTO(i) + (TO(i) - envTO(i)) \exp(-\beta(i)t) \tag{13.1}$$

In this regard, newTO(i) expresses the new position after thermal exchange and envTO(i) expresses the thermal position of the surrounding environment before thermal exchange. Also, in order to apply the parameters of time and β, we use the following suggested relationships.

$$t = \frac{NITs}{\max NITs}, \beta(i) = \frac{PFit(i)}{\max PFit(i)} \tag{13.2}$$

In these relationships, NITs and maxNITs express the number of iterations in each step and the maximum number of iterations, respectively. Also, PFit and maxPFit consider the values of the penalized objective function and its maximum value.

Step 4 Applying possible conditions: Up to this stage, the calculations are based on specific and definite relationships. In order to escape the trap of local optima and premature maturity in the algorithm, two mechanisms of search and discovery in the thermal exchange of elements should be considered. In the following relations, this possible condition is provided by the parameters C, C1, C2 and a choice of rand.

$$nevTO = (1 - c \times \boldsymbol{rand})TO \tag{13.3}$$

$$C = C1 + C2 \times (1 - t) \tag{13.4}$$

$$C1 = round(\boldsymbol{rand}), C2 = round(\boldsymbol{rand}) \tag{13.5}$$

Step 5: Elitist thermal exchange algorithm: In order to achieve the elitism of the algorithm, memories are allocated for the best results. For this purpose, the best thermal exchange (TO-M) and the objective function (Fit-M) and the corresponding penalized objective function (PFit-M) are compared with the previous periods in each period, and replaced if the results are improved.
Step 6: The termination conditions are checked and if necessary, the thermal exchange of the elements with the surrounding environment is repeated.

13.4 Chaos-Embedded Thermal Exchange Optimization (CTEO)

In order to determine the new position of the heat flow as a result of the heat exchange of each element with the surrounding environment, two exploration and exploitation strategies have been considered in Eqs. (13.3)–(13.5).

By replacing the chaos functions in the random selections related to these steps, we will have a significant improvement in the performance of the algorithm [3–5]. The suggested scenarios for this replacement are as follows:

Scenario 1 Replacing the chaos map in the exploration phase: At this stage, the first CHM1 chaos map replaces the random term in Eq. (13.3). The results of applying the chaos map will be as follows:

$$nevTO = (1 - c \times \boldsymbol{CHM1})TO \tag{13.6}$$

Scenario 2 Replacing the chaos map in the exploitation phase: In this stage, the second chaos map CHM2 is replaced to determine the values of C1 and C2 in Eq. (13.5). For this purpose, we have:

$$C1 = round(\boldsymbol{CHM2}), C2 = round(\boldsymbol{CHM2}) \tag{13.7}$$

Scenario 3 placing the chaos map in both stages: In this case, the chaos map is simultaneously applied to the random selections of both stages.

13.5 Truss Weight Optimization with Multiple Frequency Constraints

In building structures, natural frequencies and oscillation modes contain useful information about the dynamic behavior of structures, and we can easily control the dynamic behavior of a structural system by quickly accessing them. The destructive and dangerous phenomenon of resonance is one of the things that can be controlled by applying frequency constraint. Mathematical equations of frequency limit consist of implicit and complex variables, and optimization with metaheuristic algorithms in standard mode, in most cases faces premature convergence. In the Chaotic Thermal Exchange Optimization (CTEO), by embedding chaotic dynamic series, including Logistic and Gaussian series, a significant improvement in optimization results is achieved. Dynamic series can replace random distributions in the strategies of Thermal Exchange Optimization (CTEO), and by creating chaotic jumps, local optima can converge to global optima. With these conditions, a balance between exploration and exploitation has been established. In this research, 6 chaos modes are considered for each model and compared with the standard mode. To form statistical information and increase diversity in the research, each model is independently evaluated 30 times. Also, to analyze the statistical results, the best weight, the best average and the coefficient of variation are reported. With these reviews, useful information is obtained about the weakness and strength of the algorithm and how to improve the results. In the Logistic chaos map, it most likely converges from local minima to general minima. Therefore, this function is suitable for improving the exploration conditions of algorithms [6]. The Gaussian chaos map is very likely to be within the local minimums and is suitable for improving exploitation conditions [7]. Therefore, by selecting these chaos maps, the weakness of algorithms of any kind will be repaired and global optimum will not have the opportunity to escape the chaos functions.

13.6 Formulation of the Optimization Problems

In this section, the dynamic analysis of the free vibration of the structure and access to natural frequencies is performed first, and then the optimization model is presented with the formation of the objective function, constraints, penalty function and the combination of the objective function with the penalty function.

13.6.1 Free Vibration and Natural Frequencies

When a structure is affected by dynamic loads such as earthquakes or storms, in order to prevent the phenomenon of resonance, the natural frequency should be limited to a certain range [8]. In order to apply this type of constraints, the natural frequencies of the structures contain all the required information about the dynamic behavior of the structures. In low frequency vibration problems, the response of the structure depends on the basic frequencies and mode shape, and by applying the frequency constraints, the dynamic behavior of the structure can be easily controlled. In a number of optimization problems, the effects of some modes can be reduced by using these relations. For example, in the design of airplane wings, the effort is to reduce bending and torsion modes. To calculate the natural frequencies of the structure, the matrix form of the free vibration equation of the multi-degree-of-freedom system is investigated. In these relationships, M is the mass matrix, K is the stiffness matrix, Y is the displacement equation, φ_n is the shape vector of the nth mode, $q_n(t)$ is the time coordinate of the nth mode, A_n and B_n are the integration constants that are determined from the initial conditions. Dynamic relationships of free vibration are presented from Eqs. (13.8)–(13.11).

$$M \times \ddot{Y}(t) + K \times Y(t) = 0 \tag{13.8}$$

$$Y(t) = \phi_n \times q_n(t); \quad n = 1, 2, \ldots, N \tag{13.9}$$

$$q_n(t) = A_n \cos \omega_n t + B_n \sin \omega_n t \tag{13.10}$$

$$Y(t) = \phi_n \times (A_n \cos \omega_n t + B_n \sin \omega_n t) \tag{13.11}$$

In these relationships, ω_n and φ_n are unknown, which results in the placement of Y(t) in the free vibration differential equation.

$$\left[K - \omega_n^2 \times M\right] \times \phi_n = 0 \Rightarrow \left|K - \omega_n^2 \times M\right| = 0 \tag{13.12}$$

With the expansion of these determinants, the frequency equation is obtained. This equation has N real and positive roots for $\omega^2{}_n$, which determines the N natural frequency of ω_n. The roots of this characteristic equation are known as eigenvalues. Also, by reflecting on these equations, it is concluded that the design variables are not explicitly present in these equations and their presence is implicit. Therefore, if optimization is carried out with mathematical methods and implicit derivatives, one will encounter strongly non-linear and non-convex equations the Solution will be very difficult and time consuming. As a result, if we want to solve the frequency constraints in an optimization using traditional methods, we must perform a sensitivity analysis. Thus, the derivatives of eigenvalues and eigenvectors must be calculated, and this usually encounters various approximations. In addition, in some cases, we will find repetitive amounts and repetitive frequencies that are not distinguished by conventional investigations and can only be determined by directional derivatives. When analyzing the sensitivity of the structures, it is created by repetitive frequencies of particular complexity because it is mainly "unique in specific values". Another limit that greatly affect the traditional mathematical optimization methods is the choice of a good starting point. In cases where the starting point is not appropriate, these methods stop by reaching the local optimization, and there is no solution to escape these local optimists. Today, with the complication of issues and increasing the number of decision variables, the lack of accountability of classical methods is evident. Therefore, in order to overcome these challenges over the past decade, various types of powerful optimization methods have been invented, some of which have been optimized by frequency constraints. In most of these optimization methods, they are inspired by metaheuristic techniques. Meta-heuristic algorithms are widely accepted by researchers and are considered powerful tools for engineering optimization problems. The main features of these methods can be stated in the following cases:

- These are based on the population.
- These are independent of the specific problem.
- These are inspired by natural phenomena.
- These do not need any information about the gradient of the objective function and constraints.
- The quality of the final solution does not depend on the starting point.
- These are based on decisions and principles of random search.

In these algorithms, the value of the objective function itself is used instead of its derivatives, and these have global search capabilities, and are also suitable for complex, non-linear, discrete and non-convex search spaces.

13.6.2 Formulation Size and Layout Optimization with Multi Frequency Constraints

In truss design with multi-frequency constraints and optimization of size and shape, the goal is to minimize the weight of the structure while satisfying some limiting on natural frequencies of the structure. In this type of optimization, the decision variables include the cross-section of the members or the geometric coordinates of some nodes. There will be no changes in the topology of the structure, and the connection conditions remain unchanged during the optimization process. The decision variables should be chosen within the corresponding range of the upper bound and the lower bound of their search space. Optimization problems can be represented by the mathematical relationship according Eq. (13.13):

$$
\begin{array}{ll}
Find & X = [x_1, x_2, \ldots, x_n] \\
to\ minimize & P(X) = f(X) \times f_{penalty}(X) \\
subjected\ to & \\
\omega_j \le \omega_j^* & for\ some\ natural\ frequencies\ j \\
\omega_k \ge \omega_k^* & for\ some\ natural\ frequencies\ k \\
x_{i,\min} \le x_i \le x_{i,\max} &
\end{array}
\tag{13.13}
$$

In this equation, X is the vector of decision variables, including the geometric coordinates of the nodes and the cross-sectional area of the members, n is the number of variables or the dimension of the search space, which is introduced based on the grouping of members and according to the symmetry of the structure. P(X) is the penalized objective function or merit function that must reach a minimum value. f(X) is the cost function that is considered as the weight in the optimization of structures. $f_{penalty}(X)$ is the penalty function that is used to apply constraints. In cases where some constraints related to the problem are violated, this function enters the calculation process with a suitable weight. w_j expresses the jth natural frequency of the structure and w_j^* is its upper bound. W_k expresses the kth natural frequency of the structure and w_k^* is its lower bound. X_{imin} and X_{imax} are the lower and upper bound of the decision variable x_i, respectively. The cost function is expressed by the mathematical relationship according to Eq. (13.14):

$$
f(X) = \sum_{i=1}^{nm} \gamma_i \cdot L_i \cdot A_i \tag{13.14}
$$

In this equation, γ_i is the specific material density of member i, L_i is the length of member I, and A_i is the cross-sectional area of member i. The penalty function is defined by the mathematical relation according to Eq. (13.15):

$$
f_{penalty}(X) = (1 + \varepsilon_1 \cdot \nu)^{\varepsilon_2}, \quad \nu = \sum_{i=1}^{q} \nu_i \tag{13.15}
$$

where q is the number of frequency constraints and ν_i expresses the violation value which is defined from the following equation:

$$
\nu_i = \begin{cases} 0 & if\ the\ ith\ constrain\ is\ satisfied \\ \left|1 - \frac{\omega_i}{\omega_i^*}\right| & else \end{cases} \tag{13.16}
$$

The parameters ε_1 and ε_2 are chosen based on creating a balance between exploration and exploitation in the decision variables space. In these investigations, the value of ε_1 is set as unity, and the parameter ε_2 starts from 1.5 and increases linearly to 6 in all test examples. These values penalize the unfeasible solutions more severely as the optimization process proceeds. As a result, in the early stages, the agents are free to explore the search space easily, but in the final iterations, the conditions of exploitation prevail and among them, the best local optimum is selected as the sub global optimum.

13.7 Numerical Examples of Optimal Truss Design

To evaluate the Chaotic Thermal Exchange Optimization (CTEO), well-known examples of truss optimization with layout and size optimization are selected as benchmark problems. By creating competition, between chaotic states and triple scenarios, a significant improvement in optimization results is achieved. In each investigated model, 7 independent responses compete with each other. Due to the variety of checks, the optimal responses will be very close to the global optimal. The examples provided are as follows:

- The simply supported 37-bar planar truss
- The 72-bar spatial truss
- The 120-bar spatial dome
- 200-bar planar truss structure.

13.7.1 A Simply Supported 37-Bar Planar Truss

A 37-member planar truss with a simple abutment as shown in Fig. 13.1 is a well-known benchmark problem for optimizing the weight and shape of a truss with a frequency limit. This truss considers both optimizations of the section size and geometric coordinates of the nodes and the geometric shape of the structure is determined during the optimization process. The lower edge member is considered to be a rod member with a rectangular cross-section and an area of 40 cm^2. The other members are modeled as rod members. These decision variables are grouped according to symmetry in the geometric shape. The y-axis coordinates form the shape variables in the upper year, taking into account the symmetry. In the lower axis, a non-structural concentrated mass of 10 kg affects all free nodes. The mechanical properties of the structure are: The density of the materials is 7800 kg/m^3, the modulus of elasticity is 21,000 MPa, the frequency limits of the structure in the first, second and third modes are greater than 20, 40 and 60 Hz, respectively. For the cross section of the members, the lower limit is 1 cm^2 and the upper limit is 10 cm^2. The total size and shape decision variables are 19 variables, of which 14 variables are related to the size and cross-sectional area of the members and 5 variables are related to the shape of the structure and the coordinates of the nodes.

There are also three frequency limits. In order to ensure the performance of turbulent functions and algorithms, as well as to increase the accuracy and sensitivity of calculations, each of the independent modes has been performed 30 times and the results related to the best response and the average value of responses. Is presented in statistical Table 13.1. Also, the coefficient of variation of responses, which is a

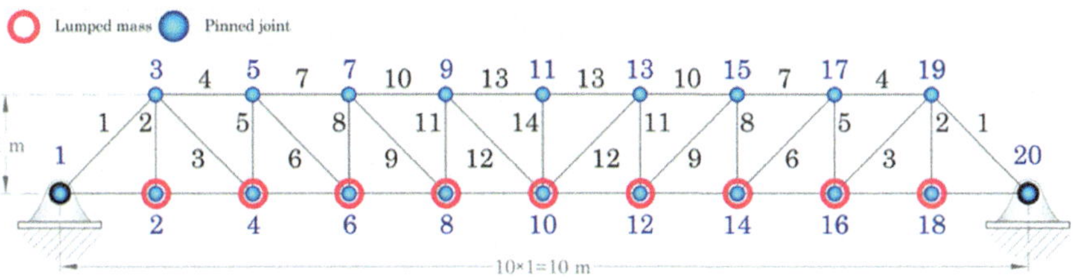

Fig. 13.1 Schematic of a simply supported 37-bar planar truss

measure of the robustness and robustness of responses, has been calculated and used to compare the efficiency of turbulence functions and algorithms. For quick access to optimization information, a bar chart of each component is shown in Fig. 13.2. Examining the optimization results for different combinations of algorithms with turbulence functions and comparing it with the standard mode, shows a significant improvement in reducing the weight of A simply supported 37-bar planar truss. The enhanced results for Chaotic Thermal Exchange Optimization (CTEO), by Logistic chaos map and scenario 1, a weight of 357.271 kg has the optimal response.

13.7.2 A 72-Bar Spatial Truss

The 72-bar spatial truss as shown in Fig. 13.3 is a well-known benchmark problem with weight-limit optimization. This truss only considers "optimizing the size of the sections" and the geometric shape of the structure is constant during the optimization process.

The decision variables related to the section size are classified into 16 groups according to the symmetry in the geometric shape. In nodes 1–4, the concentrated non-structural mass of 2270 kg has affected the truss.

The mechanical characteristics of the structure are: material density 2770 kg/m^3, modulus of elasticity 69,800 MPa, frequency limits of the structure in the first mode are equal to 4 Hz and in the second mode are greater than 6 Hz. For the cross section of the members, the lower limit is 0.645 cm^2 and the upper limit is 20 cm^2. In order to ensure the performance of chaotic functions and algorithms, as well as to increase the accuracy and sensitivity of calculations, each of the modes has been performed independently 30 times and the results related to the best response and the average value of responses are presented in Table 13.2. Also, the coefficient of variation of responses, which is a measure of the robustness and robustness of responses, has been calculated and used to compare the efficiency of turbulence functions and algorithms. For quick access to optimization information, a bar chart of each component is shown in Fig. 13.4. Examining the optimization results for different combinations of algorithms with turbulence functions and comparing it with the standard mode, shows a significant and significant improvement in reducing the weight of the 72-bar spatial truss. The enhanced results for Chaotic Thermal

Table 13.1 Optimal design comparison for the simply supported 37-bar truss

Decision variable	Standard	Logistic map			Gauss map		
	TEO	CTEO 21	CTEO 22	CTEO 23	CTEO 31	CTEO 32	CTEO 33
Y_3, Y_{19} (m)	0.85531	0.93454	0.91565	0.9032	0.96312	0.91608	0.94764
Y_5, Y_{17} (m)	1.2288	1.2877	1.2722	1.267	1.3479	1.2831	1.3242
Y_7, Y_{15} (m)	1.4109	1.4371	1.4328	1.4377	1.5254	1.4563	1.4984
Y_9, Y_{13} (m)	1.5255	1.5447	1.5471	1.5579	1.6556	1.5742	1.6227
Y_{11} (m)	1.586	1.6142	1.6176	1.6272	1.7297	1.6412	1.6953
A_1, A_{27}	3.1169	2.8424	2.8927	2.9646	2.9232	2.8572	2.9206
A_2, A_{26}	1.0462	1	1	1	1.0005	1.0005	1
A_3, A_{24}	1.0016	1	1	1	1	1	1
A_4, A_{25}	2.8745	2.5496	2.4964	2.748	2.5232	2.598	2.6112
A_5, A_{23}	1.164	1.1221	1.128	1.1257	1.1981	1.1363	1.1528
A_6, A_{21}	1.2144	1.1845	1.2212	1.2137	1.2296	1.2376	1.2377
A_7, A_{22}	2.7355	2.4032	2.5238	2.6569	2.5652	2.6596	2.4725
A_8, A_{20}	1.3642	1.3043	1.3267	1.3304	1.3441	1.3651	1.3657
A_9, A_{18}	1.4535	1.528	1.4925	1.4969	1.5159	1.5051	1.4904
A_{10}, A_{19}	2.9452	2.3894	2.59	2.6099	2.5346	2.608	2.4981
A_{11}, A_{17}	1.2398	1.157	1.1766	1.1727	1.2443	1.2057	1.2002
A_{12}, A_{15}	1.258	1.3045	1.3283	1.3092	1.2899	1.3025	1.3123
A_{13}, A_{16}	3.1511	2.3626	2.4132	2.5956	2.4391	2.5671	2.4123
A_{14}	1.001	1	1	1	1	1	1
Best weight (kg)	360.499	**357.271**	357.898	358.924	359.791	358.923	359.102
Mean weight (kg)	361.663	**358.068**	358.593	359.253	359.834	359.668	359.302
Coefficient variation	0.2821	0.2185	0.1849	0.0972	**0.0122**	0.1165	0.0366
ω_1 (HZ)	20	20.07	20.001	20.012	20	20	20.003
ω_2 (HZ)	40	40	40	40.034	40.051	40	40.042
ω_3 (HZ)	60	60	60.03	60.001	60.024	60	60.011

Exchange Optimization (CTEO), by Gauss chaos map and scenario 1, a weight of 319.199 kg has the optimal response.

13.7.3 A 120-Bar Spatial Dome

The 120-bar spatial dome as shown in Fig. 13.5 is a well-known benchmark problem with weight-limit optimization. This truss only considers "optimizing the size of the sections" and the geometric shape of the structure is constant during the optimization process. The decision variables related to the size of the members' sections and

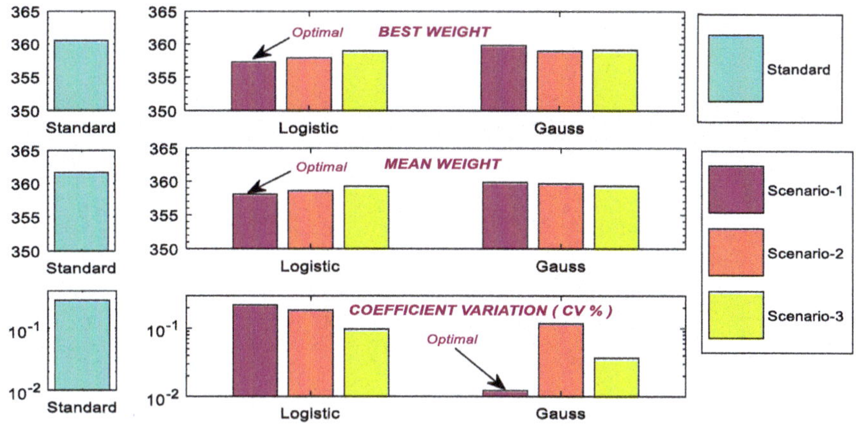

Fig. 13.2 Optimization results in standard mode and selection of chaos map for the simply supported 37-bar planer truss

according to the symmetry in the geometric shape of the dome along the X and Y axes, are classified into 7 groups. Non-structural concentrated mass in all free nodes affects the structure. Their values are 3 kg in node 1500 kg in nodes 2–13 and 100 kg in other nodes. The mechanical characteristics of the structure are: material density 7971.81 kg/m^3, modulus of elasticity 210,000 MPa, frequency limits of the structure in the first and second modes are greater than 9 and 11 Hz, respectively.

For the cross section of the members, the range of the lower limit is 1 cm^2 and the upper limit is 129.3 cm^2. In order to ensure the performance of chaos map and algorithms, as well as to increase the accuracy and sensitivity of calculations, each of the modes has been performed independently 30 times and the results related to the best response and the average value of responses are presented in Statistical Table 13.3. Also, the coefficient of change of responses, which is a measure of the robustness and robustness of responses, has been calculated and used to compare the efficiency of turbulence functions and algorithms. For quick access to optimization information, the bar chart of each component is shown in Fig. 13.6. Examining the optimization results for different combinations of algorithms with turbulence functions and comparing it with the standard mode, shows a significant and significant improvement in reducing the weight of the 120-bar spatial dome. The enhanced results for Chaotic Thermal Exchange Optimization (CTEO), by Gauss chaos map and scenario 1, a weight of 8708.667 kg has the optimal response.

13.7.4 A 200-Bar Planar Truss Structure

The 200-bar planar truss structure as shown in Fig. 13.7 is a well-known benchmark problem for weight optimization with frequency limits. This truss only considers

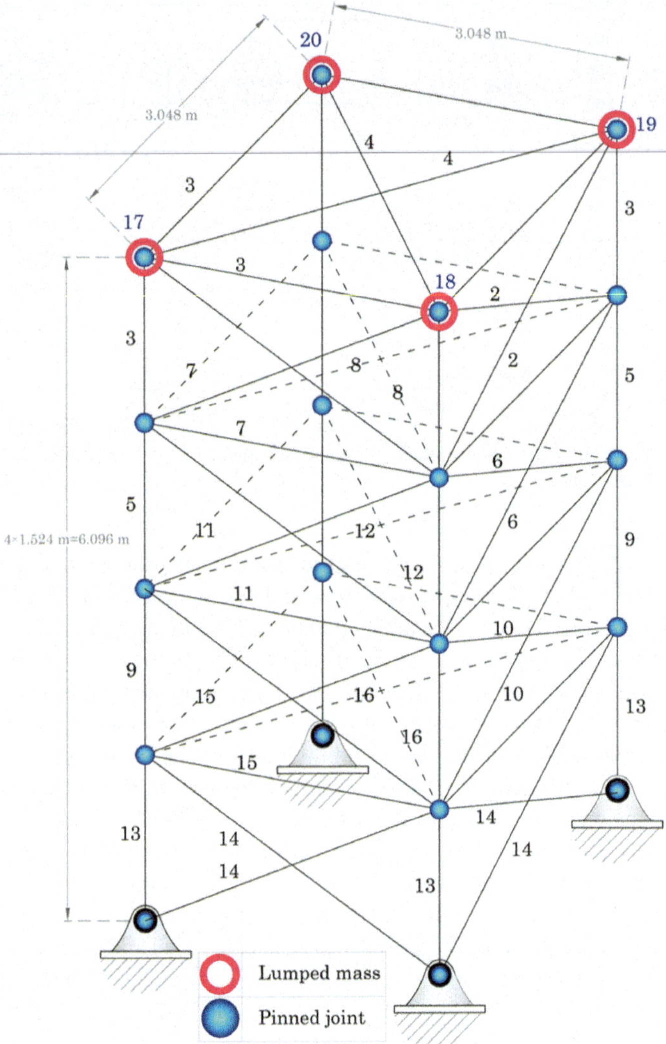

Fig. 13.3 Schematic of a 72-bar spatial truss

"optimizing the size of the sections" and the geometric shape of the structure is constant during the optimization process. The decision variables related to the sections size of the members are classified into 29 groups according Table 13.4. Non-structural concentrated mass in nodes 1–5 and in the amount of 100 kg affect the structure. The mechanical characteristics of the structure are: material density 7860 kg/m^3, modulus of elasticity 210,000 MPa, frequency limits of the structure in the first, second and third modes are greater than 5, 10 and 15 Hz, respectively. For the cross section of the members, the lower limit range is 0.1 cm^2 and the upper

Table 13.2 Optimal design comparison for the 72-bar spatial truss

Element group	Standard	Logistic map			Gauss map		
	TEO	CTEO 21	CTEO 22	CTEO 23	CTEO 31	CTEO 32	CTEO 33
A_1–A_4	4.61434	3.66343	3.20694	3.696	3.76825	4.32925	3.65463
A_5–A_{12}	7.98497	7.82537	7.75812	7.78388	7.5837	7.75208	7.68404
A_{13}–A_{16}	0.5	0.5	0.5	0.5	0.5	0.5	0.5
A_{17}–A_{18}	0.634986	0.555035	0.553503	0.534577	0.522802	0.53568	0.53865
A_{19}–A_{22}	8.30318	8.75774	7.99876	8.79415	9.78592	8.1105	8.49768
A_{23}–A_{30}	8.13245	7.86553	7.8706	7.75671	7.79707	7.88343	7.80368
A_{31}–A_{34}	0.5	0.5	0.5	0.5	0.5	0.5	0.5
A_{35}–A_{36}	0.562028	0.5	0.5	0.5	0.5	0.5	0.5
A_{37}–A_{40}	14.0369	12.7063	13.8413	13.0739	13.6794	13.4288	13.6968
A_{41}–A_{48}	8.12052	7.85697	7.75688	7.78508	7.80532	7.79064	7.80151
A_{49}–A_{52}	0.508733	0.5	0.5	0.5	0.5	0.5	0.5
A_{53}–A_{54}	0.547082	0.5	0.5	0.5	0.5	0.5	0.5
A_{55}–A_{58}	15.6536	17.7626	18.2878	17.8123	16.5606	17.4624	17.4677
A_{59}–A_{66}	8.15442	7.98117	7.77755	7.72004	7.75611	7.76839	7.8084
A_{67}–A_{70}	0.5	0.5	0.5	0.5	0.5	0.5	0.5
A_{71}–A_{72}	0.5	0.5	0.5	0.5	0.5	0.5	0.5
Best weight (kg)	328.705	322.181	320.165	319.303	**319.199**	320.353	319.605
Mean weight (kg)	330.453	323.527	322.07	322.249	323.33	**321.887**	322.34
Coefficient variation	0.5589	**0.3438**	0.416	0.7815	0.7993	0.4232	0.6921
ω_1 (HZ)	4.001	4.011	4	4.037	4	4.001	4.031
ω_3 (HZ)	6.002	6.001	6	6.005	6	6.024	6.028

limit is 25 cm^2. In order to ensure the performance of turbulent functions and algorithms, as well as to increase the accuracy and sensitivity of calculations, each of the modes has been performed independently 30 times and the results related to the best response and the average value of responses are presented in Statistical Table 13.5.

Also, the coefficient of change of responses, which is a measure of the robustness and robustness of responses, has been calculated and used to compare the efficiency of turbulence functions and algorithms. For quick access to optimization information, a bar chart of each component is shown in Fig. 13.8.

Examining the optimization results for different combinations of algorithms with turbulence functions and comparing it with the standard mode, shows a significant and significant improvement in reducing the weight of the 200-bar planar truss structure.

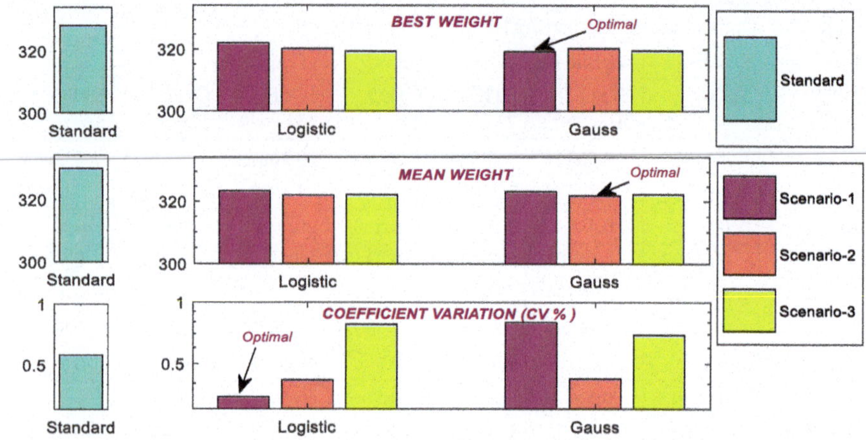

Fig. 13.4 Optimization results in standard mode and selection of chaos map for the 72-bar spatial truss

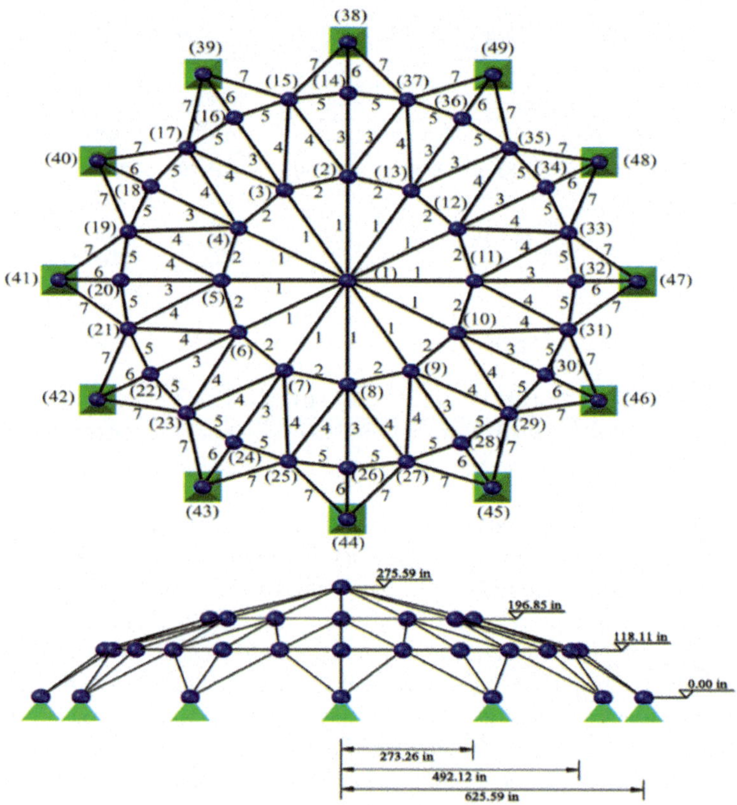

Fig. 13.5 Schematic of a 120-bar spatial dome

Table 13.3 Optimal design comparison for the 120-bar spatial dome

Decision	Standard	Logistic map			Gauss map		
Variable	TEO	CTEO 21	CTEO 22	CTEO 23	CTEO 31	CTEO 32	CTEO 33
A_1	19.6213	19.5724	19.4179	19.3379	19.5155	19.4437	19.7925
A_2	39.892	40.5617	40.7361	41.3432	40.512	41.4023	39.4558
A_3	10.8336	10.898	10.5987	10.5229	10.6373	10.5794	10.597
A_4	21.0565	21.0768	21.0937	21.1395	21.0685	20.8655	21.1274
A_5	9.85169	9.57706	9.83072	9.6574	9.74735	9.64452	9.83936
A_6	11.4955	11.4376	11.853	12.0545	11.746	11.8524	11.725
A_7	15.0211	14.973	14.7909	14.7197	14.8939	15.0546	14.9793
Best weight (kg)	8713.355	8714.335	8708.737	8716.217	**8708.667**	8724.145	8710.345
Mean weight (kg)	8715.417	8721.072	8710.306	8720.087	**8709.095**	8729.086	8711.405
Coefficient variation	0.0222	0.0559	0.0199	0.0633	**0.0061**	0.0495	0.009
ω_1 (HZ)	9.001	9.028	9.002	9	9	9.006	9.052
ω_2 (HZ)	11.002	11.015	11.023	11	11	11.027	11.074

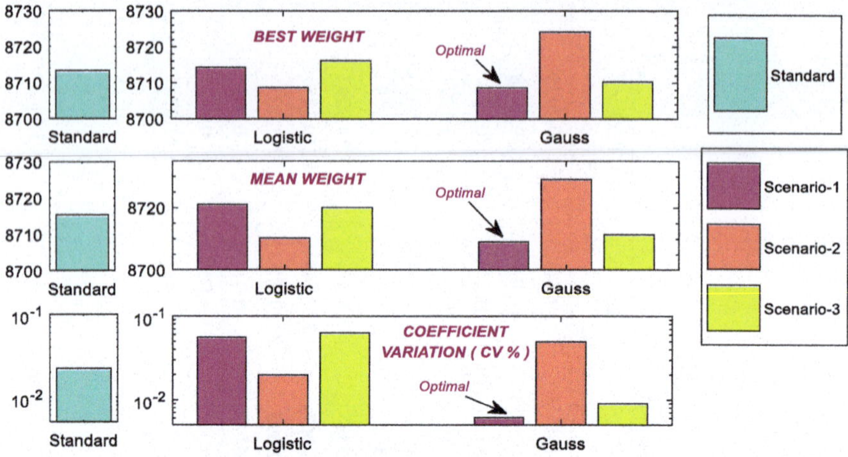

Fig. 13.6 Optimization results in standard mode and selection of chaos map for the 120-bar spatial dome

The enhanced results for Chaotic Thermal Exchange Optimization (CTEO), by Gauss chaos map and scenario 1, a weight of 2005.036 kg has the optimal response.

In the biogeography-based optimization, the Liebovitch chaos map with scenario 2 with a weight of 215/5657 kg has the optimal answer. In the optimization algorithm based on teaching–learning-based optimization, the Piecewise chaos map with scenario 1 with a weight of 2156/8604 kg has the optimal answer, and finally, in the particle swarm optimization, the Piecewise chaos map with scenario 2 weighs 2156/8582 kg has an optimal response.

13.8 Concluding Remarks

Thermal Exchange Optimization (TEO), is in the branch of physical algorithms and imitates Newton's law of cooling. In Newton's law of cooling, the rate of heat loss of an object is proportional to the temperature difference between the object and its surroundings. In this algorithm, each particle can be considered as a cooling or heating object, and assuming another agent as the environment, heat transfer is done between the two. After thermal exchange, the new temperature of the object determines its new position in the search space. In the original version of the algorithm, two phases of exploration and exploitation are considered to escape from the local optimal trap and early maturity. Each of these phases is associated with a random distribution. Although the existence of random distributions is useful for creating diversity in the search space, in most cases, it is not able to create a balance between the phases of exploration and exploitation, and it causes premature convergence and stopping at local optima. By embedding chaos functions, it is possible to jump from local

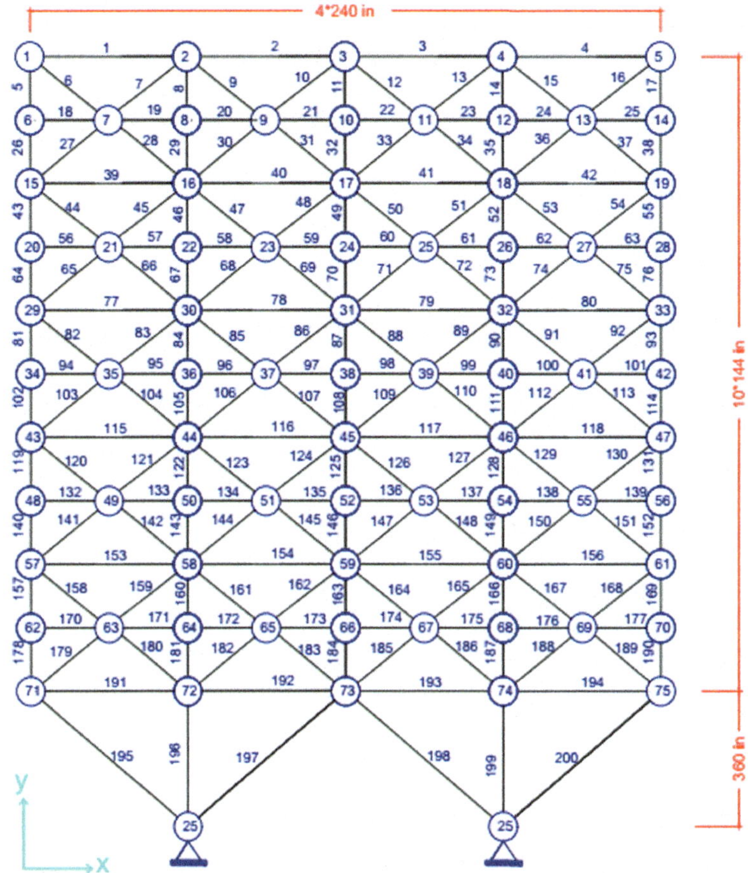

Fig. 13.7 Schematic of a 200-bar planar truss structure

optima and converge towards global optima. In this study, the Logistic and Gaussian chaos functions have saved the embedding and algorithm from premature convergence with three scenarios. In chaotic models, 6 different chaotic modes compete with each other, and the best response is compared with the standard mode. For problems related to multi-frequency, size, and layout constraints, the standard mode of the algorithm faces extensive stagnation, which can be significantly improved by embedding chaos functions and chaotic mutations. With the formation of chaotic models, the weaknesses and strengths of algorithms in dealing with optimization problems are determined.

Table 13.4 Classification of members in the 200-bar planar truss structure

Group	Element group	Group	Element group
A_1	1, 2, 3, 4	A_{16}	82, 83, 85, 86, 88, 89, 91, 92, 103, 104, 106, 107, 109, 110, 112, 113
A_2	5, 8, 11, 14, 17	A_{17}	115, 116, 117, 118
A_3	19, 20, 21, 22, 23, 24	A_{18}	119, 122, 125, 128, 131
A_4	18, 25, 56, 63, 94, 101, 132, 139, 170, 177	A_{19}	133, 134, 135, 136, 137, 138
A_5	26, 29, 32, 35, 38	A_{20}	140, 143, 146, 149, 152
A_6	6, 7, 9, 10, 12, 13, 15, 16, 27, 28, 30, 31, 33, 34, 36, 37	A_{21}	120, 121, 123, 124, 126, 127, 129, 130, 141, 142, 144, 145, 147, 148, 150, 151
A_7	39, 40, 41, 42	A_{22}	153, 154, 155, 156
A_8	43, 46, 49, 52, 55	A_{23}	157, 160, 163, 166, 169
A_9	57, 58, 59, 60, 61, 62	A_{24}	171, 172, 173, 174, 175, 176
A_{10}	64, 67, 70, 73, 76	A_{25}	178, 181, 184, 187, 190
A_{11}	44, 45, 47, 48, 50, 51, 53, 54, 65, 66, 68, 69, 71, 72, 74, 75	A_{26}	158, 159, 161, 162, 164, 165, 167, 168, 179, 180, 182, 183, 185, 186, 188, 189
A_{12}	77, 78, 79, 80	A_{27}	191, 192, 193, 194
A_{13}	81, 84, 87, 90, 93	A_{28}	195, 197, 198, 200
A_{14}	95, 96, 97, 98, 99, 100	A_{29}	196, 199
A_{15}	102, 105, 108, 111, 114		

Table 13.5 Optimal design comparison for the 200-bar planar truss structure

Number group	Standard	Logistic map			Gauss map		
	TEO	CTEO 21	CTEO 22	CTEO 23	CTEO 31	CTEO 32	CTEO 33
A_1	0.316476	0.358162	0.266191	0.304292	0.322936	0.356361	0.344772
A_2	0.438582	0.64412	0.485938	0.458607	0.485501	0.585633	0.627738
A_3	0.100013	0.1	0.1	0.1	0.1	0.1	0.1
A_4	0.100463	0.115526	0.1	0.1	0.1	0.109412	0.1
A_5	0.514046	0.744162	1.00092	0.532835	0.564392	0.677304	0.697711
A_6	0.821685	0.775408	0.735358	0.823283	0.72473	0.777914	0.767488
A_7	0.1001	0.1	0.1	0.100001	0.1	0.1	0.1
A_8	1.4306	1.41708	1.30456	1.51472	3.01913	1.37533	1.39033
A_9	0.1	0.1	0.1	0.1	1.63994	0.1	0.1
A_{10}	1.59322	1.66522	1.51327	1.56038	2.13908	1.58865	1.55512
A_{11}	1.14013	1.00104	1.0027	1.17907	0.995565	1.02585	1.00589
A_{12}	0.130731	0.100031	0.1	0.125024	0.1	0.1	0.1
A_{13}	2.97971	2.58601	3.12112	2.98134	2.79988	2.87513	4.10786
A_{14}	0.101518	0.104348	0.1	0.102846	1.2911	0.1	4.29646
A_{15}	3.24446	5.02345	3.25387	3.34815	2.89037	4.69201	3.19324
A_{16}	1.59978	1.23661	1.33548	1.59997	1.309	1.30046	1.35506
A_{17}	0.297173	0.100141	0.117071	0.218013	0.379946	0.427063	0.1
A_{18}	5.14395	5.10409	6.04685	5.08849	6.2622	5.6211	5.34706
A_{19}	0.101492	0.437744	0.1	0.100014	3.51731	0.100903	0.202362
A_{20}	5.61043	5.14377	5.61061	5.42652	4.776	6.12266	4.28548
A_{21}	2.0813	1.68835	1.69341	2.11369	1.81578	1.7807	1.78931
A_{22}	0.799689	0.16936	4.53037	0.77989	0.386963	0.20124	0.505963
A_{23}	7.71675	10.412	6.24621	7.55768	6.4546	11.1711	9.20475
A_{24}	0.166691	9.29393	0.743405	0.102427	0.398491	0.177485	3.21431
A_{25}	7.79151	10.5338	6.35396	8.09054	7.63253	13.9788	8.58691
A_{26}	2.82696	3.35933	3.34244	2.88339	2.45499	2.41314	3.18086
A_{27}	9.98003	4.87951	6.43945	10.5285	7.17695	11.9935	6.55212
A_{28}	20.9605	14.3365	15.3283	20.8348	17.182	17.4125	16.2798
A_{29}	11.7726	19.6991	18.1482	10.9007	12.352	10.1053	18.7961
Best weight	2158.103	2139.6	2018.5	2157.699	**2005.036**	2147.228	2141.909
Mean weight	2161.8	2285.482	2411.276	**2160.289**	2179.738	2384.314	2317.831
Coefficient variation	0.1871	3.899	13.534	**0.1165**	5.0617	8.676	4.662
ω_1 (HZ)	5	5.002	5	5.006	5	5	5.038

(continued)

Table 13.5 (continued)

Number group	Standard	Logistic map			Gauss map		
	TEO	CTEO 21	CTEO 22	CTEO 23	CTEO 31	CTEO 32	CTEO 33
ω_2 (HZ)	12.315	13.195	12.281	12.506	12.819	11.967	12.001
ω_3 (HZ)	15.016	15.317	15.375	15.162	15	15.259	15.382

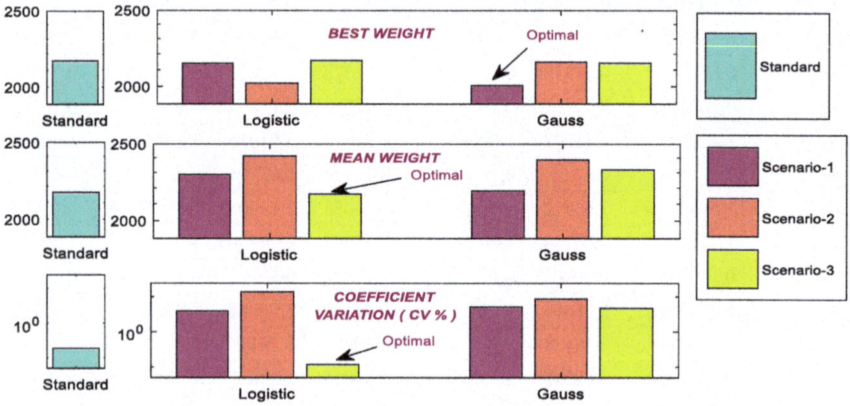

Fig. 13.8 Optimization results in standard mode and selection of chaos map for the 200-bar planar truss structure

References

1. Kaveh, A.: Advances in Metaheuristic Algorithms for Optimal Design of Structures, 2nd edn. Springer, Cham, Switzerland (2021). https://doi.org/10.1007/978-3-319-05549-7
2. Kaveh, A., Dadras, A.: A novel metaheuristic optimization algorithm: thermal exchange optimization. Adv. Eng. Softw. **110**, 69–84 (2017)
3. Kaveh, A., Yousefpoor, H.: Chaos-embedded meta-heuristic algorithms for optimal design of truss structures. Sci. Iranica **29**(6), 2868–2885 (2022). https://doi.org/10.24200/sci.2022.59812.6441
4. Kaveh, A., Yosefpoor, H.: Chaotically enhanced metaheuristic algorithms for optimal design of truss structures with frequency constraints. Periodica Polytech. Civ. Eng. **66**(3), 900–921 (2022). https://doi.org/10.3311/PPci.20220
5. Kaveh, A., Zarfam, P.,Aziminejad, A., Yosefpoor, H.: Comparison of four chaotic metaheuristic algorithms for optimal design of large-scale truss structures. Iran. J. Sci. Technol. Trans. Civ. Eng. https://doi.org/10.1007/s40996-022-00908-8
6. Ott, E.: Chaos in Dynamical Systems. Cambridge University Press, Cambridge (2002)
7. Bucolo, M., Caponetto, R., Fortuna, L., Frasca, M., Rizzo, A.: Does chaos work better than noise? IEEE Circuit Syst. Mag. **2**(3), 4–19 (2002)
8. Kaveh, A., Sheikholeslami, R., Talatahari, S., Keshvari-Ilkhichi, M.: Chaotic swarming of particles: a new method for size optimization of truss structures. Adv. Eng. Softw. **67**, 136–147 (2014)

SPRINGER NATURE

GPSR Compliance

The European Union's (EU) General Product Safety Regulation (GPSR) is a set of rules that requires consumer products to be safe and our obligations to ensure this.

If you have any concerns about our products, you can contact us on ProductSafety@springernature.com

In case Publisher is established outside the EU, the EU authorized representative is:

Springer Nature Customer Service Center GmbH
Europaplatz 3
69115 Heidelberg, Germany

The manufacturer's authorised representative in the EU is Springer Nature Customer Service Centre GmbH, Europaplatz 3, 69115 Heidelberg, Germany. If you have any concerns regarding our products, please contact ProductSafety@springernature.com

Printed and bound by CPI Group (UK) Ltd, Croydon, CR0 4YY

28/09/2026

02224085-0009